D0935001

ELECTRICAL TRANSFORMERS AND POWER EQUIPMENT

Third Edition

ELECTRICAL TRANSFORMERS AND POWER EQUIPMENT

Third Edition

Anthony J. Pansini, EE, PE
Life Fellow, IEEE
Sr. Member ASTM

Published by
THE FAIRMONT PRESS, INC.
700 Indian Trail
Lilburn, GA 30047

Library of Congress Cataloging-in-Publication Data

Pansini, Anthony J.
Electrical transformers and power equipment / Anthony J. Pansini.--3rd. ed.
p. cm.
Includes index.
ISBN 0-88173-311-3
1. Electric transformers. 2. Electric apparatus and appliances--Protection.
I. Title.
TK2551.P286 1998 621.31'4--dc21 98-31729
CIP

Electrical transformers and power equipment by Anthony J. Pansini.
--Third edition.

Published by The Fairmont Press, Inc.
700 Indian Trail
Lilburn, GA 30047

This book was previously published by Prentice-Hall, Inc.

Printed in the United States of America

10 9 8 7 6 5 4 3 2 1

ISBN 0-88173-311-3 FP

ISBN 0-13-012967-4 PH

Distributed by Prentice Hall PTR
Prentice-Hall, Inc.
A Simon & Schuster Company
Upper Saddle River, NJ 07458

Prentice-Hall International (UK) Limited, London
Prentice-Hall of Australia Pty. Limited, Sydney
Prentice-Hall Canada Inc., Toronto
Prentice-Hall Hispanoamericana, S.A., Mexico
Prentice-Hall of India Private Limited, New Delhi
Prentice-Hall of Japan, Inc., Tokyo
Simon & Schuster Asia Pte. Ltd., Singapore
Editora Prentice-Hall do Brasil, Ltda., Rio de Janeiro

CONTENTS

PREFACE TO THE THIRD EDITION

Events in the electric utility industry in the fast few decades have made knowledge of transformers and power equipment assume even greater importance. In general, the trend has been toward squeezing out every ounce of capacity to achieve a greater efficiency, all increasing the potential for decreased reliability.

Earlier efforts to reduce the demand and consumption of electric energy through load management programs (brought about by the increased cost and difficulty in providing generating facilities), largely through measures affecting consumers:

1. Retrofitting consumers loads with more efficient units; e.g., replacement of incandescent lamps with fluorescent ones.

2. Peak suppression by manipulating consumers' "schedules" of operation of specific loads to avoid their coincidence; e.g., clothes washer-dryers not operate at the same time as ranges or air conditioners.

3. Encourage development of cogeneration by large consumers.

Cash bonuses, favorable rates, attractive financing, and other incentives employed to accomplish these objectives. Deregulation of utilities that ostensibly replace monopolies with free market competition between suppliers of electric energy further increased pressures for efficiency improvement. Here the tactics employed, mainly by utility managements:

1. Intensification of load management programs.

2. Increase the use and capacity of transmission facilities to deliver low cost power from distant, not necessarily contiguous, sources; e.g., increase line capacities by converting to higher voltages.

3. Eliminate, physically or legally, less efficient generating units and other invested capital; e.g., land, buildings, other "stranded" facilities, no longer necessary or desirable.

In the quest for greater efficiency, efforts are directed principally toward transformers and transmission systems, where the potential for failure is perhaps greatest, where overloads and exposure play a large role. Development of transformers that cannot be overloaded, plastic insulation to replace frangible porcelain, and improved protection from electronic relays, all tend to mitigate the potentially negative effects on service reliability. These are more fully developed in the accompanying text.

At a time when advances in the fields of computers, communication, and automation have materially affected the daily activities of almost everyone, and at a time when the demand for greater reliability and quality of electric service is more critical than ever, the potential for degrading service reliability (especially because of deregulating processes) makes imperative that the design, manufacture, construction, installation, operation and maintenance of transformers and power equipment be thoroughly examined, that prudence and caution be exercised in the implementation of load management and deregulation programs.

Anthony J. Pansini
Waco, Texas
1998

Part I

ELECTRICAL TRANSFORMERS

1

THEORY OF OPERATION

INTRODUCTION: THE TRANSFORMER

The invention of the transformer in 1884, overcoming the technical and economic limitations associated with the original dc (direct current) commercial system, made practical the eventual availability of electric power to almost every home, farm, office, store, and factory. The first commercial ac (alternating current) distribution system using transformers was put in operation two years later in Great Barrington, Mass. In the same year, 1886, ac power was transmitted at 2000 volts over 30 kilometers of line built at Cerchi, Italy. From these small beginnings, the electric power industry has grown to the giant, almost universal, force for the good of mankind the world has ever seen. For all practical purposes, this industry today is entirely a supplier of alternating current; and its development is entirely due to the transformer.

Figure 1-1 (on page 4), showing typical facilities required to supply electric service, vividly indicates the many places in the supply system where transformers are used.

What It Does

The transformer is a device, having no moving parts, which transfers electric power form one circuit to another by electromagnetic means, usually with changes in values of both voltage and current. A step-up transformer receives electrical power at one voltage and delivers it at a higher voltage; conversely, a step-down transformer receives electrical power at one voltage and delivers it at a lower voltage.

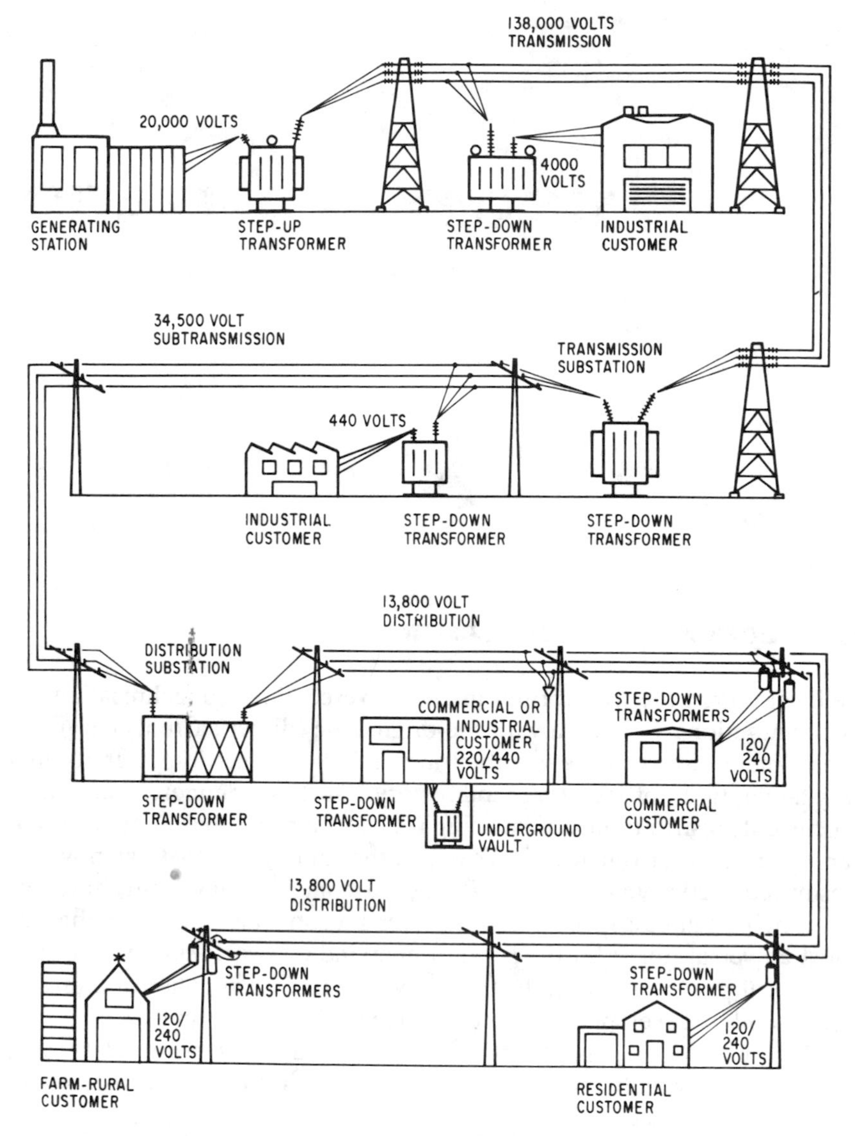

Figure 1-1 Electrical supply from generator to customer showing transformer applications and typical operating voltages.

The diagram of the hydraulic jack in Fig. 1-2 shows that a total force of 10 lbs. applied to one side of the jack can lift a weight of 100 lbs. on the other. The first piston, however, must be pushed down 10 in. in order that the weight of 100 lbs. be lifted a distance of 1 in. Thus a force of 10 lbs. has been transformed into one of 100 lbs.

In a transformer, the forces exerted correspond to the voltage while the

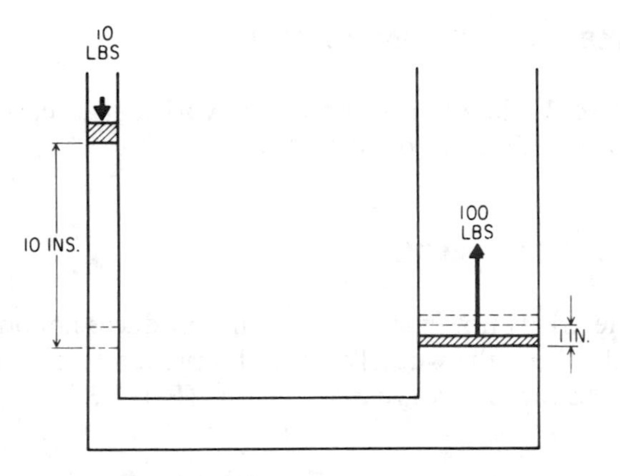

Figure 1-2 Hydraulic (oil or water jack) analogy of a transformer.

distances traveled correspond to the amperes. The total power on both sides, however, is the same. Thus, if the power input on one side was 1 ampere at a pressure or voltage of 100 volts, the same power would be obtained on the other side, but this time 10 amperes would be received at a voltage of 10 volts. It will be noticed that the product of the volts and amperes is the same for both the input and output. This product then is a means of measuring the capacity of the transformer, and is termed volt-ampere (VA). This unit is usually too small for practical purposes and so the unit called kilovoltampere (kVA) is used. The prefix kilo means 1000, so that 1 kVA equals 1000 VA (also a kilovolt is equal to 1000 volts). Power transformers range in capacity from ½ kVA to about 5000 kVA for supply to distribution customers, and from about 100 kVA to some 500,000 kVA for bulk or wholesale power supply.

Referring to the hydraulic jack again, if a continuous motion were substituted for the up and down pressure applied to the piston of one of the cylinders, power would be transmitted only during that time that the piston moved to its ultimate position. There would be no way to have the piston move back so that it would be able to transmit power on another stroke. The same effect is true in the transformer. Hence, it will be seen that the transformer will not work for direct or continuous currents.

The transformer is a highly efficient device so that practically the full electrical power received from one circuit is delivered to the other. Since electric power (P) may be considered the product of voltage (E) and current (I) (that is, $P = E \times I$), it is evident that the values of current in a transformer change in opposite fashion to those of the voltage. That is, in a step-up transformer where the voltage delivered is higher than that received, the current delivered is lower than that received; in a step-down transformer, where the voltage delivered is lower than that received, the current delivered is higher than that received.

HOW IT WORKS: ELECTROMAGNETISM

In order to describe how the transformer works, it is necessary to review briefly some simple elements of electricity.

ELECTROMAGNETIC FIELDS

A current of electricity flowing through a wire produces not only heat, but also a magnetic field about the wire. This may be proved by placing a compass in the vicinity of the current-carrying conductor. (Fig. 1-3.)

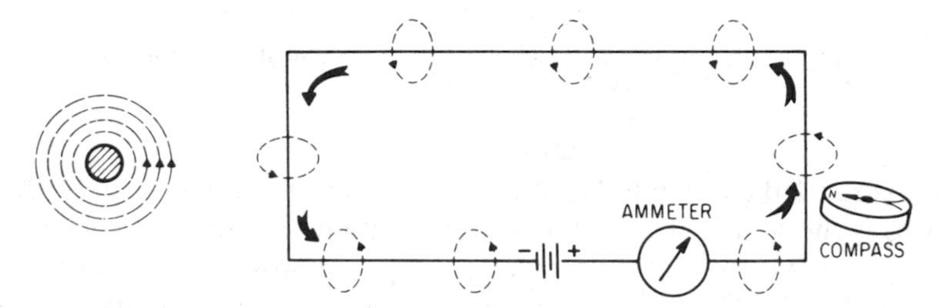

Figure 1-3 Magnetic field around a conductor.

The magnetic field around a single wire carrying a current may be rather weak. By winding the wire into a ring, the magnetic lines of force (sometimes called magnetic flux), are concentrated in the small space inside the coil and the magnetic effect is much increased. (Fig. 1-4.) The grouping of the lines of force is known as a magnetic field.

A coil of wire is nothing but a succession of these rings stacked one after the other (Fig. 1-5A). Each adds its quota to the magnetic field. Most of the magnetic lines of force pass straight through the coil. Each line makes a complete circuit, returning by a path outside the coil. A coil carrying a current is in fact a magnet. Where the lines come out is referred to (for identification) as the "north" (N) pole, where they enter as the "south" (S) pole (Fig. 1-5B).

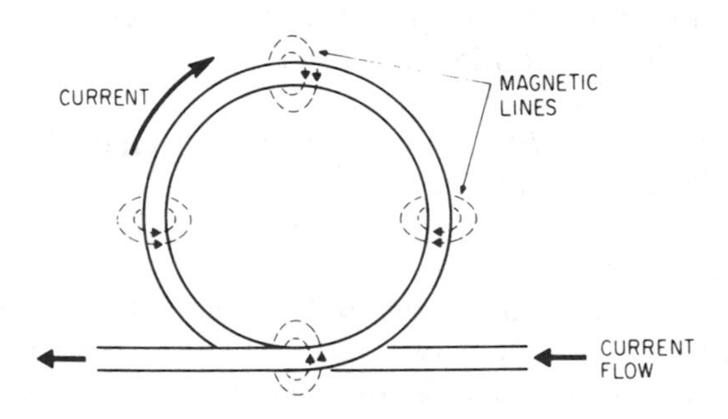

Figure 1-4 Magnetic field about a wire loop.

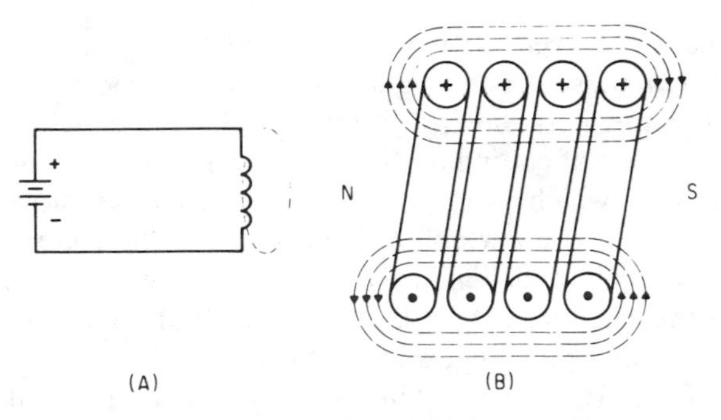

Figure 1-5 Magnetic field about a coil.

The strength of the magnetic field inside a coil depends on the strength of the current flowing and the number of turns. Its strength is, therefore, expressed in "ampere-turns," that is, amperes multiplied by the numbers of turns. Thus, a single turn carrying a very large current may produce the same effect as a great many turns carrying a small current.

A coil with an air core, however, produces a comparatively weak field. The strength is enormously increased by putting in a core of iron. (Fig. 1-6.) This is generally referred to as an electromagnet.

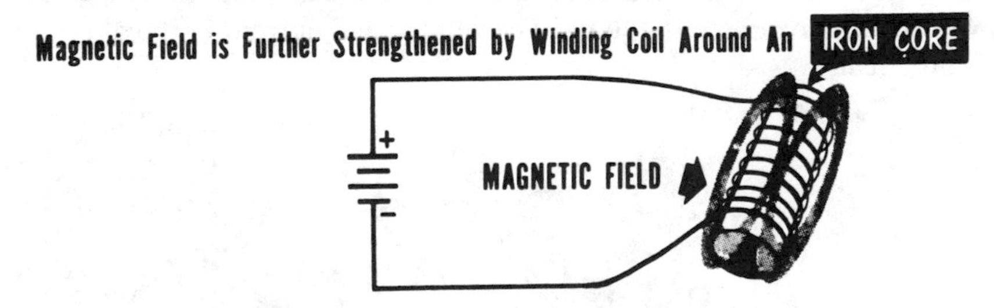

Figure 1-6 Coil with iron core.

ELECTROMAGNETIC INDUCTION

One relation between electricity and magnetism has already been demonstrated, that of producing magnetism with the aid of electric current. There is another important relation and that is the production of electricity with the aid of magnetism.

When a conductor is moved through a magnetic field an electrical pressure—or voltage—is produced in the conductor. If the strength of the magnetic field is increased (say, by replacing a simple magnet with a more powerful electromagnet), it will be found that a greater electrical pressure is now induced in the conductor cutting the lines of force. Similarly, the greater the length of the conductor, the more voltage is produced because more lines of force are cut. Also, if the speed at which the conductor is moved through the magnetic field is increased, it is found that the voltage induced in the conductor is also increased. The magnitude of the electrical pressure induced in a conductor while it is moving through a magnetic field, therefore, is determined by the *rate* of cutting of the lines of force of the magnetic field.

If a conductor which is part of a closed circuit is moved through a magnetic field, an electric current will flow in the conductor. The current results from what is called the *"induced"* electromotive force (EMF). But this happens only if the directions (1) of the current, (2) of the magnetic field, and (3) of the motion are at right angles to each other.

If a loop of wire (as illustrated in Fig. 1-7) is rotated at uniform speed in a magnetic field, a voltage is induced in the conductor that makes up the loop. If the voltage for different positions of the loop is measured and the angular position of the loop, (in Fig. 1-7) the waveform of the induced voltage can be obtained. This waveform is called a sine wave. Note that the instantaneous values of the voltage changes continuously from zero to maximum, from maximum back to zero, from zero to a maximum in the opposite direction, and then back to zero, thus completing one cycle. The maximums, in either direction of the voltage wave are called the "amplitude" of the voltage, and the number of cycles per second the "frequency" of the wave. The part of the curve from zero-to-maximum-back-to-zero is called an "alternation." Such a waveform is an example of the form of voltage and current in the alternating current circuit.

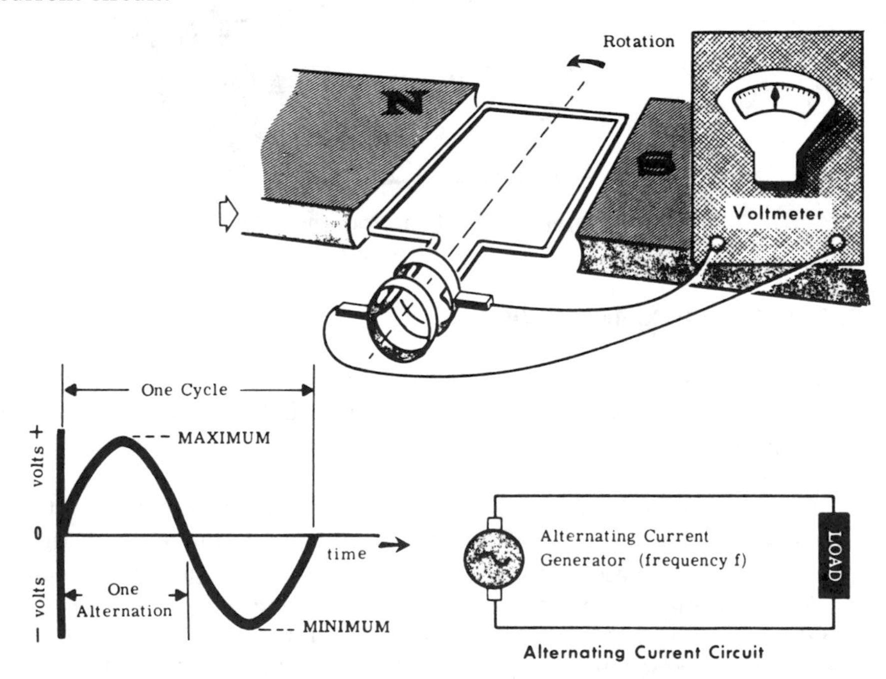

Figure 1-7 Sine-wave voltage induced in coil as it rotates in magnetic field.

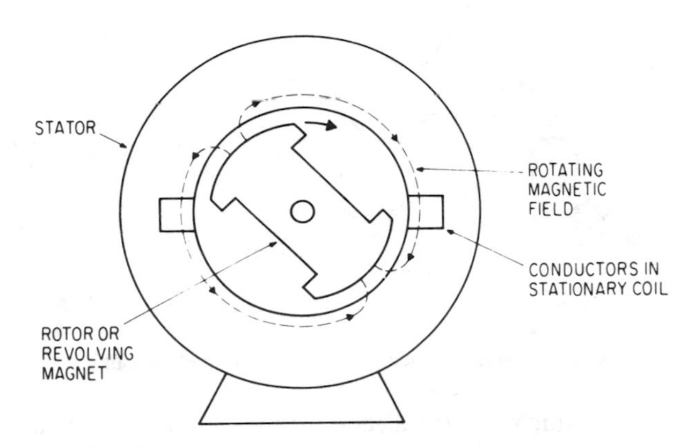

Figure 1-8 A basic alternator.

In a 60-cycle system, each cycle or two alternations takes place in 1/60 second.

An electrical pressure can be induced in a conductor by moving it through a magnetic field as described above. A voltage can also be induced electromagnetically by moving a magnetic field across the conductor. (Fig. 1-8.)

It makes no difference whether the conductor is moved across the magnetic field or if the magnetic field is moved across the conductor. A stationary conductor which has a magnetic field sweeping across it is cutting a magnetic field just the same as though the conductor was moving across the magnetic field.

Inductance

A conductor carrying alternating current has a magnetic field around it which alternates its characteristics in accordance with the alternations of the current flowing through the conductor. Following the sine wave characteristic of the alternating current, the magnetic field produced is zero at the start, builds up to a maximum in one direction at the first-quarter cycle, as shown in Fig. 1-9A, B, C, and D (on page 10) respectively, and reduces to zero at the second-quarter cycle, as shown in Fig. 1-9D, C, B, and A respectively, builds up again to a maximum in the third-quarter cycle but in a direction opposite to the previous maximum, and reduces to zero again to complete the cycle.

Self-Inductance

The expansion and contraction of the magnetic field constitutes a moving magnetic field which cuts the conductor carrying the current producing it. Such action will produce a voltage in the conductor distinct from, and tending to oppose that causing the original current to flow. This second voltage also

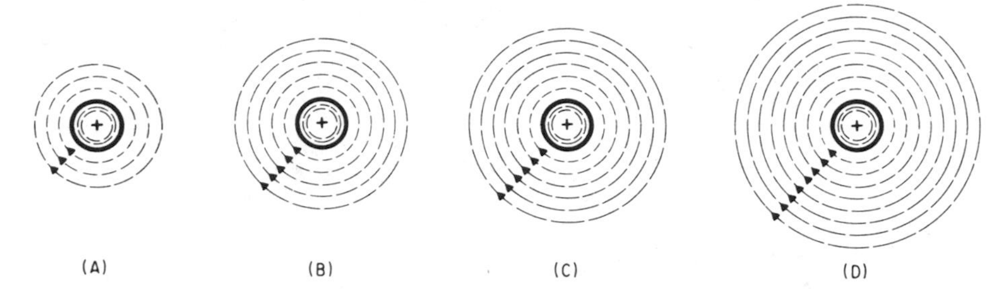

Figure 1-9 Expansion and contraction of a magnetic field about a conductor.

produces a current, which in turn affects the original current, and which in turn affects the magnetic field around the conductor, thus affecting the whole set up; this finally stabilizes at some point.

If the two voltages in the conductor are now combined (as actually they cannot exist separately), the resulting voltage will be different in magnitude and the voltage wave will be displaced somewhat when referred to the wave representing the current flow, or, put another way, it would appear that the current wave now "lags" the voltage wave (Fig. 1-10). This action prevents the

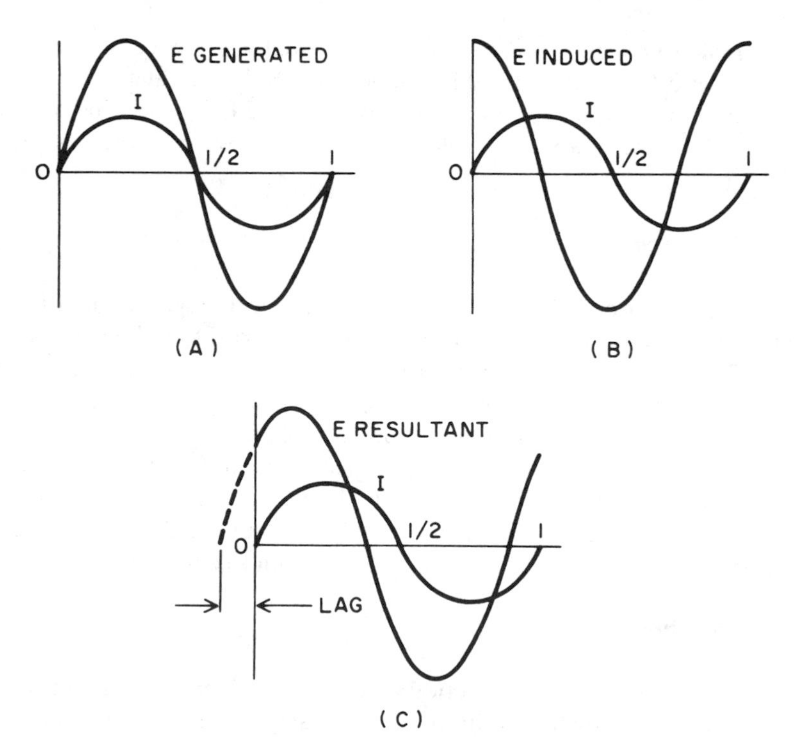

Figure 1-10 Effect of inductance on voltage and current in a conductor (not drawn to scale).

current and voltage (which produce power) from acting together for each point of the cycle, with the net effect being a reduced production of power. The net power produced compared to the 100 percent produced if both current and voltage acted together fully is known as "power factor." Further, the action can be looked at as something which prevents the free flow of electricity in the circuit, somewhat like the resistance produced by the friction of the water flow in a pipe. This magnetic reaction in a conductor is called "self-inductance" since it is caused by the magnetic field about itself.

Mutual Inductance

This same reaction may be caused by the magnetic fields of adjacent conductors, in which case it is called "mutual inductance," since both conductors affect each other (Fig. 1-11).

Inductive Reactance

The obstructive effect of these phenomena of induction within a conductor and between conductors carrying alternating current is called "inductive reactance" and, like resistance in a circuit, is measured in ohms.

CAPACITANCE–CAPACITIVE REACTANCE

When two conductors are near each other, there is also another reaction between them, but it is not due to the magnetic field. As an alternating current flows through a conductor, it will cause that conductor to have a scarcity of electrons during the first half cycle and an excess during the second half cycle. In attempting to restore a balance, it will therefore tend to attract and repel electrons from the adjacent conductor. This to-and-fro motion of electrons sets up an alternating current in the second conductor. If both conductors are carrying alternating currents, they will react upon each other.

The amount of this reaction, called "capacitance" will depend on the

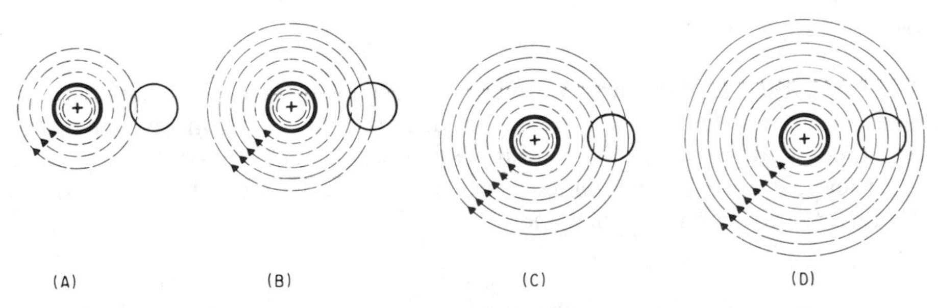

Figure 1-11 The effect of the magnetic field about a conductor on an adjacent conductor.

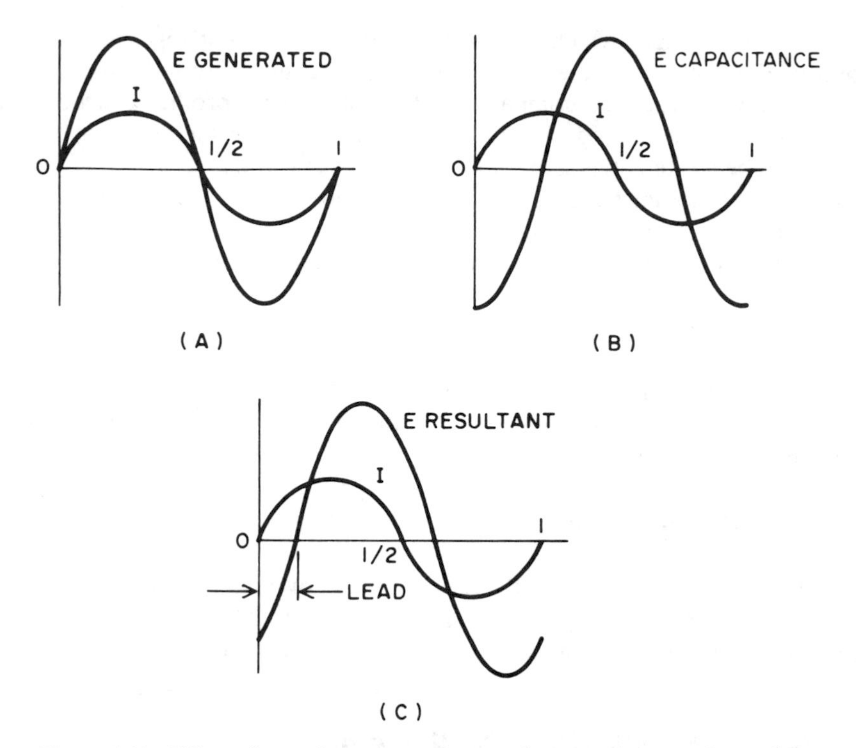

Figure 1-12 Effect of capacitance on voltage and current in a conductor (not drawn to scale).

area of the conductors exposed to each other, the distance between them, the kind of insulating material between them, and the voltage (and current) in the conductors. Like the inductance in a circuit, the current set up in a conductor because of this capacitive effect will tend to cause the resultant current to be displaced from the voltage wave (Fig. 1-12). The obstructive effect of this capacitance phenomenon between conductors is called "capacitive reactance," and is also measured in ohms.

IMPEDANCE

The total obstruction to the flow of current in an ac circuit may, therefore, be caused by resistance, inductance, and capacitance or more generally, resistance and reactance. This total obstruction, which impedes the flow of electricity, is called "impedance" (usually denoted by the letter Z). In ac circuits, then, Ohm's Law becomes

$$I = \frac{E}{Z}$$

where I is the current, E the voltage, and Z the impedance.

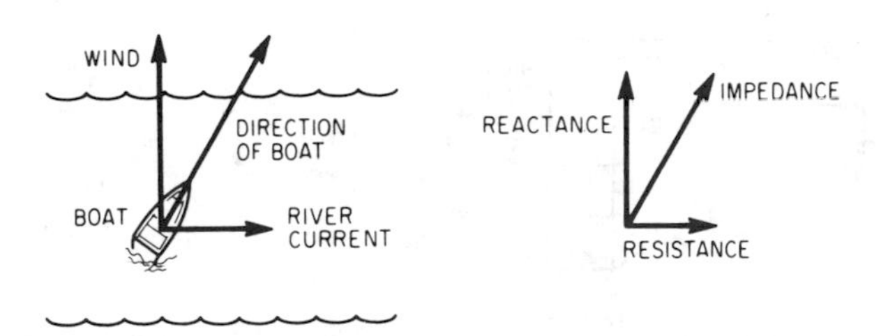

Figure 1-13 Water analogy of resistance and reactance relationship to impedance.

Resistance and reactance are not added directly to arrive at the impedance of the circuit. Rather, their relationship may be compared to a boat traveling in a body of water affected by both current and wind (Fig. 1-13); it is pushed by the current (resistance) of the water, and by the wind (reactance). If the wind in one direction is at right angles to the boat, it may be referred to as inductive reactance; if in the other direction, it may be referred to as capacitive reactance; the net effect of the two is the overall reactance. The relative strengths of these three components will determine the direction the boat would take; similarly, the relative position of the current wave with respect to the voltage wave will be determined by the relative values of the three components of impedance: resistance, inductance, and capacitance of the circuit. It is obvious that if the inductance and capacitance values are equal, then the net reactance is equal to zero and the only quantity left is resistance. When this condition occurs, the circuit is said to be in "resonance."

GENERATION OF VOLTAGE IN A TRANSFORMER COIL

It has been demonstrated how an electrical pressure or voltage may be generated in a conductor adjacent to another one carrying an alternating current. This process is called induction, and, in the conductors it acted to obstruct the normal flow of current. This same effect of induction, however, can prove useful in another way.

Instead of two wires adjacent to each other, assume there are two coils of wire adjacent to each other and an alternating current of electricity flows in one of them. As was shown previously, the voltage induced in the second coil will depend on the length of the conductor, the relative speed between conductor and the magnetic field, and the strength of the magnetic field. The entire magnetic field set up by the first coil can be assumed to cut the turns of the second coil if both coils are wound on an iron core (Fig. 1-14 on page 14). The relative speed between the conductor and magnetic field is fixed, being dependent on the frequency of the alternating current flowing through the first coil. The voltage in the second coil will therefore depend on the length of the conductor, or on the number of turns. If the magnetic field and the rate of

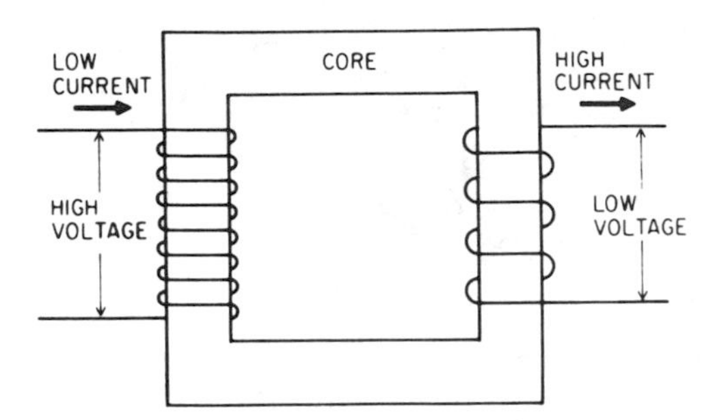

Figure 1-14 Diagram of transformer having two windings insulated from each other and wound on a common iron core.

cutting are the same for both coils, then each turn of each coil will have the same voltage produced in it. Therefore, to obtain the desired voltage in the second coil, the volts per turn are determined from the first coil. If a voltage of 1000 volts is applied to a coil of 1000 turns, then 1000 volts will be generated in the entire coil, each turn generating 1 volt. Now, if only a voltage of 100 volts is desired in the second coil, only 100 turns will be required as each turn generates 1 volt.

It must be remembered that the voltage in the second coil is an induced voltage and will therefore be displaced from the voltage in the first coil by a half cycle (Fig. 1-10). The currents will also be displaced by a half cycle as shown in Fig. 1-15.

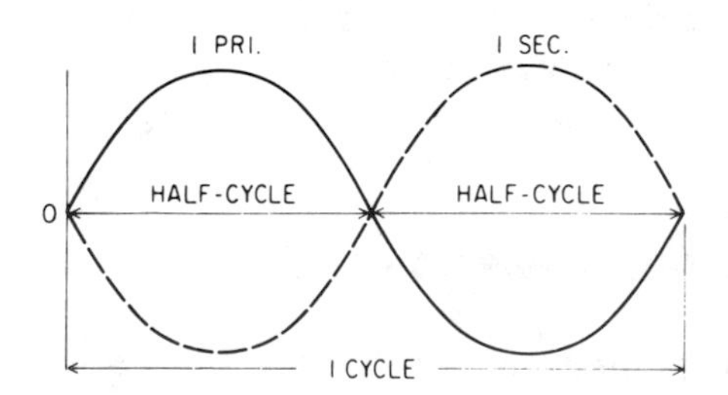

Figure 1-15 Current in secondary of transformer displaced from current in primary of transformer by one-half cycle.

RATIO OF TRANSFORMATION

The usual type of power transformer consists of two electrical windings placed on a common iron core, the number of turns in each winding being dependent on the desired ratio of transformation as described above. One winding, which is called the primary winding, is connected to the source from which power is supplied. The circuit to which power is delivered is connected to the other or secondary winding of the transformer. The primary and secondary windings are insulated from each other and from the iron core. Hence the transformer not only is a device by which the voltage of an ac system may be changed, but it also serves as a safety device for effectively isolating the lower voltage circuit from the higher voltage circuit.

The magnetic field produced by the first coil is determined by the number of amperes flowing in the turns of the coil, that is, on the ampere-turns as mentioned earlier. Since the same magnetic field is responsible for the current induced in the second coil, then the ampere-turns of the second coil must match the ampere-turns of the first coil. In the coils mentioned above, therefore, the current in the first coil will be one tenth that of the second; for instance, the current in the secondary winding (or second coil) will be 10 amperes.

Another way to prove the same thing is to measure the power input and compare it to the output. Remembering that power is represented by the product of the voltage and current, then, neglecting losses,

$$I_{PRI} \times E_{PRI} = I_{SEC} \times E_{SEC}$$

$$\text{or: } 1 \text{ amp}_P \times 1000 \text{ volts}_P = 10 \text{ amps}_S \times 100 \text{ volts}_S$$

All of the above relations may be expressed simply as follows:

$$\frac{N_P}{N_S} = \frac{E_P}{E_S} = \frac{I_S}{I_P}$$

N_P = Number of turns in primary winding.
N_S = Number of turns in secondary winding.
E_P = Voltage induced in primary winding.
E_S = Voltage induced in secondary winding.
I_P = Current flowing in primary winding.
I_S = Current flowing in secondary winding.

Turns Ratio

Whether a transformer is of a step-up or step-down type, the power in the primary is equal to the power in the secondary. Thus, if the load draws 1000 watts, the product of voltage and current in the primary is also equal to 1000 watts. Another important principle is the fact that the primary and secondary voltages are in the same ratio as their turns. If the secondary has twice the turns of the primary, the secondary voltage will be twice as great as that of the primary. Figure 1-16 (on page 16) illustrates the significance of turns ratio.

ACTION OF TRANSFORMER UNDER LOAD

The transformer also acts automatically to regulate the flow of energy in the primary as it is demanded by the load connected to the secondary. When no circuit is connected to the secondary coil, no current will flow in it. However, a current will flow in the primary coil because it presents a continuous path. This current will set up an alternating magnetic field which will cause a voltage to be self-induced in the primary coil. If there are no losses considered in the transformer, that is, if a "perfect" transformer is assumed, then the self-

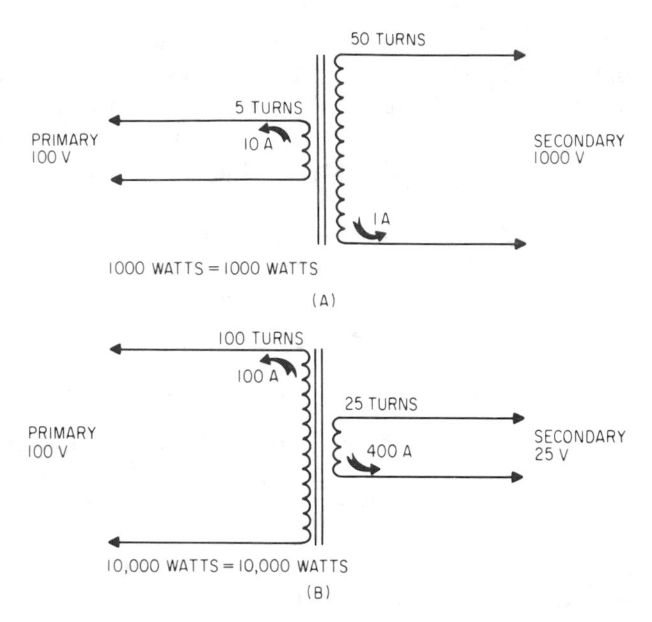

Figure 1-16 Turns ratio of transformers. (A) Step-up transformer, (B) Step-down transformer.

induced voltage will exactly balance the original voltage in the primary coil; the net result will be a zero voltage and consequently, a zero current. If a load is connected to the secondary, the following will occur:

1. A continuous circuit will be established and current will therefore flow in the secondary coil.
2. This secondary current in turn creates an alternating magnetic field about it, which will tend to induce within itself a voltage of such value as to stop the current from flowing.
3. At the same time, this secondary magnetic field will act on the primary in such a way as to reduce the effect of self-induction in the primary coil.
4. This will allow more current to flow in the primary coil, which in turn will set up a stronger magnetic field.
5. This stronger magnetic field will induce a greater voltage in the secondary coil, which will in turn tend to reduce the effect of self-induction in the secondary coil.

Thus, a state of equalization will be reached where the magnetic fields of the primary and secondary are so balanced that the ampere-turns in one match the ampere-turns in the other.

As more load is connected to the secondary coil, more paths will be provided for the current to flow. The impedance of the circuit connected to the secondary will be reduced and consequently more current will flow. The previously mentioned cycle of events will then reoccur until a new state of equalization will be reached. This new balance, it is obvious, can only be obtained by increasing the primary current.

The same chain of events is repeated in the reverse sequence when the load connected to the secondary coil is reduced.

TRANSFORMER LOSSES

All that has been said refers to an ideal or perfect transformer. In any actual instrument, the output is less than the input because of the transformer losses (heat developed in the wires and in the core).

Copper Losses

The heat losses developed in the wires are those caused by the resistance to the flow of current, which has been likened to the friction of water in a pipe, or friction between electrons moving in the conductor. This loss is determined by multiplying the square of the value of the current by the resistance (I^2R). It must be emphasized that only the resistance, R, causes this effect. The inductance (usually denoted by the letter L) does not resist the flow of current, it prevents it from coming into being. Therefore, in determining the heat losses, only the resistance, R, is to be considered; the inductance, L (or inductive reactance, X_L) and the impedance, Z, are not to be considered. It is obvious, too, that the greater the current flowing in the transformer (whether primary or secondary) the greater will be the heat (I^2R) losses in the conductors of the coils. These losses are usually referred to as the "copper losses" in a transformer.

Hysteresis

If each molecule of the iron comprising the core of the transformer is considered as a minute magnet, then, as the magnetic field set up by an alternating current changes direction, these small magnets will reverse their position to accommodate the strong magnetic field. As these molecules change their position with each alternation of the magnetic field, friction between them is produced and energy is used up in overcoming it. This loss of energy due to friction between the molecules is given off as heat, and is called "hysteresis."

Eddy Currents

The alternating magnetic field set up in a transformer not only induces voltages and currents in the coils through which it passes, but does so also in the core of iron. The currents thus induced in the iron core swirl around like eddies in a pool of water, and are therefore appropriately named "eddy currents." These currents in the core also produce an I^2R heat loss.

Iron Losses

The sum of both the hysteresis and eddy current losses is usually referred to as the "iron losses" in a transformer. Since there is only a relatively small difference in the magnetic field in the iron core at any time, as explained above, these losses will present little variation as the load on the transformer is increased or decreased.

No-Load Losses

It will be observed that when the secondary of a transformer is open, that is, it has no load connected to it, a small current will nevertheless flow in the primary. As mentioned previously, the large self-induced voltage in the primary practically counterbalances the original voltage in that coil. The current which the primary takes under these circumstances (sometimes referred to as the "exciting" current) serves to supply the hysteresis and eddy current (or iron) losses as well as a small I^2R loss in the primary coil itself. The sum of these losses is usually referred to as the "no-load loss" in a transformer.

IMPEDANCE AND REGULATION OF TRANSFORMER

In determining the performance of a transformer, the impedance of the transformer caused by its core and coils, must be considered. The effect of this impedance is to cause a drop in the secondary voltage of the transformer, depending on the current flow. Transformers are designed to give a predetermined secondary voltage drop at full load current. The nameplate on the transformer gives its impedance, expressed as a percentage of its rated voltage. For example, a transformer connected for 240-volt output, with a 2.2 percent impedance, will deliver 2.2 percent less than 240 volts or 234.72 volts at full load.

The regulation of a transformer is the ratio of the change in secondary terminal voltage from no load to full load (at constant primary impressed terminal voltage) to the rated secondary voltage, usually at some specified power factor. This value is usually small, which allows the ordinary power transformer to be known sometimes as a constant potential transformer.

Despite the low values, however, the regulation and impedance of transformers is important when two or more transformers are connected in parallel (as in a network illustrated in Fig. 1-17), as the load currents they supply will be different if the regulation and impedance of the units are different; that is, the transformers will not share the load equitably.

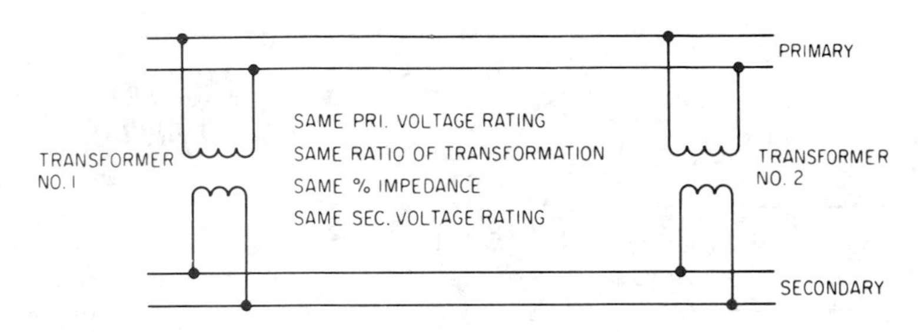

Figure 1-17 Networking or banking of transformers.

The conventional power transformer is designed to operate with the primary connected across a constant potential source and to provide a secondary voltage that is substantially constant from no load to full load.

Like any other piece of equipment, the efficiency of a transformer is given by:

$$\frac{\text{Output}}{\text{Input}} = \frac{\text{Output}}{\text{Output} + \text{copper losses and iron losses}}$$

TRANSFORMER RATINGS

The capacity of a transformer is limited by the permissible temperature rise. Both the current and the voltage contribute to determine the heat generated in a transformer. As has been shown above, because of reactance, the current and the voltage waves do not always act in concert, and the power supplied (as expressed in watts) is not always a measure of the heat generating processes. Hence, the rating of a transformer is expressed as a product of the volts and amperes, or volt-amperes (for practical purposes, kVA is used). For a given power factor of load, these values often are used interchangeably.

Figure 1-18 (on page 20) illustrates the nameplate on a transformer giving all the pertinent information required for the proper operation and maintenance of the unit. The capacity of a transformer (or any other piece of electrical equipment) is limited by the permissible temperature rise during operation. The heat generated in a transformer is determined by both the current and the voltage. Of more importance is the kilovolt-ampere rating of the transformer. This indicates the maximum power on which the transformer is designed to operate under normal conditions (when the current and voltage are "in phase"). Other information generally given is the phase (single-phase, three-phase, etc.), the primary and secondary voltages, frequency, the permitted temperature rise, the cooling requirements—which include the number of gallons of fluid that the cooling tank may hold. Primary and secondary currents may be stated at full load.

Depending upon the type of transformer and its special applications,

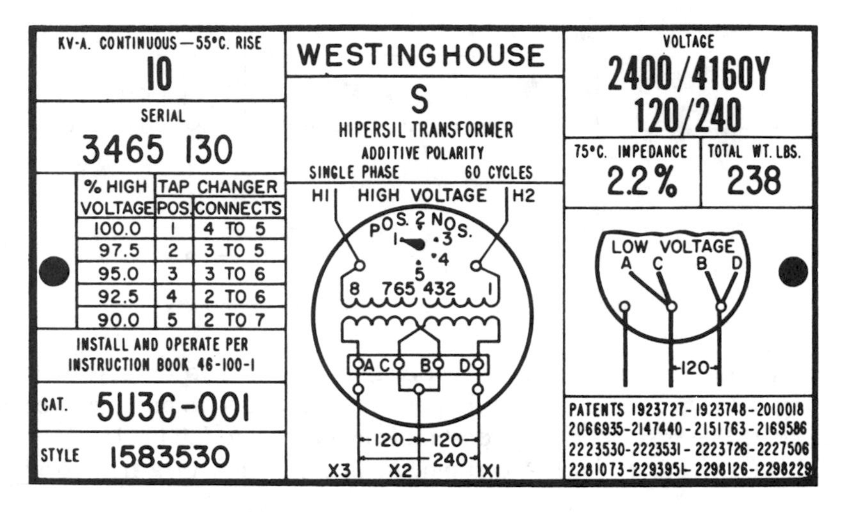

Figure 1-18 Transformer nameplate.

there may be other types of identifications for various gauges, temperature indication, pressure, drains, and various valves.

Thus, it can be seen that while the transformer consists primarily of a primary and a secondary winding, there are many other factors to take into consideration when selecting a transformer for a particular use. Most power supply transformers are designed to operate at a frequency of 50 or 60 cycles per second. Aircraft and other special application transformers are designed for a frequency range from 400 to 4000 cycles per second. The higher frequencies permit a saving in size and weight of transformers and associated equipment; further, where separate circuits prove economical, fluorescent lighting at the higher frequencies has been found to be more efficient.

REVIEW

1. The transformer is a device, having no moving parts, which transfers electric power from one circuit to another by electromagnetic means, usually with changes in values of both current and voltage. They may be used to step up or step down input voltage. Its operation depends on the relationship between magnetism and electricity.
2. An electric current flowing in a conductor produces a magnetic field around that conductor (Fig. 1-3). The magnetic field can be concentrated and increased in strength if the conductor is made into a loop (Fig. 1-4), and further increased if several loops are made into a coil (Fig. 1-5). The magnetic field can be increased appreciably if an iron or steel core is inserted inside the coil (Fig. 1-6).

3. If a conductor that is part of a closed circuit is moved through a magnetic field, an electric current will flow in the conductor. The current results from what is called an induced electromotive force (EMF). The EMF depends on the strength of the magnetic field, the length of the conductor, and the speed with which the conductor cuts the magnetic field.
4. When a coil or loop is turned within a magnetic field, a voltage will be produced which will rise from zero to maximum, back to zero, then to a maximum in the opposite direction and back to zero (Fig. 1-7). The voltage thus induced will produce an alternating current (ac) in the coil or loop (Fig. 1-8).
5. A conductor carrying alternating current will produce an alternating magnetic field around it; if another conductor parallels it closely, another ac voltage will be produced in it by the alternating magnetic field of the first conductor cutting it (Fig. 1-11).
6. If two coils are wrapped around a common iron or steel core, an ac voltage in one will produce an ac voltage in the other. The ratio of these voltages will be the same as the ratio of the number of turns in the coils (Fig. 1-16). This assembly of coils and core constitutes a transformer (Fig. 1-14). The input coil is usually called the primary and the output coil the secondary.
7. Losses developed in the transformer are given off as heat. The rating of a transformer is usually limited by how hot it is allowed to become. It is rated in kilovolt-amperes or kVA (Fig. 1-18).

STUDY QUESTIONS

1. What is a transformer and what does it do?
2. What general relationship exists between the input and output of a transformer?
3. What two effects are produced when a current of electricity flows through a wire?
4. How may a voltage be induced in a conductor? What factors affect its magnitude?
5. How is a voltage generated in a coil of the transformer? What is the relationship between the input or primary voltage and the output or secondary voltage?
6. What is meant by the ratio of transformation or turns ratio?
7. Describe the action of a transformer when a load is connected to the secondary.
8. What are the energy losses in a transformer? What form do they take?
9. What is meant by the regulation of a transformer? What factors affect it?
10. How are transformers' ratings expressed? What limits their capacity?

2

TRANSFORMER CONSTRUCTION

PRINCIPAL PARTS

The ordinary power supply transformer basically consists of two windings, insulated from each other, wound on a common core of magnetically suitable material such as iron or steel. One winding receives the energy from an ac source and is known as the primary winding. The other, which receives the energy by mutual inductance from the primary, delivers that energy to the load and is known as the secondary winding. The core provides a circuit or path for the easy flow of the magnetic lines of force or magentic flux. In addition to these, there are several other important parts which make up the practical power supply transformer: the enclosure, the terminals or bushings, and the coolant or cooling medium (Fig. 2-1). The requirements of all of these determine the kind of construction for the specific power supply transformer.

CORES

The construction of a transformer starts with the core. The material for this is chosen to afford its molecules the greatest ease in reversing their position as the ac magnetic field reverses its direction. The friction developed between these magnetic molecular particles as they reverse themselves creates heat, and this action causes a loss known as "hysteresis" (Fig. 2-2). This is minimized by using a special grade of heat-treated, grain-oriented silicon steel. These reversals of molecular particles also contribute to the noise or hum of a transformer.

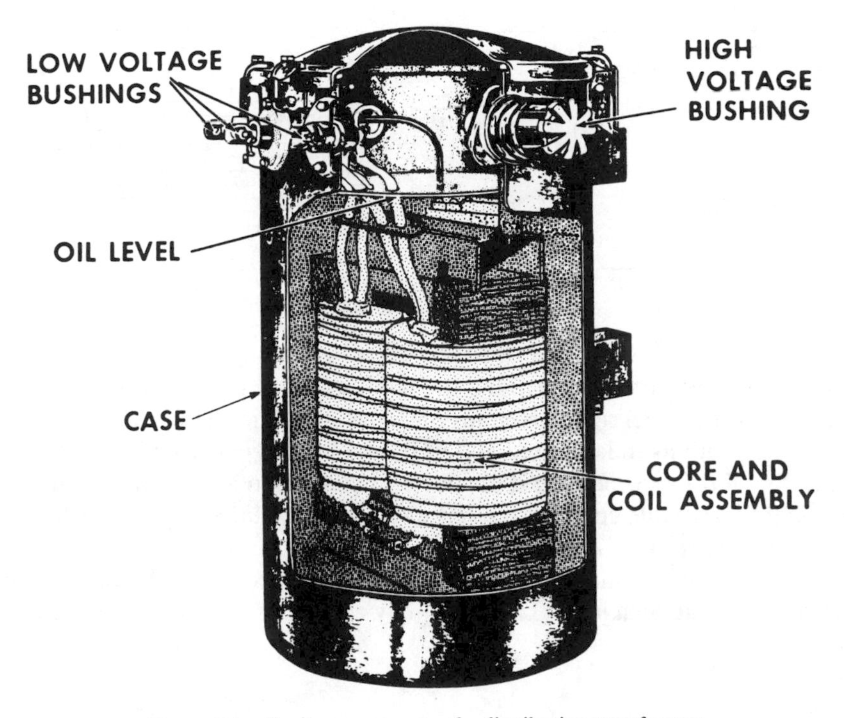

Figure 2-1 Basic components of a distribution transformer.

Currents are generated in the core by the alternating magnetic flux (or lines of force) as it cuts through the metal core. These currents, known as "eddy currents," also create heat and contribute to the losses in a transformer. These are minimized by constructing the core of thin sheets or laminations, and by insulating adjacent laminations with insulating varnish (Fig. 2-3 on page 24). The laminations and varnish tend to present a high resistance path to these eddy currents, thereby reducing their magnitude, and the consequent losses. If these laminations are not properly secured, they will tend to vibrate contributing further to the noise or hum of a transformer.

The principal types of transformer construction are known as the "core" type and the "shell" type (Fig. 2-4 on page 24); both of these refer to the arrangement of the steel core with reference to the windings. In the core-type

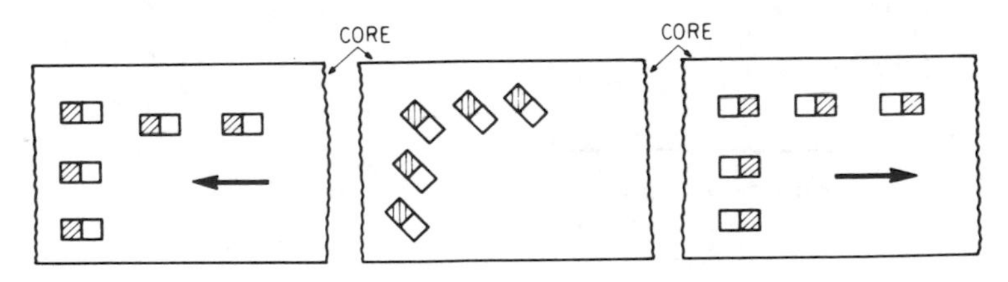

Figure 2-2 Molecular motion during magnetic field reversal (hysteresis).

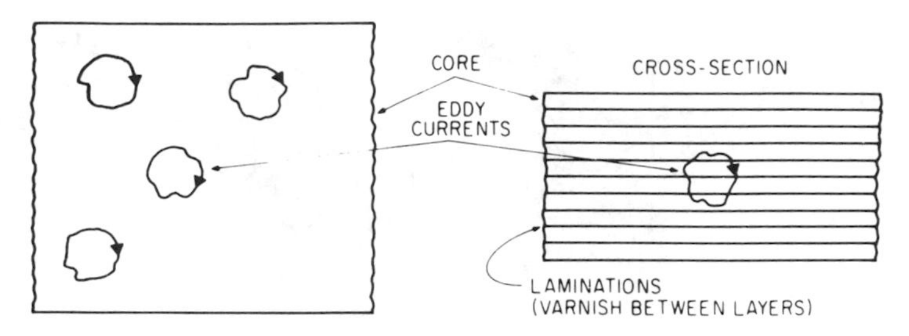

Figure 2-3 Eddy currents in core.

transformer, the windings surround the laminated steel core. In the shell-type transformer, the steel core surrounds the windings. The choice depends on the economics, both as to labor and material, involved in construction and installation. Small transformers used for power distribution are usually of the core type (Fig. 2-5), while some of the larger station or plant-power transformers are the shell type (Fig. 2-6 on page 26).

The insulated laminations which make up the core may be only several thousandths of an inch in thickness and may consist of various shapes: I, U, E,

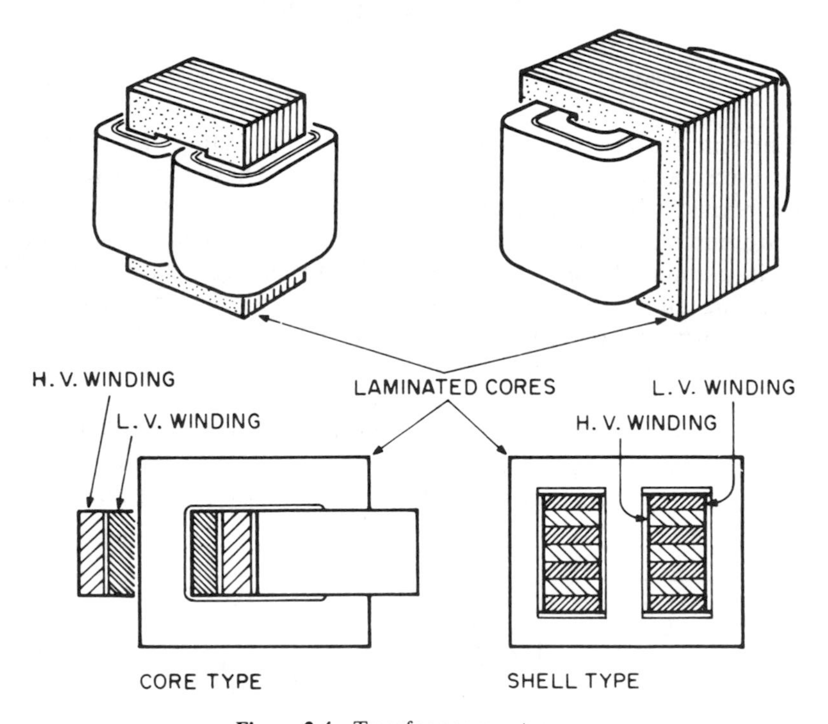

Figure 2-4 Transformer core types.

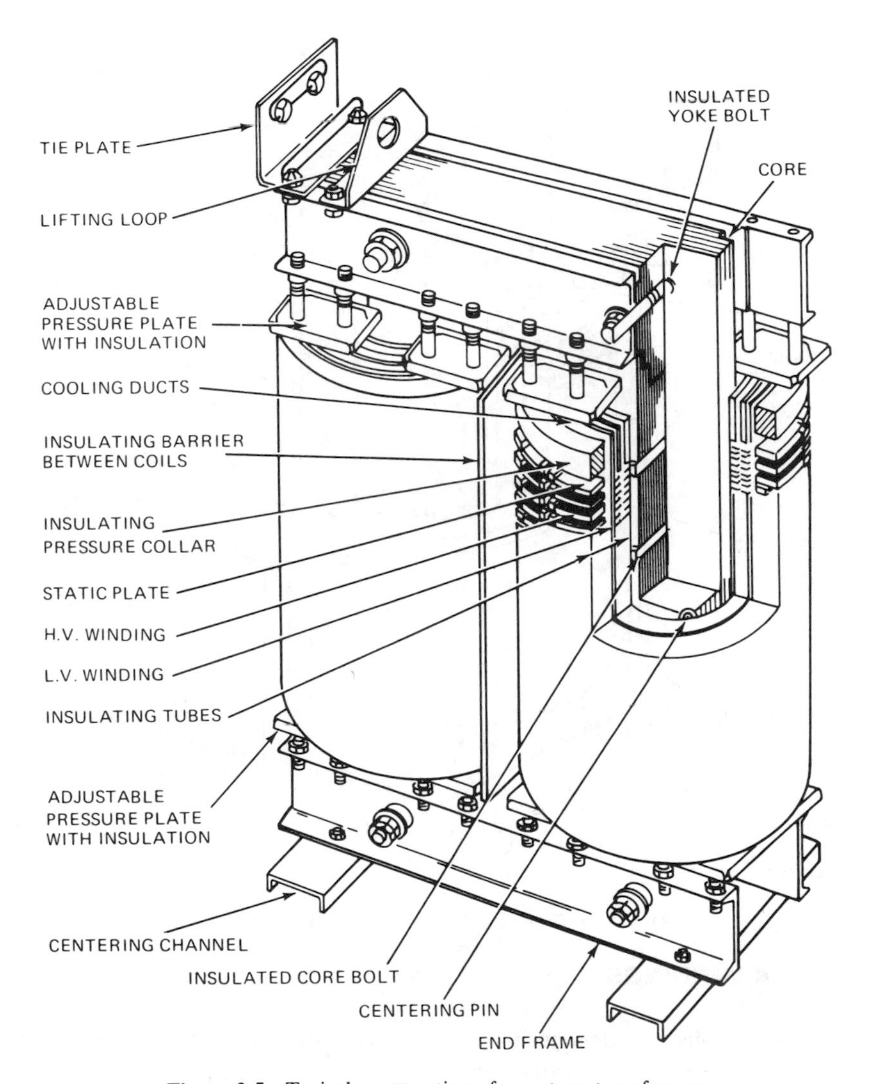

Figure 2-5 Typical construction of core-type transformer.

and L stampings, or a split-spiral-wound core. The cores are thus constructed to allow them to be inserted within the coils. The joints in the construction of the cores may overlap or butt together; the two halves of the spiral butt together. Mechanically, these joints must be rigid and strong, not only to prevent their coming apart under severe stress, but also to hold down the noise which may emanate from their vibration during the operation of the transformer.

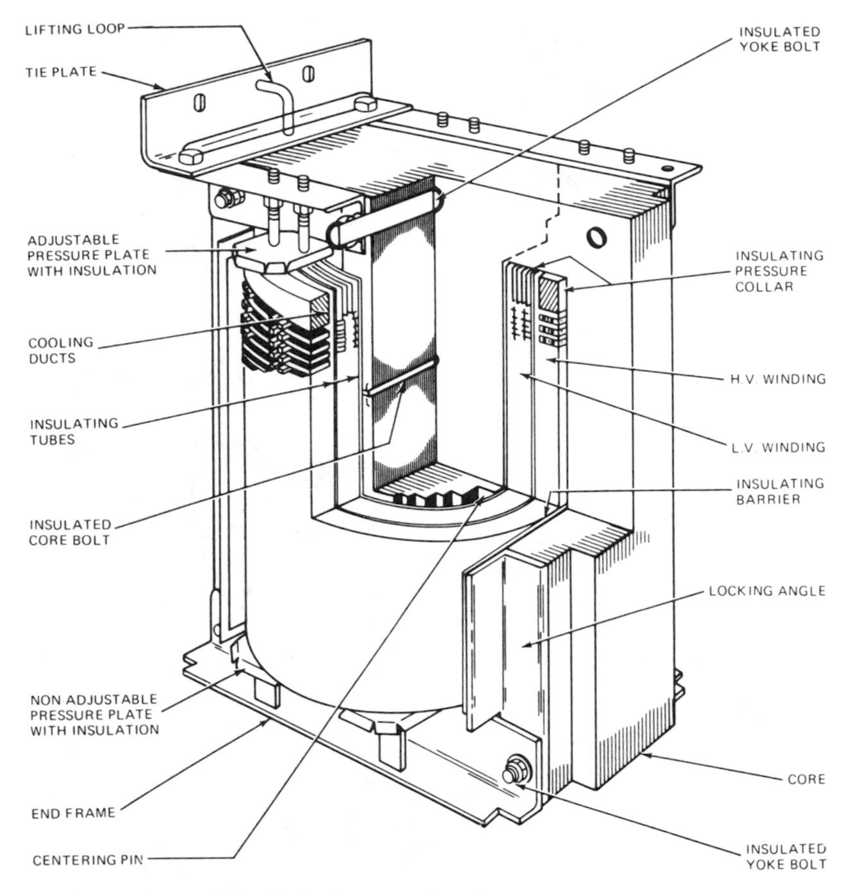

Figure 2-6 Typical construction of shell-type transformer.

COILS OR WINDINGS

Transformer coils are designed to get the required number of turns into a minimum of space. At the same time, the cross section of the conductor must be large enough to carry the current without overheating, and sufficient space must be provided for the insulation and for cooling paths, if any. These coils may be made of copper or aluminum, the choice depending on the cost to achieve the low resistance and small space requirements.

For small units, the coil wire may be round, insulated with cotton, enamel, shellac, varnish, paper, or a combination of these. For larger units, the wire may be square or rectangular ribbon, usually insulated with oil impregnated paper. Where transformer operation at high temperatures is desired, special glass or plastic insulation may be used. The insulation should provide not only for normal operating voltages, but also for surges of high voltage resulting from lightning or switching.

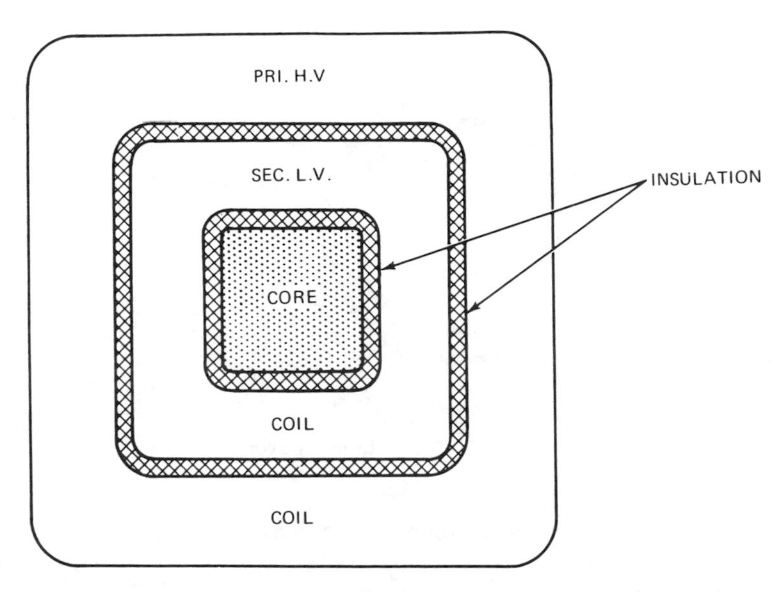

Figure 2-7 Concentric windings.

In the core-type transformer, the low-voltage coils are usually placed next to the core and the high-voltage coils are external and concentric with them. This reduces the insulation requirements of both coils; if the high-voltage winding was placed next to the core, two layers of high-voltage insulation would be required, one next to the core, and the other between the two windings (Fig. 2-7). (Sometimes, where large and heavy connections are involved, this arrangement may be reversed.) The high-voltage windings may be separated from the low-voltage windings by insulating cylinders; the high voltage may be composed of several disc-shaped coils, each disc insulated from others by insulating strips. If the windings were placed on separate legs of the core, a relatively large amount of the flux produced by the primary windings would fail to link the secondary winding, resulting in a large loss of effectiveness of the flux.

For very small sizes of transformer, a cruciform-type core (cross-shaped) is often used for economy (Fig. 2-8 on page 28). In both this type and the core-type transformers, the coils may be made cylindrical in shape, which enables the insulation to have a factor of safety since there are no sharp bends in the insulating material and a better opportunity for the radiation of heat. Both these factors permit the use of less material for a given output.

In the shell-type transformer, "pancake" construction is often employed (Fig. 2-9 on page 28). Here, the high-voltage and low-voltage coils are alternately placed around the core, with the required insulation between them, each coil having the rough appearance of a pancake. Often, space is left between coils for cooling purposes. Such an arrangement of coils reduces the reactance between the coils and improves the operation of the transformer,

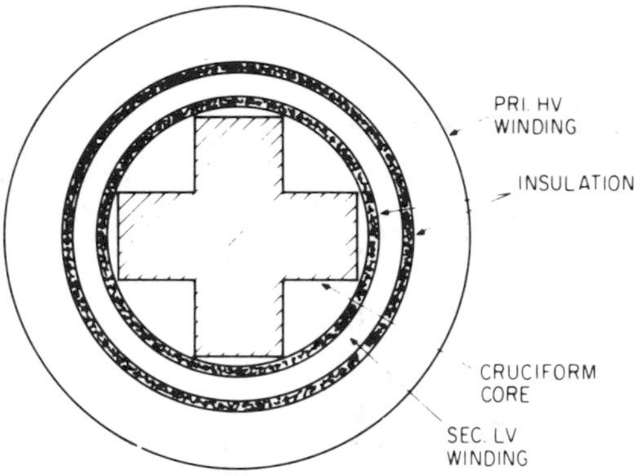

Figure 2-8 Cruciform-type core.

particularly in large-size transformers where heavy currents are experienced. The shell-type transformers also make arrangements for air cooling paths simpler and easier to provide.

In transformers, all coils which carry current in the same direction attract one another and coils which carry currents in opposite directions repel one another. Hence, all the coils of the primary attract one another as do all the coils of the secondary; however, primary coils repel the secondary coils and vice versa. Under normal operating conditions, these forces are relatively small. In case of short circuit or the carrying of very large currents, these forces may become great enough to damage the transformer if the coils are not adequately supported.

Coils and cores must be mechanically capable of withstanding these short circuit stresses. Where the coils are concentrically placed, the forces produced are radial which may tend to distort the shapes of the coils. If

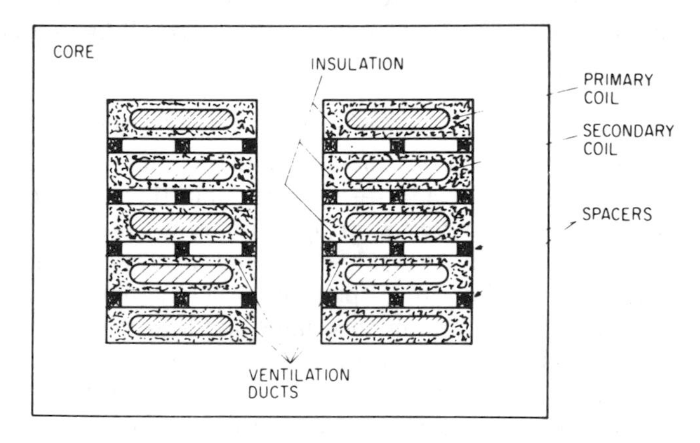

Figure 2.9 Shell-type unit ("pancake" coils).

interleaved coils are not exactly balanced, axial forces develop, also tending to distort the coils; these are usually so interleaved that forces between coils are balanced, except at the ends. Hence, when assembled, coils must be carefully centered on cores and rigidly blocked to prevent any movement, bending or distortion from normal positions under the stresses caused by heavy currents. Such bracing and blocking also tends to reduce the noise emanating from the vibration of the several elements, brought about by the effects of the alternating magnetic fields.

Precaution should also be taken so that a failure of the high voltage insulation to the low voltage side will not impose the high voltage on the windings of the low voltage coil. Additional insulation and barriers are often placed between the primary and secondary coils to lessen the chances of such occurrences.

ENCLOSURE AND COOLING SYSTEMS

The complete core and the coil assembly are placed in an enclosure, generally a steel tank, to protect them from the elements, vandalism, and, for safety purposes, from possible contact by people. In commercial transformers, the complete assembly is usually immersed in a special mineral oil to provide means of insulation and cooling. The oil covers the assembly completely and fills small voids, especially in the insulation, or pockets where air or contaminants may collect which may eventually cause failure of the transformer.

METHOD OF TRANSFORMER COOLING

Transformers are usually cooled by oil or air (Fig. 2-10 on page 30). The wasted energy in the form of heat generated in transformers due to unpreventable iron and copper losses must be carried away to prevent excessive rise of temperature and injury to the insulation about the conductors. The cooling method used must be capable of maintaining a sufficiently low average temperature. It must also be capable of preventing an excessive temperature rise in any portion of the transformer, and formation of "hot spots." This is accomplished, for example, by submerging the core and coils of the transformer in oil, and allowing free circulation for the oil. As the coil and core assembly heats up from both load and losses, it gives up its heat to the surrounding oil. The heated oil will tend to rise to the top causing the cooler oil to drop toward the bottom of the tank, thus setting up a natural circulating current of oil within the tank or enclosure (Fig. 2-11 on page 30). The oil in contact with the tank, in turn, gives up its heat to the tank walls and to the surrounding air. In this manner, the temperature of the coil and core assembly is held to safe limits.

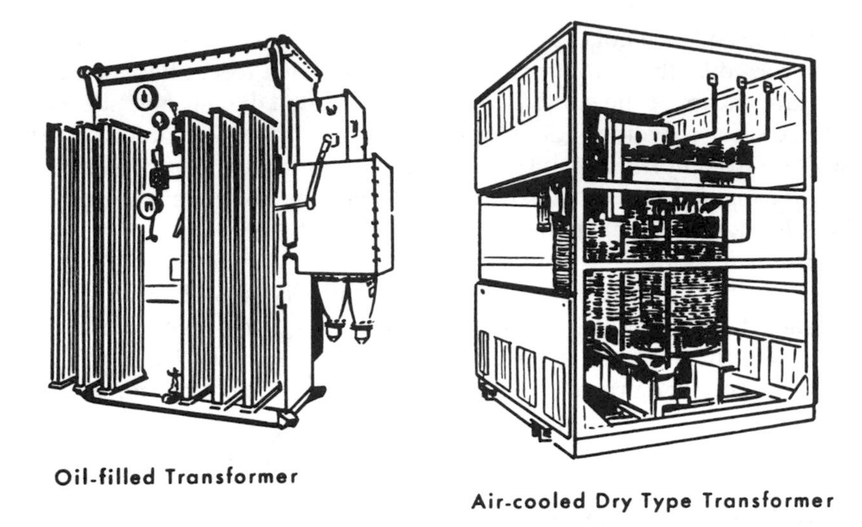

Figure 2-10 Oil- and air-cooled transformers.

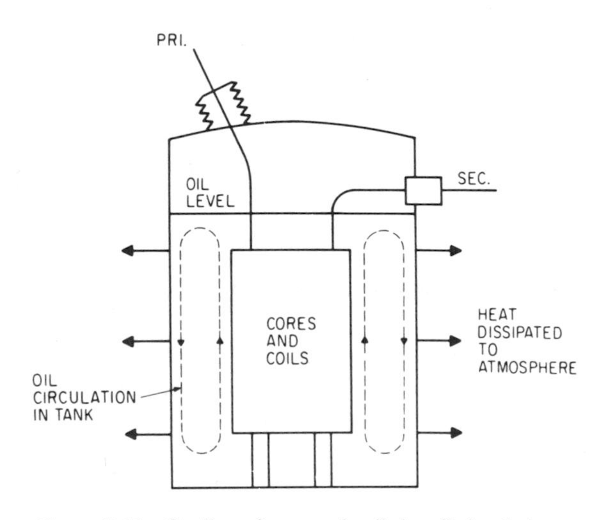

Figure 2-11 Cooling of core and coils by oil circulation.

MOISTURE AND SLUDGE

Moisture and oxygen affect the quality of the insulating oil. Moisture reduces the dielectric strength and oxygen helps form sludge.

Sludging is principally due to decomposition of oil resulting from exposure to the oxygen of the air while hot. Moisture condensation is caused by changes in the volume of air and oil in transformer tanks, due to changes in temperature of the transformer itself or that of the surrounding air. This

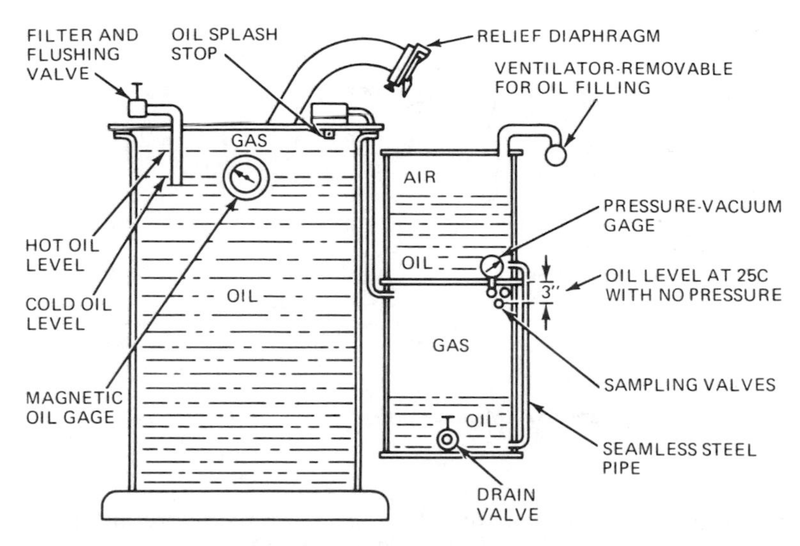

Figure 2-12 Method of gas-oil seal application to power transformers.

produces a constant breathing or interchange of air in the top of the tank and the surrounding air. This breathing may result in the lowering of the temperature of the enclosed air to a dew point, resulting in condensation of water vapor within the tank. The gradual accumulation of quantities of moisture will decrease the insulating quality (dielectric strength) of the oil. Large drops of water may collect, and, as water is heavier than oil, the globules of water will fall through the oil to the windings, which may result directly in a breakdown.

Isolation of the oil from the air by using an inert gas (Fig. 2-12), such as nitrogen, above the surface of the oil in a sealed transformer tank often eliminates this source of possible trouble.

Some outdoor transformers are provided with oil conservators. These are auxiliary tanks used for oil expansion to eliminate the air space above the transformer tank and to isolate the hot oil from the surrounding air. This is done by completely filling the transformer tank with suitable piping. The conservator is open to the surrounding air through a breathing device and is provided with a sump from which water from condensation may be drawn off (see Fig. 2-13 on page 32).

The use of oil conservators eliminates breathing and moisture condensation in the transformer tank containing the core and coils, thus preserving the original quality of the oil there, protecting it from sludging.

Large substation transformers are equipped with a diaphragm covered vent, usually at the top of the transformer. These transformers are usually sealed against the weather so that the air (or inert gas) and the oil vapor existing above the oil level may generate excessive pressures within the tank. The purpose of this diaphragm, known as a "relief" diaphragm, is to rupture when pressures exceed a predetermined amount, thus relieving the pressure in

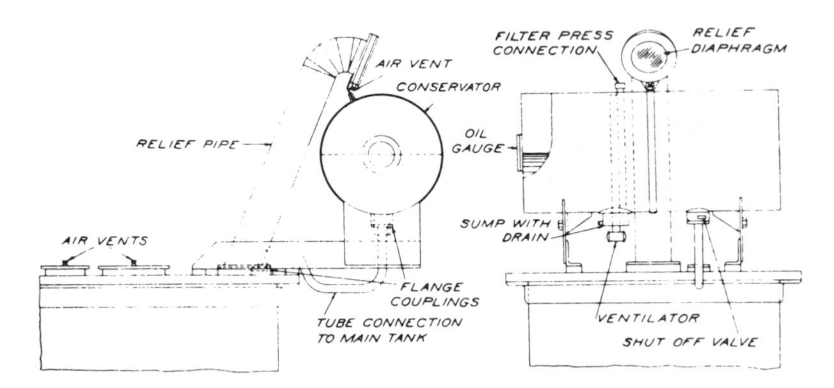

Figure 2-13 Main features of an oil conservator.

the transformer tank. In some instances, pressure relays are installed to give indication of a rise in pressure within the tank. In as much as deteriorating insulation often gives off heat and gas, such indication may also be a sign of approaching insulation failure.

INCREASED COOLING

In order to improve the cooling effect, the surface of the enclosing tank exposed to the surrounding air is increased by corrugating the flat sides or by adding fins to them. Sometimes, radiators or tubes are attached to the tanks so that the hot oil on top may circulate through them, entering from the top and returning to the tank at the bottom; in effect, these radiators add still more cooling surface and improve the circulatory effects of the oil (Fig. 2-14).

The effectiveness of the cooling system may be increased by forcing a current of air to blow away the heated air adjacent to the heated surfaces and replacing it with cool air in rapid motion. The air blast may be directed against the transformer enclosure and radiating surfaces, or may be forced through air ducts contained within the transformers (Fig. 2-15).

The effectiveness of cooling systems may also be increased by "forced" circulation of the oil by means of a pump (Fig. 2-16 on page 34). The oil may be circulated from the bottom of the tank to the top through the pipes or radiators connected to the tank. It may also be circulated from the tank to a remote and separate cooling device and back to the transformer.

Additional cooling of the transformer oil, in very large units, is sometimes obtained by water cooling pipes situated within the tank of the transformer.

Depending on circumstances and requirements, combinations of the forced air, forced oil, and water cooling processes may be used.

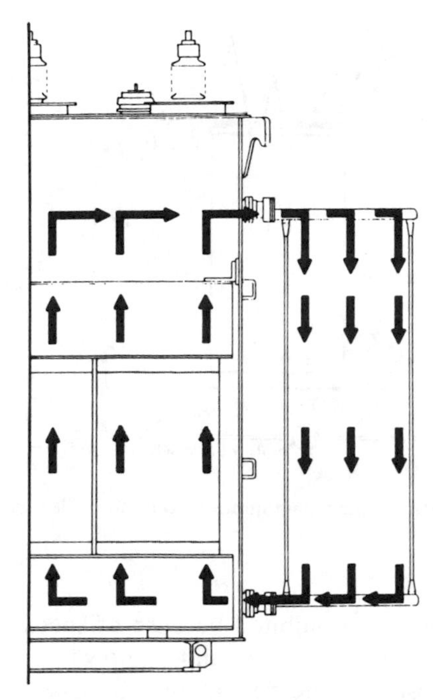

Figure 2-14 Oil-immersed, self-cooled (arrows show flow of hot oil).

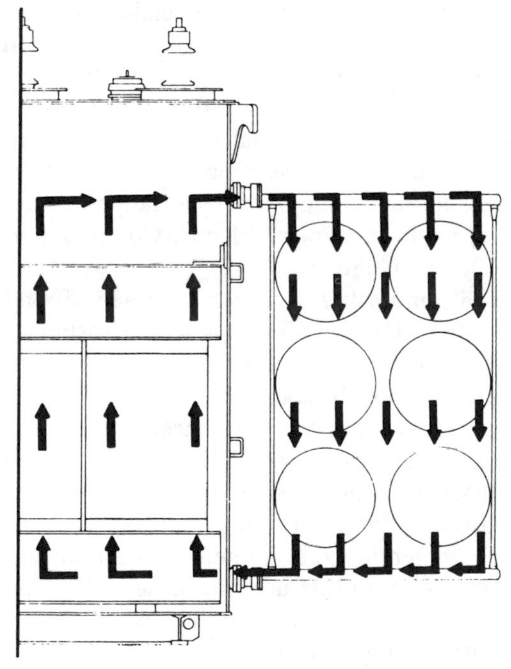

Figure 2-15 Oil-immersed, self-cooled forced-air cooled.

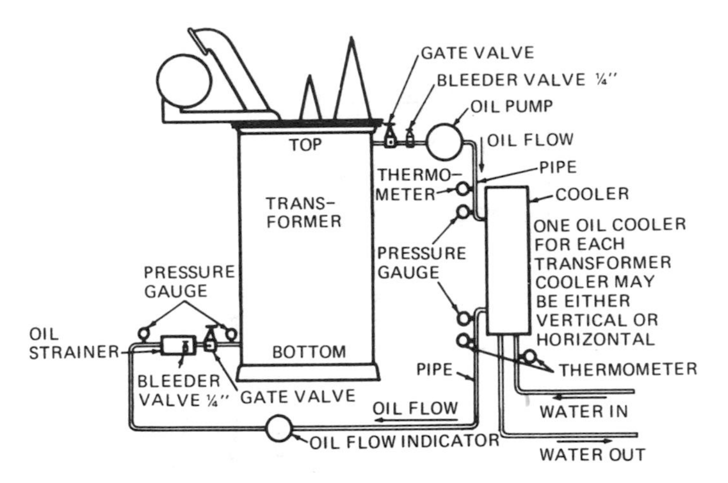

Figure 2-16 Piping diagram for forced, oil-cooled transformer.

ASKARELS

Where oil as a coolant is prohibited because of special fire hazard requirements, special nonflammable compounds, known as "askarels," are used. Even when decomposed by electric arcing, the gaseous mixture evolved is nonexplosive. These are considerably heavier than oil, almost double in weight, and considerably more expensive. Special treatment (usually shellacking) is required for coils, cores, and connections immersed in askarels, as they will attack many of the insulating varnishes generally used. The rate of heat transfer in askarels is also lower than oil, so that their cooling effectiveness is lower than oil. In addition to the usual precautions taken for handling oil, special care should be taken with askarels as they may cause irritation and other harmful effects to the human body, especially the eyes. Trade names for askarels are Pyranol made by General Electric Company, and Inerteen, made by Westinghouse Electric Company.

Some of the oils and askarels in use have been found to contain polychlorinated biphenyl (PCB), a cancer producing substance. Steps have been taken to eliminate the hazard by replacing the oils and askarels with cancer-free oils and other coolants. In some instances this is accomplished by replacing existing transformers; in other instances, the contaminating fluid may be drained at the site and, after several flushings with special contaminate-absorbing fluids that draw the PCB from the transformer core and parts, replaced with PCB-free oil or other fluids.

The rigid fire prevention requirement is sometimes met by the use of air-cooled transformers. No oil is used in these units, and the enclosure is usually designed to be drip proof, to keep rain away from the "innards" of the unit, though moisture may penetrate through vents and perforations in the enclosure. These units are usually restricted to relatively small sizes.

BUSHINGS AND TERMINALS

Leads from the windings must be brought outside the transformer enclosure or tank. Leads operating at low voltages (usually less than 5000 volts), may consist of insulated flexible wires brought out through insulating collars, made of porcelain or plastic, with a terminal lug soldered or mechanically fastened to the wire (Fig. 2-17). For higher voltages (usually up to about 25,000 volts), stiff leads or bars are used, projecting through porcelain cylinders of varying thickness; the assembly is known as a "bushing." Internally the coil is connected to the lead or bar of the bushing by suitably insulated wire or bar conductor; externally, the lead or bar contains some terminal arrangement which may be a simple thread and nut on the end of a bar, or a more complicated device involving springs and interlocks. Externally, the distance over the porcelain surface may be increased by installing "petticoats" that increase the "creepage" distance from the line terminal and the tank; this reduces the possibility of flashover over the surface due to rain and dirt.

For transformers operating at greater voltages (usually up to about 75,000 volts) (Fig. 2-18 on page 36), the conductor may be mounted inside porcelain cylindrical insulators filled with oil, and the oil space may be divided by thin insulating cylinders, with an oil gauge at the top to indicate the level of oil in the bushing. The conductor is generally a metallic rod which also serves to strengthen the construction of the bushing.

For even higher voltages (generally over 75,000 volts), the rod type conductor may be insulated with layers of oil-impregnated special paper, with metal foil inserted at several locations among the layers. The layers of insulation at the center are subjected to greater electrostatic stresses than the outer layers. The metal foils inserted between layers form a series of condensers which tend to even out and equalize the stresses among the layers. Such construction permits operation at higher voltages with less material, at lower cost, and with better performance. Refer to Appendix A for polymer insulation.

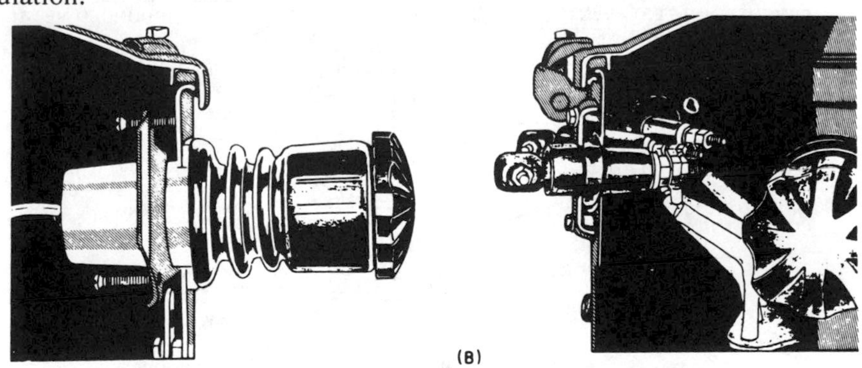

Figure 2-17 Sidewall mounted bushings. (A) Primary bushing. (B) Secondary bushing.

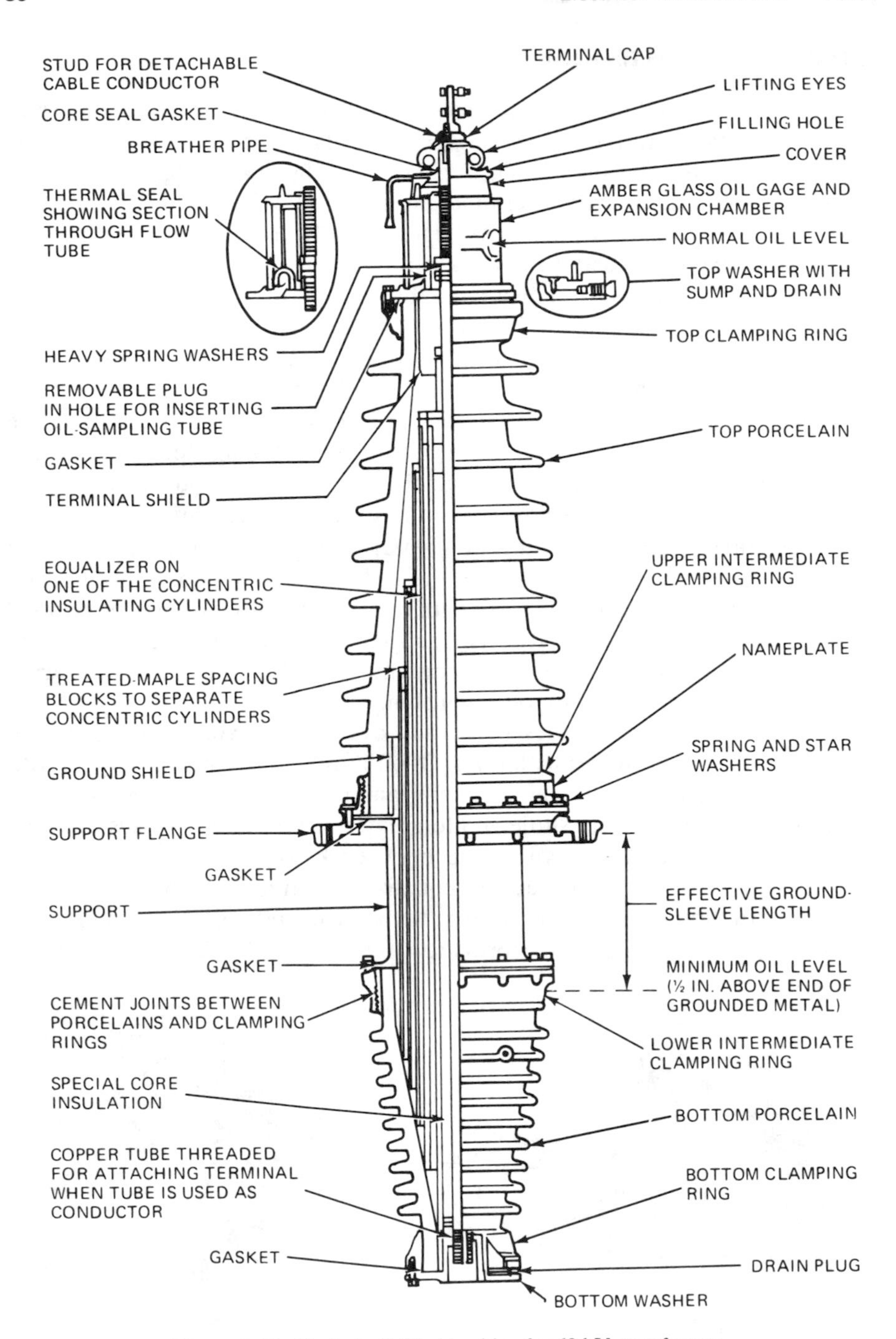

Figure 2-18 Typical oil-filled bushing for 69 kV transformer.

Connecting leads within the transformer, between coils or between coil and terminals require careful consideration because these conductors may be subjected to great stresses from the proximity of high-voltage coils and from heavy short-circuit currents which they may be called upon to carry. Such conductors often have additional insulation and bracing to minimize the possibility of breakdown.

TAPS AND TAP CHANGING

Often it may be desirable to change, by a relatively small amount, the ratio of a transformer. This may be done to compensate for the voltage drop in the supply source or to accommodate a specific characteristic of the connected load. The simplest method brings taps from one or both windings to a terminal board below the oil level to provide an arrangement of studs and links by which the connections may be changed (Fig. 2-19). Changes in these connections are made only when the transformer is deenergized. Sometimes, the tap changing equipment is designed so that the change of connections, under the oil, may be accomplished by the turn of a handle protruding above the oil level; it is still necessary that the transformer be deenergized. The connections from the coils to the terminal branch, the board itself, and the studs and links all must conform to the same requirements as those bushing leads and terminals.

Taps may be changed under loads, that is, while the transformer is energized and operating. Such applications are used where the maintenance of voltage is of paramount importance, and such units are used as voltage regu-

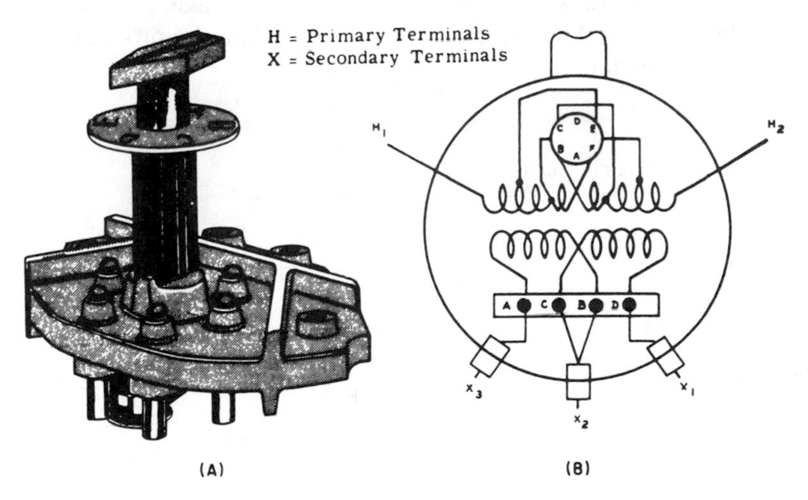

Figure 2-19 Tap changer used to adjust the turns ratio of a transformer. (A) No-load tap changer. (B) Typical internal wiring of transformer with tap changer.

lators to maintain desired outgoing supply voltages. The arrangements employ auto-transformers (these will be discussed in the section relating to the auto-transformers in Chap. 6).

INSULATION COORDINATION

The insulations associated with the several parts of the transformers must not only withstand the normal operating voltages imposed on them, but must also be able to withstand higher voltage impulses or surges that result from lightning or switch operations and that may find their way into the transformer. The insulation from coils to ground (steel cores, tanks, etc.), from coil to coil, and from turn to turn, must be strong enough to withstand the voltages produced by the impulses at these points.

To provide adequate insulation without the use of excessive and expensive insulating material, the insulation requirements of the several parts are coordinated. In general, this means that the internal parts are insulated as nearly equal as practical, but generally stronger than that of the bushings, thus insuring that any breakdowns that may occur would be outside the tank, where damage would be comparatively light. Further, the bushing is usually protected by an air gap or lightning arrester (which will be discussed later), whose insulating value under the surge conditions is even lower than that of the bushing, so that the failure will be by flashover across the bushing and the tank, rather than by puncture of the bushing insulation (Fig. 2-20). Finally, the insulation of the weakest point in the transformer is stronger, by great enough margin, than the protective device. Such a coordinated arrangement of insulation whereby the transformer is stronger against voltage impulses or surges than the bushings, and the bushings are stronger than the protective devices, tends to restrict damage.

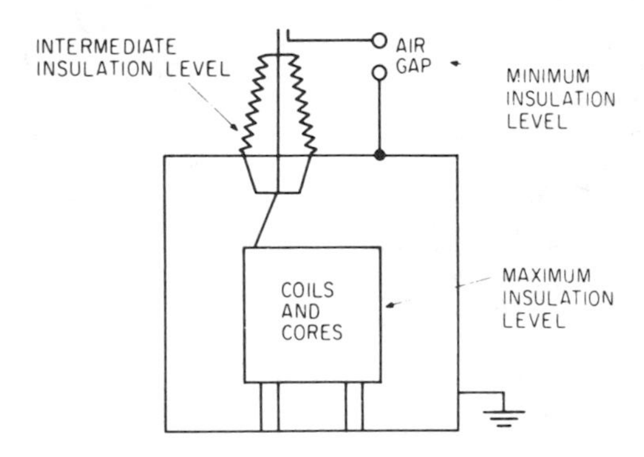

Figure 2-20 Insulation coordination.

TRANSFORMERS WITH MORE THAN TWO WINDINGS

Under special circumstances, transformers have been constructed with three or four windings, Sometimes, the coil arrangement for the normal two-winding transformer provides for two separate coils to constitute the secondary winding. The construction details described above apply equally to these multi-winding types.

SINGLE- AND THREE-PHASE TRANSFORMERS

In general, the same construction details described above for single-phase transformers also apply to three-phase transformers. These latter usually consist of three single-phase units within the same enclosure or tank. However, when advantage is taken of the combination of cores, the three separate sets of windings are wound on a three-legged core. Three-phase transformers similar to the one illustrated in Fig. 2-21 are often used in the larger sizes for substation and industrial use, principally because of the savings in space and weight, and hence cost.

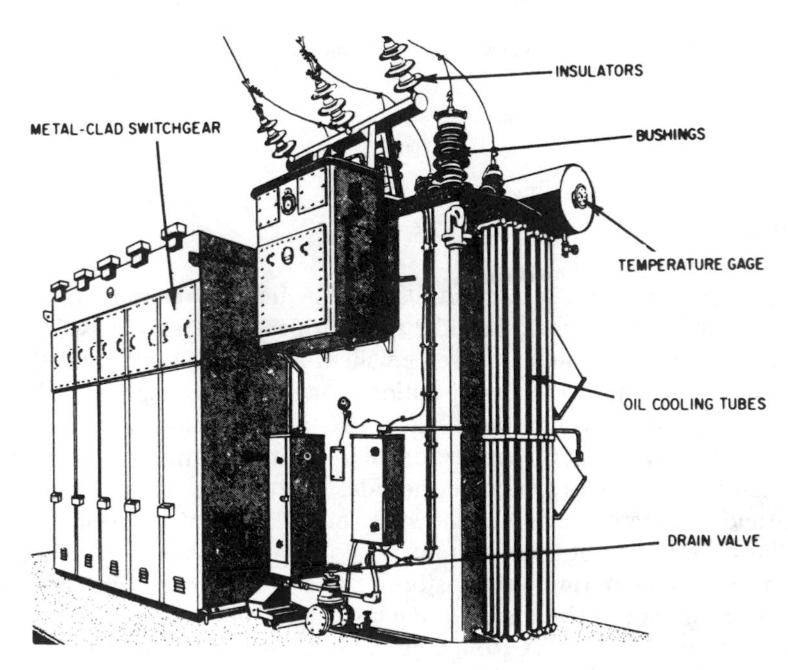

Figure 2-21 Substation transformer.

PORTABLE TRANSFORMERS

Transformers, especially larger power units, have been mounted on wheels, as shown in Fig. 2-22, for temporary operation in emergencies or for maintenance. These are equipped with protective devices and are sometimes referred to as portable or movable substations.

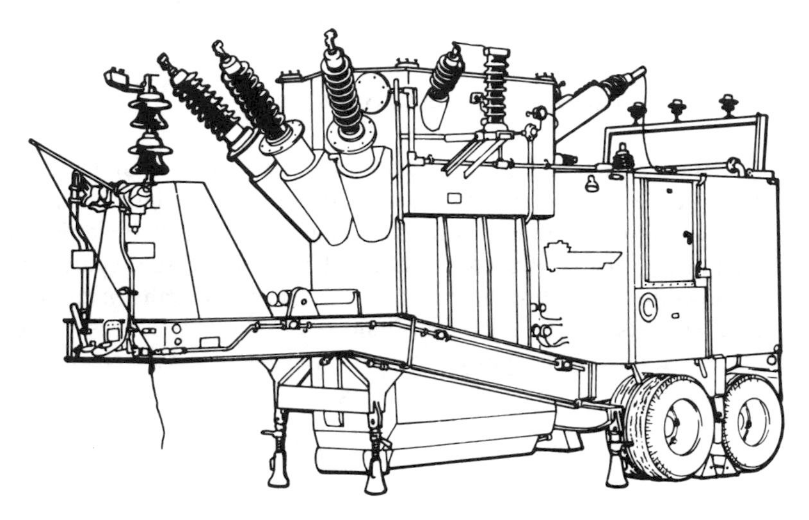

Figure 2-22 Portable transformer.

TANKS

Transformer tanks vary in size and shape depending on the manufacturer and the age of the transformer. The older transformers have heavy cast-iron cases, but the later ones are made of welded sheet steel. The tank holds the core, windings, bushings, and leads in position, and must be oil tight to hold the oil that insulates and cools the transformer. The tanks have fittings in the back, so that crossarms or hangers may be attached for erecting the transformers on a pole. Lifting lugs are provided on the sides for handling the unit. The nameplate and connection diagram showing the location of the leads from the windings is attached outside the tank. A cover and gasket to prevent foreign substances from entering the transformer and to keep rain from causing rust and corrosion, and deterioration of the insulating materials, is securely held in place by bolts or clamps. A grounding lug is usually installed near the bottom of each tank so that it can be electrically grounded for safety purposes when installed.

REVIEW

1. The basic components of a transformer are the primary coil, the secondary coil, and the iron or steel core (Fig. 2-1). Other major components include the tank, oil, and bushings. Transformer coils are designed to get the required number of turns into a minimum space. At the same time, the cross section of the conductor must be large enough to carry the current without overheating, and sufficient space must be provided for the insulation and for cooling paths, if any. These coils may be made of copper or aluminum, the choice depending on the cost to achieve the low resistance and small space requirements.
2. Special care is taken to hold losses down in the core. Heat created by hysteresis, friction from the reversing magnetic molecular particles (Fig. 2-2), is minimized by using a special grade of heat-treated, grain-oriented silicon steel. Heat created by eddy currents, created when the alternating magnetic field cuts through the metal core (Fig. 2-3), is minimized by laminating the core, and insulating the adjacent laminations with insulating varnish. Cores may be the "core" type, where the windings are outside the core (Figs. 2-4 and 2-5), or of the "shell" type, where the core surrounds the coils (Figs. 2-4 and 2-6).
3. The magnetic fields in both coils repel each other; during short circuits, these forces may become large enough to wreck the transformer. Hence, the coils must be suitably wound (Figs. 2-7 and 2-9), mounted on the core, and firmed to prevent movement, bending or distortion; this may be done by suitable bracing and blocking. (This incidentally also reduces noise from the vibration of the several parts.) Extra insulation and barriers are often used as a precaution to keep a failure on the high-voltage winding from imposing a high voltage on the low-voltage coil.
4. The coils and core are usually enclosed in a steel tank to protect them from the elements, vandalism and, for safety purposes, from possible contact by people. Oil is usually placed in the tank to cover the coils and core to provide a means of cooling (Fig. 2-11) and filling small air voids which may eventually cause failure. Moisture reduces the dielectric strength of oil, while oxygen helps form sludge. Moisture is condensed in the tank from changes in volumes of air and oil due to changes of temperature of the transformer itself or that of the surrounding air. Sludging is principally due to decomposition of oil resulting from exposure to the oxygen of the air while hot. Accessories to keep air from affecting the oil use gas seals (Fig. 2-12) or conservators (Fig. 2-13).
5. Increased cooling, and hence increased transformer capacity, may be obtained by the use of radiators which increase the area of cooling surfaces, by fans to increase the movement of air over the cooling surfaces, and by pumps to increase the circulation of oil (Figs. 2-14, 2-15,

and 2-16). Combinations of these methods are often used; in special cases, circulating water around radiators or in pipes within the transformer is sometimes employed.

6. Bushings take the terminals of the coils through the tank, insulating them from the tank. Usually, these consist of a conductor through an insulating collar, usually porcelain (Fig. 2-17). The distance from the line terminal to the tank may be increased by "petticoats," reducing the possibility of flashover from rain or dirt. For higher voltages, the porcelain cylinders may also be filled with oil (Fig. 2-18) or contain layers of insulation with metal foil inserted between them to equalize electric stresses among the layers.
7. The primary coil sometimes has taps which permit changes in the turn ratio or ratio of transformation (Fig. 2-19). Taps on the winding are brought to a terminal board and must conform to the same specifications as the bushing leads. Changes in taps are usually made with the transformer deenergized; they may also be changed while the unit is energized, as in voltage regulators.
8. For economy and to restrict damage when a voltage surge, such as lightning, is applied to the transformer, the insulation of the internal parts is usually stronger than that of the bushings, which in turn is stronger than that provided by the protective device, usually a lightning arrester. Flashover will then be apt to occur outside the transformer tank, restricting damage to more easily replaceable parts (Fig. 2-20).
9. Transformers may be constructed to include more than two windings, may incorporate several units in a poly-phase transformer for large or substation application (Fig. 2-21), and may be made mobile (Fig. 2-22) for special application.
10. So-called polymer insulators are replacing porcelain; lighter in weight, better rain sheding properties, do not fragmentize, less attractive as targets to vandals.

STUDY QUESTIONS

1. What are the three essential parts of a transformer? What are some other important parts?
2. Why are transformer cores laminated?
3. What are the principal types of core construction and what are their differences? What determines what and where they are used?
4. What factors affect the design of coils?
5. Why and how are transformers cooled?
6. What affects the quality of insulating oil in a transformer, and how may this be remedied?
7. How may the effectiveness of cooling be improved?
8. What is a bushing? What types are there and where used?
9. What are taps, and how may they be changed?
10. What is meant by insulation coordination?

3

INSTALLATION

Power-supply transformers may be installed almost anywhere; on poles, on the ground, under the ground, in the home, on the farm, in industrial plants, shops, offices, in aircraft, trains, ships, and vehicles. The usual precautions used in installing any piece of equipment are also valid in the installation of transformers; these may include: adequate mechanical and structural support; space for access and maintenance; sufficient clearances; barriers and signs for safety to the public; and, means of holding down noise levels. For transformers, one additional precaution is necessary, and that is proper means for dissipating the heat generated by these transformers.

DISTRIBUTION TRANSFORMERS: OVERHEAD

Distribution transformers are very often located outdoors where they are hung from crossarms, mounted on poles directly, or placed on platforms. In general, transformers up to 100 kVA size are mounted directly to the pole or on a crossarm. Larger size transformers (or groups of several transformers) are placed on platforms or mounted on poles in banks or clusters (Fig. 3-1 on page 44).

How a transformer is mounted is a matter of considerable importance; the transformer or transformers must stay put and continue functioning even in the midst of violent winds, pouring rain, freezing cold, sleet and snow. Besides weather, there is the danger of the pole itself being hit by a carelessly-driven vehicle.

Modern pole-mounted transformers as shown in Fig. 3-2 (on page 45)

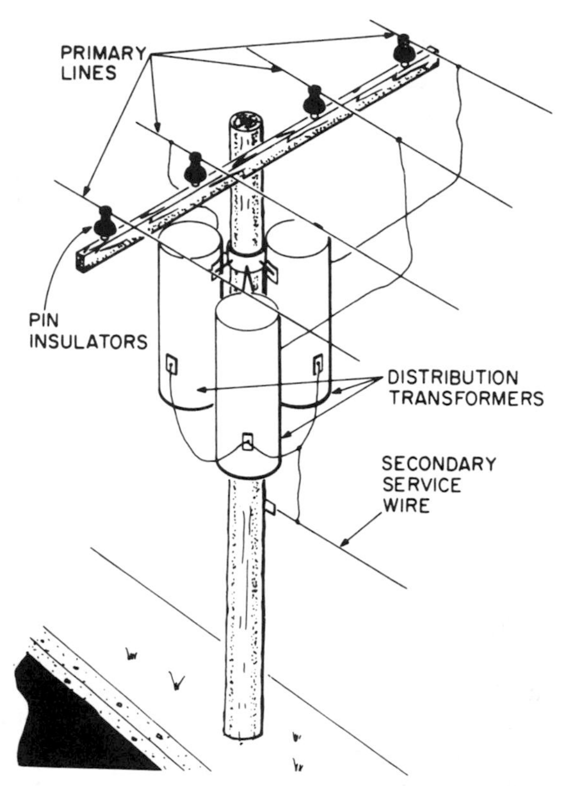

Figure 3-1 Modern cluster mounting of distribution transformers on a pole.

have two lugs welded directly on the case; these lugs engage two bolts on the pole from which the whole apparatus hangs securely. This method, which is known as "direct mounting" eliminates the need for cross arms and hanger irons (as was done in the past), thus saving a considerable amount of material and labor.

In the past, smaller transformers were hung from crossarms by means of hanger irons, which were two flat pieces of steel with their top ends bent into hooks with squared sides (Fig. 3-3A). The transformer was bolted to these pieces of steel, the assembly was raised slightly above the crossarm and then lowered so that the hooks on the hanger irons would engage the crossarm. Many such installations still exist. Installing larger transformers is illustrated in Fig. 3-3B.

Transformers may be raised to their position by means of block and tackle as shown in Fig. 3-4 (on page 46), with one set of blocks attached to a stub extension attached to the top of the pole and the other to a sling attached to the transformer; a guide line attached to the transformer keeps it under control during the lifting process. Smaller units are lifted by hand, whereas larger units make use of the truck power winch. In some instances, smaller

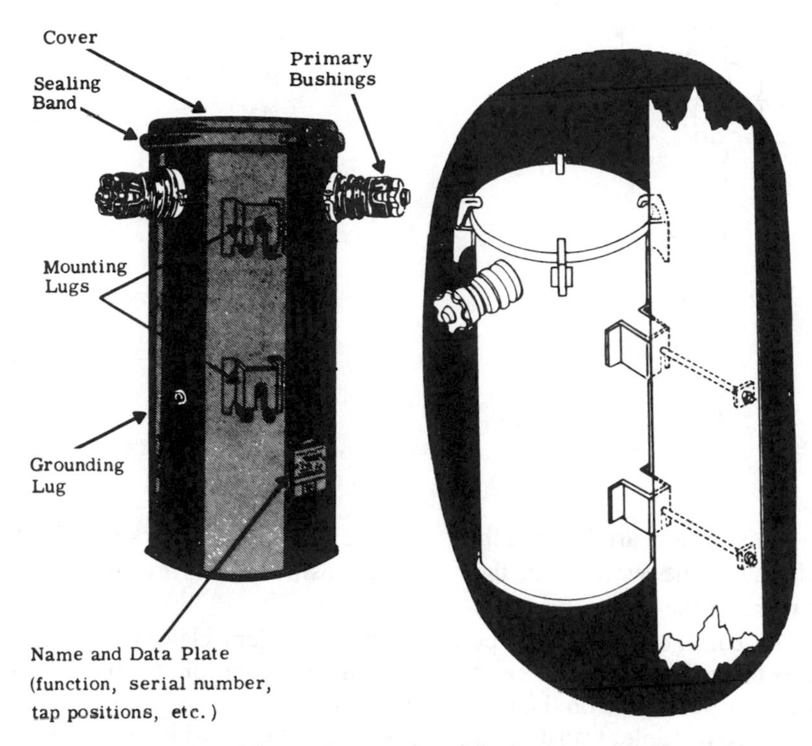

Figure 3-2 Direct pole-mounting of distribution transformers.

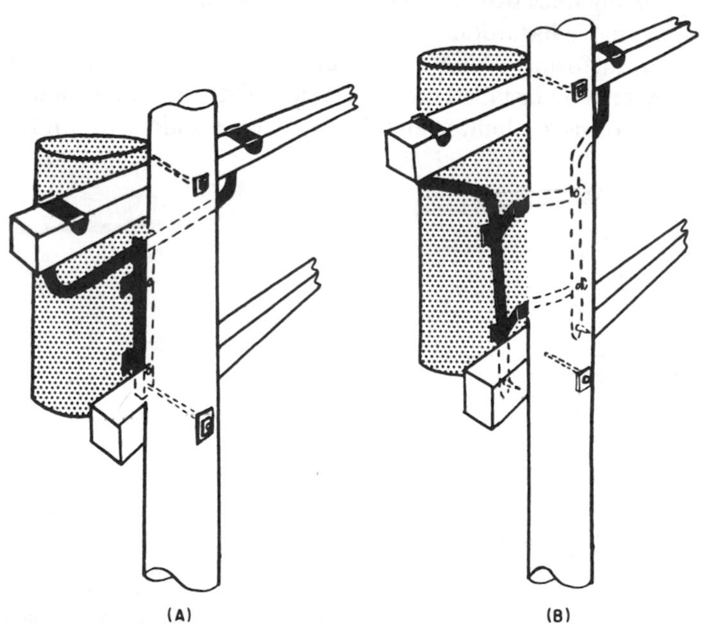

Figure 3-3 Hanger-iron method of mounting distribution transformers on poles.

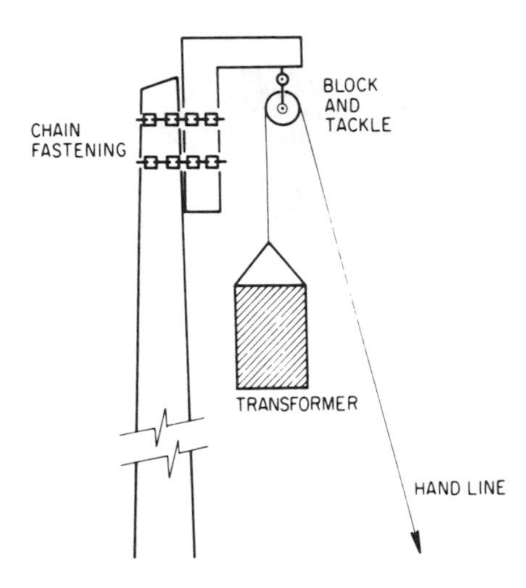

Figure 3-4 Pole top stub for raising transformer.

sized transformers are lifted at the end of the boom of the bucket type trucks which raises the workers to the working position. Platform-lifts are also used to raise the transformer.

The configuration of the pole provides sufficient clearances to permit workers to work safely whether operating from a bucket, platform truck, from climbers, or pole steps on the pole. (Fig. 3-5.)

Where possible, transformers should not be mounted on a junction pole (a pole supporting lines from three or more directions), as this makes working on such a pole more hazardous for the worker.

Where transformers cannot be mounted on poles because of the size or number, they may be installed on an elevated platform or a ground-level pad (Fig. 3-6). Platforms are built in any shape or size required to suit the particu-

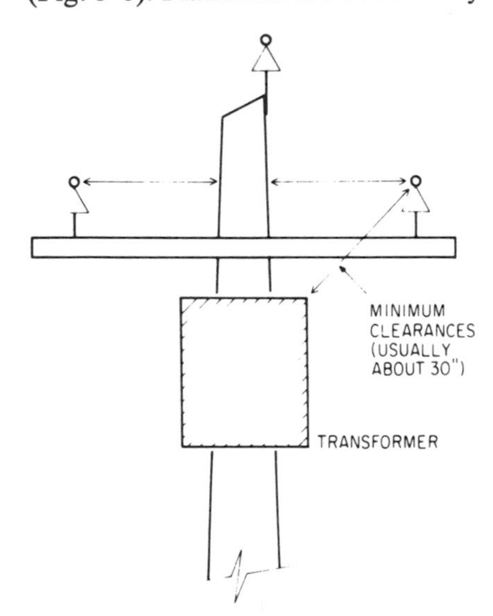

Figure 3-5 Pole configuration showing minimum clearances for workers when installing transformers.

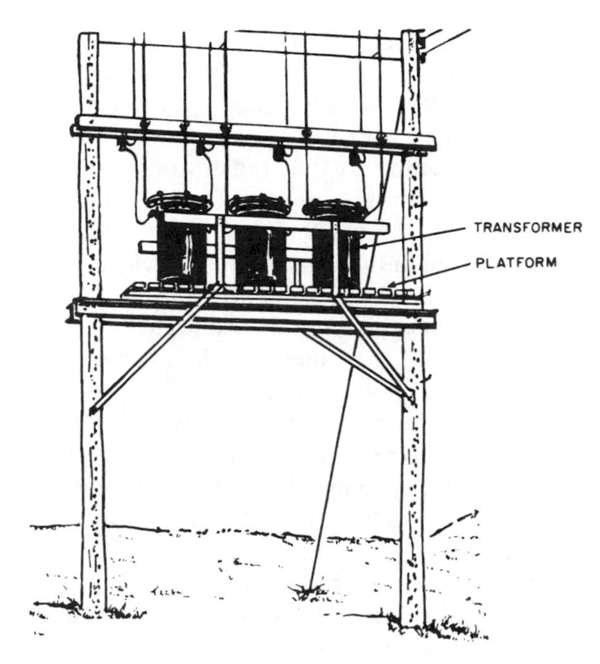

Figure 3-6 Platform mounting of transformers.

lar need. They are usually constructed of wood, though steel is often used for some of the members.

DISTRIBUTION TRANSFORMERS: PAD-MOUNTED

Ground-level pads are usually made of concrete or reinforced concrete and have provisions for enclosing the transformers within a fence, wall, or other enclosure for safety reasons. Ground-level pads are very useful when appearance is a major consideration.

DISTRIBUTION TRANSFORMERS: UNDERGROUND

In urban areas, transformers are usually placed in manholes or vaults under the street or sidewalk, and those serving large loads in vaults located within the building; in these instances, subway-type transformers are usually installed. These may be either of the single- or three-phase types.

In the subway-type transformer, the core and coils are essentially the same as for the overhead or line type, but the construction of the tank insures its watertightness when submerged for extended periods of time. The tanks have thicker walls, bottom and cover than the line type, and are designed to withstand the pressure of watertightness tests. Between the cover and the tank, there is a gasket to make the seal watertight; in larger sizes, it may be necessary to space closely a relatively large number of bolts around the periphery of the cover and cover plates.

One method of connecting to the transformer is by bringing a lead-covered cable into the transformer, thus making a wiped joint between a bushing of tinned brass and the cable sheath where it passes through the bushing. A barrier is inserted in the connector or sleeve connecting the internal and external cables to prevent the transformer oil being siphoned into the cable. In another type, the wiped joint to the cable sheath is made at the outer end of a wiping bushing, and the inner end is attached to the tank by a pipe-union type or bolted-flange type joint with a gasket provided to insure watertightness. In some instances, to simplify the process, the connections are made in a chamber attached to the transformer case into which the terminals of the windings are brought by means of water and oil-tight bushings, which sometimes may be of the porcelain type. In special instances, where such transformers supply low-voltage networks, these chambers may also contain high-voltage grounding or disconnecting switches.

In the supply to large loads by means of low-voltage networks (Fig. 3-7), the transformers may also be equipped with a network protector, a relay-activated secondary breaker. This breaker or protector may be installed in a separate tank or may be in a compartment attached to the transformer tank.

For economy reasons, line-type transformers are occasionally used in manholes under sidewalks or streets where there is little possibility of their being submerged.

The ventilation of transformer manholes and vaults is of special importance. Not only should the temperature of the transformer be kept within safe limits even under extreme operating conditions, but the design must consider the possibility of the formation of explosive mixtures of oil vapor or gasses and air in the event of a transformer failure. This is achieved by insuring sufficient vent area outlet to the volume of the manhole or vault; the smaller the ratio of the volume of air in the manhole or vault to the total net ventilating area (openings), the less will be the pressure developed in the manhole or vault, and the lighter the construction necessary to withstand the force of the possible explosion.

In some instances, the transformers have been installed in vaults located on the upper floors in tall buildings or to supply specialized equipment. The transformers may be of the subway type or the line type depending on the application and overall economy.

In suburban residential or rural areas, transformers may be buried directly in the ground, or installed in commercially manufactured enclosures of steel or concrete which may be partially or completely located below ground level; in some instances these installations are made in sections of concrete pipe or other makeshift enclosures. The transformers installed may be of the

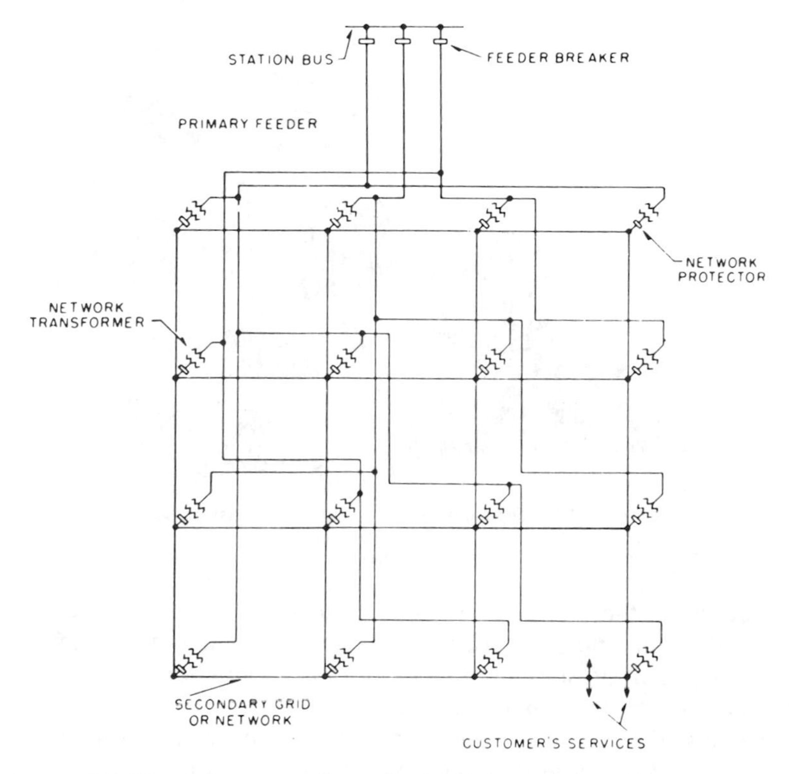

Figure 3-7 Network protector for application on low voltage ac network systems.

subway type described above, but more often are conventional line-type transformers modified with insulated bayonet-type connections to make them somewhat waterproof and to enable rapid disconnection in the event of fault.

For subway-type transformers, usually no lightning arresters are installed. Oil fuse cut-outs are used for smaller transformers, while oil-filled switches or circuit breakers are used for larger sizes; the oil switches generally may not operate under load. In larger units, the switch or disconnecting device is often attached to the transformer case, and may be so arranged to act also as a grounding switch.

SUBSTATION POWER TRANSFORMERS

Substation transformers may be installed either indoors or outdoors. When installed indoors, connections to the units may be made by cables in ducts, by cables carried in trays, or to open busses usually located above the units. Reinforced concrete barriers are installed between units to prevent explosion and fire in one unit spreading to other units. Provision is usually made to take care of oil leakage by means of drains (to safe holding areas) installed in the

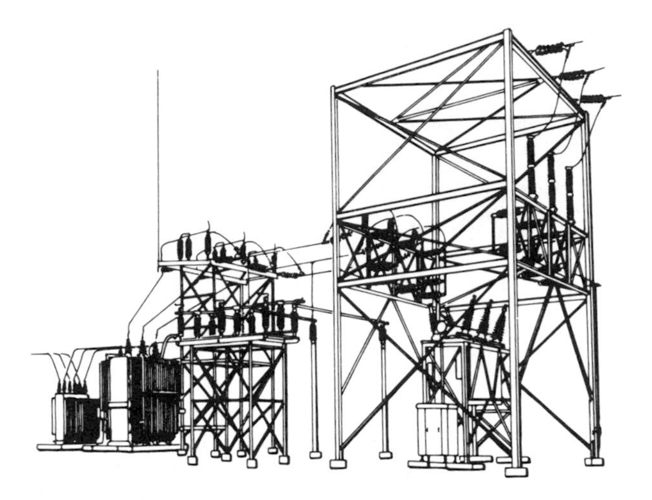

Figure 3-8 Typical distribution substation installation.

floor. Because of the barriers, ventilation is difficult and usually accomplished with the aid of fans and often also with air ducts. Such indoor installations may be necessary when substations are located in residential or other areas where appearance and noise are paramount considerations.

Perhaps the greater number of substation installations are made outdoors (Fig. 3-8). Here the units are mounted on concrete pads, with the larger units sometimes equipped with wheels and the unit mounted on rails for ease in handling. Barriers to prevent the spread of fire are not very common in this type installation, but the area containing the transformer is encompassed by earthen dams or concrete curbs forming a basin to contain all the oil which may flow from the unit in trouble. Acoustic barriers (Fig. 3-9), to keep the

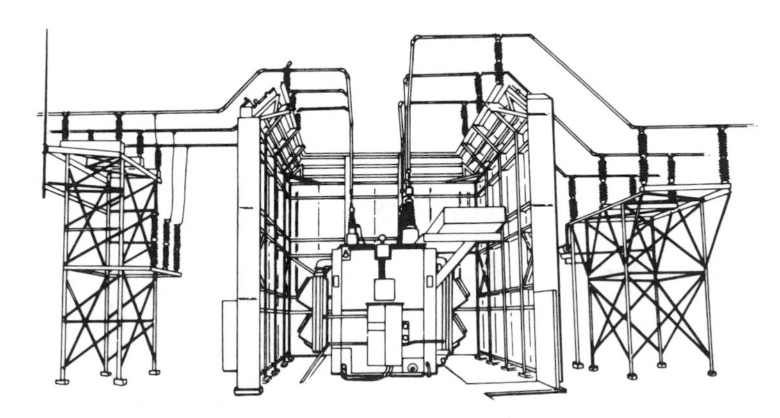

Figure 3-9 Substation power transformer with acoustical barrier.

sound levels to satisfactory levels in certain communities, are sometimes installed.

Connections most often are made by means of open bus work usually located above the units. The impressive but expensive and hazardous steel racks to support these connections and associated switches have given way to sleek streamlined busses mounted on unobtrusive supports of steel or concrete. Even though installed outdoors, these units often have provisions for forced air and forced oil cooling. The transformers may also be painted with light colors (often silver) to allow the unit to reflect heat from the sun's rays and to improve appearances generally.

PROTECTION

Practically all transformer installations include some form of electrical protection for the transformer. This generally includes protection of the unit from lightning and switching surges, deenergization to prevent damage under fault or overload conditions, and means for providing safety when worked upon. The solid grounding of system neutrals and transformer cases, sometimes using two or more separate leads for this purpose, is an absolute necessity.

Overhead line transformers of the conventional type are protected by a lightning arrester and a primary fuse cutout, mounted separately on the pole or crossarm.

Arresters designed to protect substation installations generally operate on the same principle as those protecting line installations. They are specifically designed to handle larger currents and higher voltages. See Chaps. 11 and 12.

SELF-PROTECTED DISTRIBUTION TRANSFORMERS

In the self-protected transformer (Fig. 3-10 on page 52), a weak link or primary protective fuse link is mounted inside the tank with the transformer unit as also are two circuit breakers for protection on the secondary side of the transformer. A simple thermal device causes the breakers to open when a predetermined safe value of current is exceeded. The lightning arrester is mounted on the outside of the tank. It is apparent that the self-protected transformer makes for simpler, more economic mounting and neater appearance. What's more, it is of particular advantage for higher voltage (13.8 kV) primary distribution systems where connections and disconnections are made with the unit energized.

Similar protection is provided to pad-mounted transformers installed at ground level. For larger size units, station-type lightning arresters and circuit breakers may be substituted for the pole-type arrester and fused cutout. Large

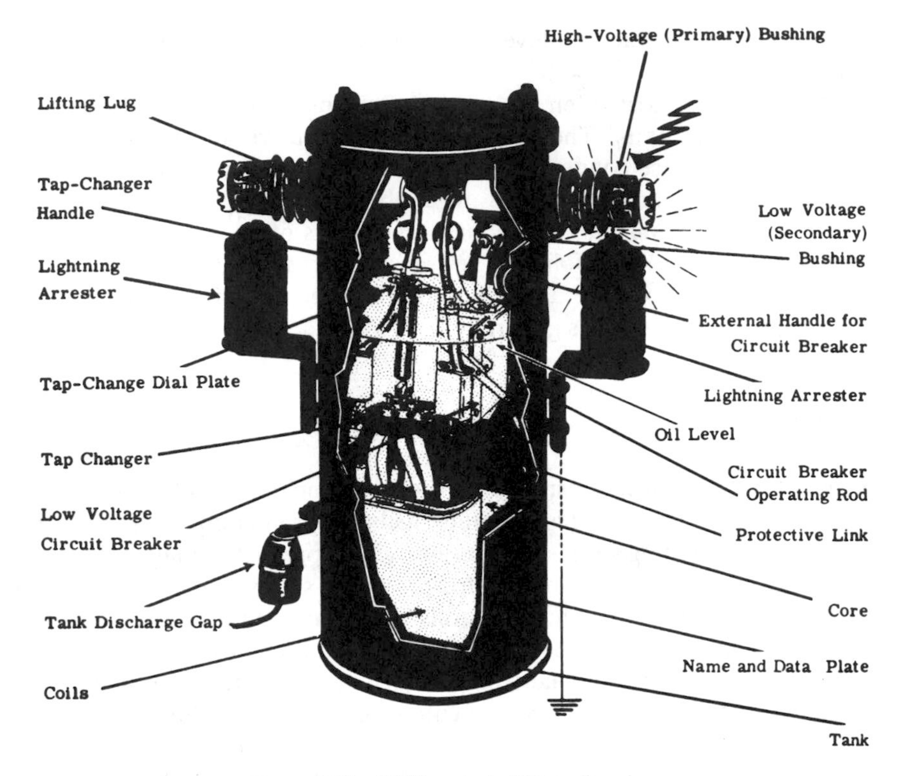

Figure 3-10 "Self-protected" transformer.

size transformers, for industrial or similar installations are essentially substations and are treated as such.

For transformers installed underground, lightning arresters are usually not necessary. However, if the primary supply is from a nearby overhead line which is subject to lightning (Fig. 3-11), the arresters are installed at the transformer, or at the pole, or at both locations. Submersible type oil filled fused cut-outs are used. Leads connecting the several pieces of apparatus are waterproof lead or plastic sheathed wires. Where self-protected transformers are used, no external connections or protective devices are used.

In all installations, special care should be taken with the installation of neutral or ground wires. These must not only be continous with special attention paid to splices and connectors, but they should also be of ample size to allow for abnormal fault currents, or for the flow of a sum of the neutral or ground currents from several units.

In this regard, it should be noted that waterpipe grounds should be supplemented by driven grounds as the use of nonconducting pipe in water supply systems no longer provides a good trustworthy ground source.

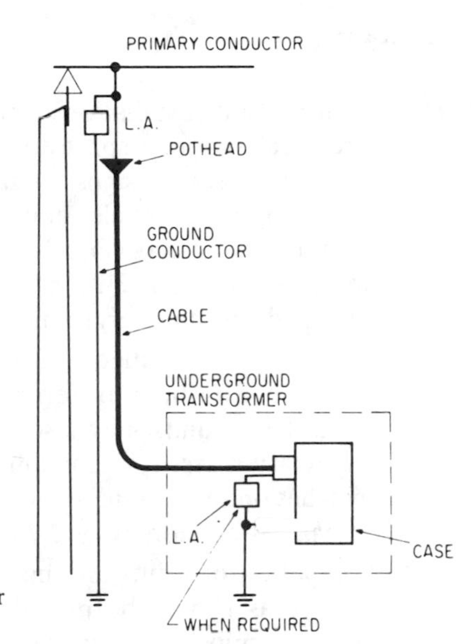

Figure 3-11 Lightning arrester on riser connected to underground transformer.

CUSTOMER INSTALLATIONS

Transformers for specific customer installations have a wide range: from small bell ringing and sign lighting where the high voltage may be 120 or 240 volts and amounts limited to a few amperes to x-ray equipment where high voltages range into the several thousands and welding and electric furnaces where values of current are measured in thousands of amperes. Their installations are covered in the National Electric Code and in local codes and regulations. In general, however, the safety precautions are similar to those for power installations described above. Where fire hazards are of paramount importance, dry air-cooled transformers are installed for modest requirements, air blast or askarel cooled units for large loads.

In all cases. special attention is given to the proper ventilation of these units and the safe grounding of transformers and appliance enclosures and neutrals.

AUTOTRANSFORMERS

Autotransformers are often used in place of two-winding transformers. These will be described in Chap. 6. Their installation, however, follows closely the same requirements as two-winding transformers.

REVIEW

1. Transformer installations should provide adequate mechanical and structural support; space for access and maintenance; sufficient clearances, barriers, or signs for safety to the public and workers; means of restricting noise levels; and means for dissipating heat. They may be mounted in clusters (Fig. 3-1), directly on poles (Fig. 3-2), or on hangers and crossarms (Fig. 3-3); also on elevated platforms (Fig. 3-6), or ground level pads or in underground manholes or vaults.
2. Transformers installed underground are watertight and equipped with waterproof bushings or terminal chambers filled with oil and are watertight. These units may also contain switches on either the primary or secondary sides, for grounding or disconnecting purposes; where such switches on the secondary are operated automatically, they are known as network protectors (Fig. 3-7).
3. Ventilation of vaults and manholes is of special importance. Sufficient vent areas should be provided, not only for heat dissipation, but to prevent explosive mixtures of oil vapor or gasses and air to form in event of transformer failure. The greater the vent-volume ratio, the less pressure will develop in the manhole or vault and the lighter the construction necessary to withstand the force of possible explosion.
4. Substation transformers may be installed indoors or outdoors (Fig. 3-8). Precautions are taken to keep failure of one unit affecting adjacent equipment, usually by installing barriers. Sumps around the transformers are often provided to collect its oil should leaks develop or the unit experience trouble. Facilities are also supplied to hold down noise (Fig. 3-9), and to provide for fans, pumps and other accessories.
5. Electrical protection of transformers include lightning arresters and switches or cut-outs to disconnect them from the line either automatically or manually. These may be installed with the unit on poles, crossarm, pad, manhole or vault, or mounted on the tank or case. Solid grounding of system neutrals and transformer cases, sometimes using two or more separate leads, is an absolute necessity.
6. Completely self-protected (CSP) transformer (Fig. 3-10) provide surge protection and disconnecting devices within the transformer itself, simplifying installation as only the connection from the unit to the line is necessary.

STUDY QUESTIONS

1. What factors should be considered in the installation of overhead distribution transformers? What are some of the methods of doing so?

2. Why should mounting of distribution transformers on junction poles be avoided where possible?
3. What factors should be considered in the installation of underground distribution transformers? What are some methods of doing so?
4. What is meant by vent-volume ratio for transformer manholes or vaults?
5. What factors should be considered in the installation of substation or plant transformers?
6. What protective devices are usually installed with overhead, underground, and substation or plant transformers?
7. What is the function of a lightning arrester?
8. What is meant by a completely self-protected distribution transformer?
9. What special precaution should be taken with neutral or ground wires?
10. Who covers installation codes for specific consumer's transformers?

4

OPERATION AND MAINTENANCE

Because there are no moving parts, permitting simple and durable rugged construction, the transformer requires little attention in the way of operation and maintenance. Prudence, however, requires that certain steps be taken to insure failure-free operation, including the operation of its auxiliary accessory equipment.

The most important consideration in transformer operation is the prevention of overheating. Overheating may result from the loads served by the transformer, from poor ventilation, from failure of cooling systems, and from incipient faults (progressive failure of the insulation) within the transformer.

HEAT INDICATORS

Devices are incorporated in some transformers to give automatic indication of excessive heating. In the small, self-protected distribution transformer, heat sensitive device turns on a small red light located on the transformer. In larger units, a "hot-spot" indicator is installed in the insulation at a point where maximum temperatures may be expected. The indication of this condition may be located at the transformer itself, or at some remote monitoring location. Smaller units may be monitored by special crayon marks on the case which change color with heat; others will have receptacles into which are placed small vials of waxes which melt at different temperatures. They will be observed during routine inspections.

AMMETERS

Other means of monitoring the loads carried by a transformer include measurement of either primary or secondary currents by ammeters and by the measurement of the collective energy of the loads connected to it.

Ammeters may be of the instant-reading type, the recording type, or the maximum-indicating type. The first type indicates the current carried at the particular times of reading. If the time chosen corresponds to well-known periods of peak load, the reading may suffice to give some indication of the load condition of the transformer. The recording ammeter provides a graphic record of the load carried by the transformer over a period of time, from which peak conditions as to magnitude and duration may be observed. The maximum-indicating ammeter is installed over a period of time and essentially is an indicating ammeter with a second free-floating needle which, being pushed by the indicating needle, is left in the maximum position to which it is pushed; it is a more economical device than a recording ammeter. Records of ammeter readings serve to indicate progressive changes in transformer loadings and call attention to possible approaching critical conditions.

TRANSFORMER LOAD MONITORING

Computers have made possible the reading of the energy consumed by the several loads (consumers) connected to the transformer. The individual meter readings used for billing purposes are added together in the computer to give an approximate total amount of energy supplied by the transformer to which they are connected. From test patterns developed separately, a factor is determined which, applied to the total energy consumption, enables the approximate calculation of the load current carried by the transformers. This operation is known as "transformer load monitoring."

AMBIENTS

As the safe operation of a transformer depends on how hot its insulation is permitted to become, the readings of load amperes provide only an approximate method of determining the adequacy of the transformer to supply its load safely. Both the ambient temperatures and the time and duration of the peak loads supplied have an effect on the internal temperature of the transformer. Evening periods are usually cooler than day time; winter is cooler than the other seasons. Summertime ampere readings may tend to give deceptive indications of the real transformer condition since the ambient temperature may be high; air-conditioning loads may impose peak loads of very long duration, sometimes approaching twenty-four hours.

For small transformers, usually pole-mounted, the load supervision is only a very approximate operation.

LARGER INSTALLATIONS

For larger, industrial or station-type units, such supervision may consist of more complex, permanently installed meter installations and associated relay and alarm circuitry (Fig. 4-1). As mentioned previously, internal pressure relays are sometimes installed on the larger, more critically important and expensive transformers.

When such supervisory measures fail to anticipate transformer troubles, protective devices such as fuses and circuit breakers operate to disconnect the transformer, usually before extensive damage occurs.

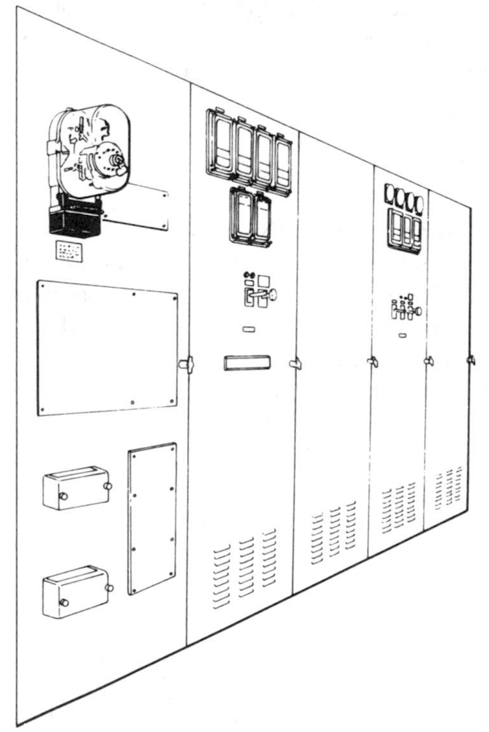

Figure 4-1 Switchboard for larger installation.

EXTERNAL INSPECTION

Much can be learned of the condition of the transformer from an external inspection. The condition of the bushings, if broken, cracked or chipped, may not only be caused by vandalism, but may be a sign of flashover caused by overload or by voltage surges (lightning or switching), and may be an indication of potential insulation failure inside the transformer. Discoloration of the bushings, and of the associated lightning arrester, may also provide evidence of such flashovers.

Discoloration of terminals and leads may be evidence of heating, which may be evidence of loose connections. Connections to the ground should receive special attention for, as a safety measure, they prevent possible energization of the case.

Excessive rust or corrosion of the tank may also indicate excessive transformer heating, although it may be caused be corrosive elements in the ambient. Signs of oil overflow on the tank may indicate excessive heating or the occurrence of a severe internal stress during a temporary fault which may have cleared itself; of course, it may also be due to having initially overfilled the transformer tank with oil, or accidental spill during installation.

Where they exist, in large, station-type transformers, pressure relief diaphragms should be inspected and replaced if any signs of imminent rupture appear.

Unusual noises or unusual heat during periods of normal loads may indicate that the unit may have been subjected to severe operating conditions, such as overloads or frequent short circuits.

The level of oil in the transformer may be observed externally in some instances where such oil gauges are installed, usually on the larger, station-type transformers. In addition, the color and the clarity of the oil may also indicate the existence of moisture or carbon or sludge; moisture may indicate leaks from improperly tightened bushings, cover or ports in the transformer; carbon or sludge may indicate the existence of excessive internal temperatures.

INTERNAL INSPECTION

Where oil gauges do not exist, the oil level may be visually inspected by opening the handhole (if any) or by removing the tank cover.

The condition of the gaskets should be observed for signs of wear. If worn, they should be replaced with new ones so that the cover may be securely fastened when replaced. If they are even slightly loose, rain or snow may be driven into the transformer ultimately causing transformer failure; loose covers also may cause noise problems.

The condition of the oil in the transformer should also be part of the routine inspection. A very small quantity of moisture will spoil the insulating qualities of the oil; a fraction of a teaspoonful of water in the tank will often necessitate changing the oil. A sample may be removed by a simple glass or metal suction tube (rubber should never be used since it contains sulphur which will dissolve in the oil and attack the copper windings and terminals) of a suitable length to take a sample from the bottom of the tank where sludge and water usually settle.

Tap changers (if they exist) should be checked for proper position and for the condition of the contacts. Discoloration of the contact may indicate excessive heat. A tap changer set between or on two tap contact points may prevent the transformer from operating or may cause excessive heating in portions of the winding between the two tap contacts.

Core, coils, terminal boards, and connections should be inspected for signs of heat discoloration. These may be caused by overloading, from loose connections, or from loose core laminations due to improper bracing.

In large transformers, water cooling coils should be carefully inspected and tested. Lagging or insulation of the water pipes above the oil level should be intact or condensation may form, affecting the insulating quality of the oil. The rate of water flow should also be checked; if found to have diminished, the cause should be looked for and remedied. Scale formed inside the cooling pipes may be removed by circulating a solution of hydrochloric acid through the pipes. Scale formed outside the cooling pipes may be removed by wire brushings.

Radiators should be inspected for dents and holes which may cause restriction of the flow of coolant or leakage, eventually causing the transformer to overheat.

MAINTENANCE

Ordinary mechanical maintenance should be performed in the normal manner. Rusted and corroded areas should be cleaned before repainting. Connections, similarly, should be cleaned before remaking or retightening.

OIL

Usually, the first thing checked in determining the condition of a transformer is the oil. In a transformer, oil not only acts as insulation but also provides a medium which carries heat away from the windings. Hence, it should not only have high dielectric strength, but should have low viscosity to provide good heat transfer. It should also be free of sulphur and other corrosive agents which might injure the insulation, conductors, and other metallic parts, it should not permit moisture particles to remain in suspension but allow them to settle quickly to the bottom, and it should be free from sludging under normal operating (heat) conditions.

The condition of the oil may be determined in the field by two simple tests. The first simply compares the color of a sample of the oil from the transformer with a panel of several standard colors which give some indication of the emulsification that may have taken place. The second is a voltage breakdown test. In both tests, care should be exercised in taking the samples. The test receptacles should be rinsed first with some of the oil to be tested. The oil sample should be taken from the bottom, after the oil has been allowed to rest for a period of time. Several samples should be taken.

When using the color test set shown in Fig. 4-2, the oil sample is placed in the clear tube which is then inserted in the holder so that the oil may be viewed through the left center port. The circular color wheel is then inserted in the holder so that the rotation of the wheel permits the various colors on the wheel to be compared with the oil sample color at the right center port. When the color on the wheel is similar to the oil sample color, the numerical designation of the oil sample color appears in the port at the lower right side of the holder.

In the electric test using the portable dielectric test set shown in Fig. 4-3, the oil is tested between two, one-inch diameter circular electrodes set in a standard test cup. The electrodes are set, usually with a spacing between them of 0.1 (one-tenth) inch. The oil sample is put in the cup, rocked, and left to stand a few minutes to allow any bubbles that may have formed during the filling to dissipate. A voltage is applied and gradually raised until breakdown occurs. The process is repeated up to about five breakdowns for each cup of oil, and tests made on about 3 cups of oil—fifteen readings in all. The breakdown values for good oil should never be less than 20,000 volts, and should average above that value.

Oil testing less than 20,000 volts should be removed from the tank and filtered through a paper press or a centrifuge before being put back in the tank. If there is evidence of sludge, the core, coils, and terminal board should be washed down with clean, hot oil under pressure, using the filter press pump and hose, before the tank is refilled with good clean filtered oil.

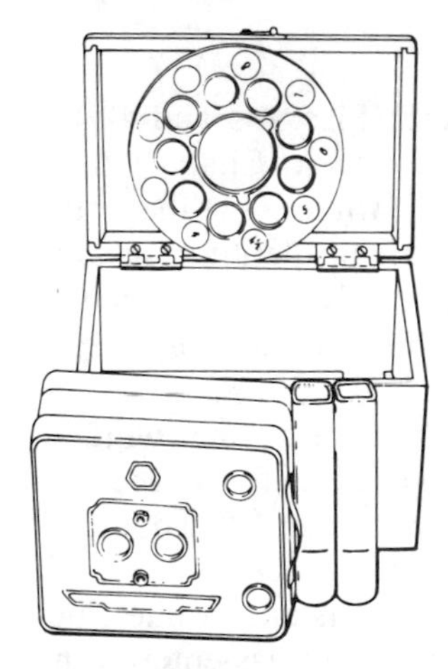

Figure 4-2 Oil color test set.

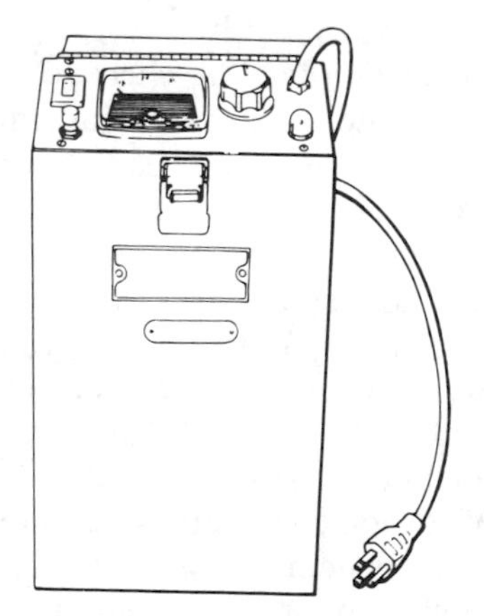

Figure 4-3 Typical portable dielectric test set.

ASKARELS

The same general procedures for testing and maintaining oils are also used with askarels. The main exception is that samples are taken from the top rather than the bottom of the transformer tank. Askarel may have its moisture removed and filtered by means of a filter press using activated clay (Fuller's earth) rather than filter paper, which is inadequate for decontaminating askarels.

Oil and askarels mix readily, and it is practically impossible to separate them once they are mixed. It is extremely important that they be kept apart, because oil changes the characteristics of askarels, making them flammable and explosive. The equipment used in decontaminating askarels should be used solely for askarels.

GASKETS

New gaskets properly cemented to cleaned surfaces should be used whenever any opening in the transformer case is made or the cover removed. Care should be used in preparing and placing the gaskets, making sure that the nuts and bolts are tightened uniformly. Gaskets for oil-filled transformers may be of cork or neoprene and other plastics. Those of cork should be compressed to one-half their thickness while those of neoprene should be compressed to two-thirds. Only cork should be used in askarel-filled transformers as askarels tend to dissolve plastic materials.

WINDING INSULATION

Winding insulation may be checked with a "Megger" (Figs. 4-4 and 4-5), a small dc generator used for measuring resistance. Refer to Chap. 7 for a description of a "Megger." Readings are taken between windings at the terminals (Fig. 4-6 on page 64), and the ground. Minimum safe insulation resistance values (in ohms or megohms) at different temperatures are usually specified for different voltage ratings of the transformers by the manufacturers. Each winding is tested separately to the ground.

If one winding under test indicates a very low resistance, that winding may be faulted to ground. The terminals should be checked to ascertain whether the fault is in the terminal board or in the winding itself. The board or one coil (if several constitute the winding) may need replacement.

If all the windings indicate lower than acceptable resistance values, but approximately proportionally equal, it may be that moisture has entered the insulation of the windings. The windings may require drying out.

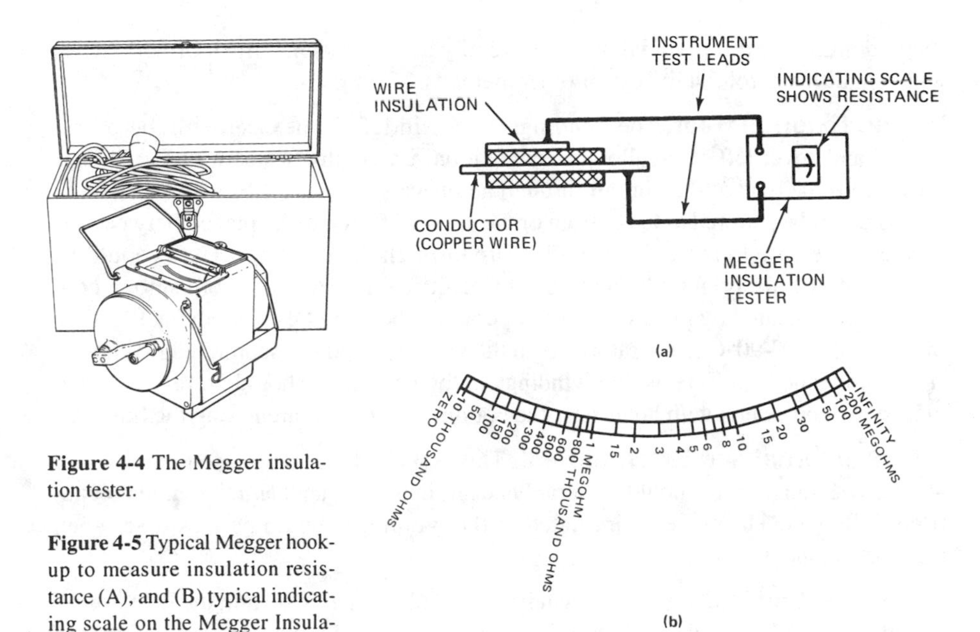

Figure 4-4 The Megger insulation tester.

Figure 4-5 Typical Megger hookup to measure insulation resistance (A), and (B) typical indicating scale on the Megger Insulation Tester. (Courtesy, James G. Biddle Co.)

Figure 4-6 Megger testing of transformer winding.

TRANSFORMER

CONNECTION TO WINDING UNDER TEST

MEGGER INSULATION TESTER

CONNECTION TO CASE OR OTHER GROUND

DRYING OUT TRANSFORMERS

These are four methods that may be used in drying out transformers:

1. Internal Heat. The windings are left in the case with the oil removed and the cover left off to allow circulation of air. The low voltage winding is short circuited. A rheostat to control current is connected in series with the high voltage winding and sufficient voltage impressed to circulate enough current through the coils to maintain a temperature between 80° to 90°C. The end terminal of the winding is used, not taps, so that the current will circulate through the total winding. This method is slow and superficial, and is generally used with small transformers, when local conditions prohibit the use of other methods.

2. External Heat. The windings are again left in the case with the oil removed and cover left off to allow air circulation. Externally heated air, (Fig. 4-7) is blown into the tank at the bottom through an oil valve for larger transformers or by means of a tube lowered to the bottom of the case. Baffles may be placed between the core and the case to force the heated air up through the windings. Care should be taken with the air, preferably heating it by electrical resistors to avoid smoke contamination if heated by open fire. Its temperature should not exceed about 125°C to maintain the 80°-90°C temperature about the windings and not scorch the insulation. If it is not expedient to leave the windings in the case, they should be removed and placed in another box with holes near the bottom and top for circulating the hot oil.

3. Internal and External Heat. This is a combination of methods 1 and 2 above. The transformer should preferably be left in its own tank and the same precautions followed. The current circulated in the windings should be less than when method 1 alone is used.

4. Heating in Oil. Here the windings are submerged in oil and the heat is generated as in method 1. The moisture is driven from the windings and insulation into the oil, and is removed from the oil by evaporation into the air or by circulating the oil through a filter, or both. The temperature of the windings should be maintained at 80°-90°C and the oil at as high a temperature as possible to keep a safe temperature in the windings. To bring the oil to a temperature sufficient to evaporate the moisture into the air, it may be necessary to reduce the cooling efforts either by insulating the case, by stopping or reducing the oil circulation, by lowering the oil level below the top radiator connections, or by reducing or cutting off the water in a water-cooled transformer. As much ventilation as possible should be provided, especially in the large transformers. Manhole covers may be raised and forced ventilation used. If ventilation is not sufficient, moisture will condense on the cover and drip into the oil, temperature of the oil should then be reduced and ventilation increased.

In all these methods, care should be taken to see that the windings and insula-

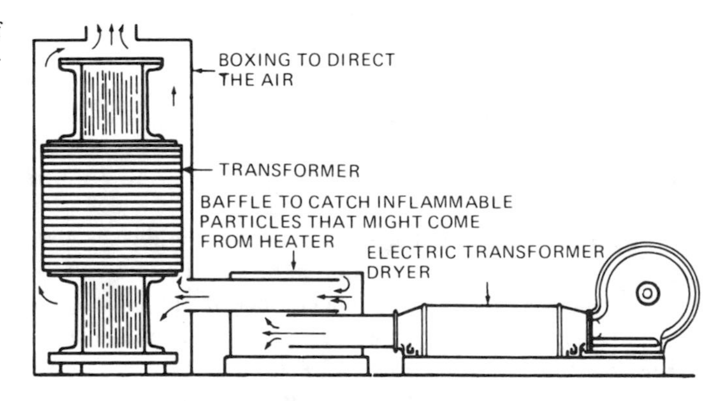

Figure 4-7 Method of using electric transformer dryer.

tion do not reach dangerous temperatures. The oil, when used for drying, may be so hot as to be flammable and easily ignited. The transformer may be considered dry when a series of Megger readings remain constant for a 24-hour period. The time periods required may be long, in the nature of 3 to 4 weeks. Method 4 may take longer than the others.

RADIATORS

Radiators may be removed for cleaning and repairs. The oil in the transformers may have to be removed before these can be unbolted from the tank.

In some designs, there are flanges and valves so arranged that with the closing of the valves, the radiators may be peened over and caulked; larger holes may be welded. Dents deep enough to affect oil circulation may have to be cut out and a patch welded in place.

ELECTRICAL GROUNDS

Electrical grounds associated with transformers are of the greatest importance. By providing low resistance paths to unbalanced or neutral currents, they help maintain output voltages reasonably balanced and within safe limits. Under fault or lightning conditions, they are essential to drain off surge voltages, holding voltages within the transformer to safe values. This is especially true of substation installations, where such grounds not only protect the transformers, but also help insure the positive operation of relays, further assuring the correct operation and safety of electrical systems. Such grounds should be checked periodically to ascertain not only that their circuit connections are unbroken, but also that their resistance be reasonably low. The Megger is often used to check the resistance of these grounds (Fig. 4-8).

CAUTIONS

Though some items of inspection may be accomplished without deenergizing the transformer, thorough inspection and maintenance, particularly of larger units, require that they be deenergized when work is done on them.

To provide maximum safety to the workers, both the high voltage and the low-voltage leads should be disconnected from the transformer. Preferably, there should be a visual break between the transformer terminals and the high and low-voltage lines. If this is not possible, grounds should be applied to both the high and low-voltage terminals of the transformer. It should be kept in mind that the transformer may be energized not only from the high voltage (usually the primary) side, but also the high-voltage terminals or bushings even though disconnected, will continue to receive full voltage

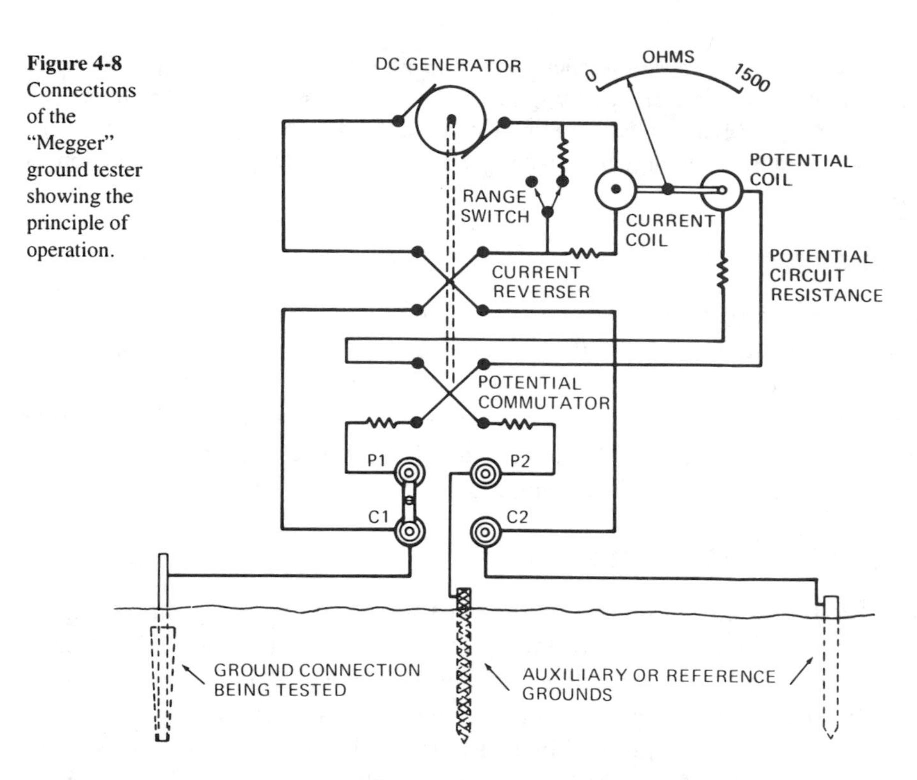

Figure 4-8 Connections of the "Megger" ground tester showing the principle of operation.

from the transformer from the low side, and thus constitute a serious risk.

Transformers should never be moved or lifted by the bushings or other attachments. Lifting lugs are provided for this purpose.

When working on transformers, particularly the larger ones, care should be taken that nothing falls into the open tank. Objects, such as tools, may not only cause failure if lodged in critical places, but may be the source of puzzling and unexplainable noises emanating from the transformer when returned to service.

CRYOGENIC TRANSFORMERS

Present-day transformer capacity ratings are based on holding the temperature of the hottest part of the insulation to about 100°C. The heat developed in the transformer stems from the I^2R losses, the greatest part in the windings, and from eddy current and hysteresis losses.

Developing technology that calls for the use of high temperature superconductors (HTS) in the windings and cooled by liquid nitrogen, has promises of meeting the needs for meeting the economic challenges of the future.

Major advantages of HTS transformers are:

1. Substantial reduction in size and weight. (Figure 4-9a)
2. Ability to operate at high overloads for long periods of time without insulation deterioration.
3. Elimination of fire hazard and potential contamination from failure, an environmental advantage (nitrogen is non-flammable).

The HTS conductors, capable of being made into windings, presently being employed consist of alloys of copper oxide and calcium with such exotic (and expensive) materials as barium, bismuth, titanium, and yttrium. Cooling is accomplished by 'cryocoolers' that operate at very low temperatures where common gases liquefy. (Figure 4-9b)

The HTS transformer windings and insulation operate in ultra cold ranges of temperature (–253°C to –196°C) where insulation will not degrade. At this temperature, the resistance of HTS windings decreases markedly approaching zero, hence I^2R losses in the windings also approach zero. Eddy current and hysteresis losses and other a-c losses still remain.

The savings in losses and the ability to overload are matched against the cost of the cryocooler and associated piping and controls. Overloading of present oil cooled transformers substantially hastens the degradation of the winding insulation, decreasing the life of the transformer. In the HTS unit, the overloading is controlled by increasing the refrigeration load (the use of liquid nitrogen), decreasing the overall efficiency of the unit during the period of overload. The nitrogen coolant may be dissipated harmlessly into the atmosphere, or may be recycled (at additional cost) to liquid form and reused.

HTS transformers that are safer, operationally superior, more environmentally friendly, and high in efficiency for longer lifetime operation, presently in the development stage, may very well meet the demand for replacement of the aging inventory of oil filled transformers. (Figure 4-9c)

REVIEW

1. The most important consideration in transformer operation is the prevention of overheating. Causes generally are overloads, poor ventilation, failure of cooling systems, and incipient faults; also failure of auxiliary accessory equipment. These may be uncovered through heat indicators, internal to give hot spot indication in the windings, or external through case temperature indicators. Ammeters, instant reading, maximum indicating, or recording, may give indications of transformer loading. Analysis of readings of watt-hour meters supplied from the transformer and the effect of ambient temperatures also give indications. Large substation-type installations may include more complex relay and alarm systems (Fig. 4-1).
2. Periodic inspections, with the unit energized, are often limited to external conditions. Broken or cracked bushings, discoloration of bushings, terminals and leads, rust or corrosion of tank, signs of oil overflow, broken or ruptured diaphragms on conservators, unusual noises, level as well as color and clarity of oil in tank oil gauges, all serve to indicate the need for

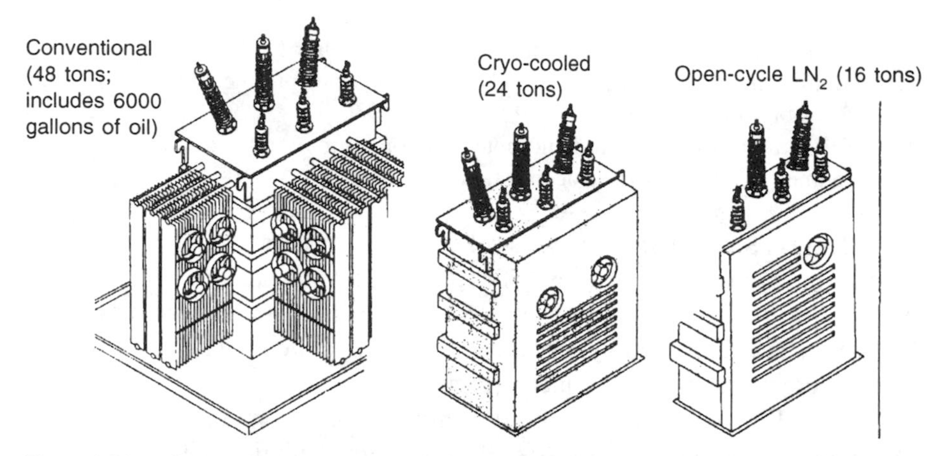

Figure 4-9A Superconductors. (1) Transformers using high-temperature superconducting (HTS) windings can operate in closed or open cycle. The cryocooled HTS transformer operates closed cycle and all refrigeration is self-contained (middle). The open-cycle unit (right) has an internal supply of liquid nitrogen refrigerant, which is automatically replenished, periodically, from remotely located liquefiers or storage vessels. Oil-filled transformers (left) in urban locations, which superconducting transformers may replace, are often surrounded by sprinkler systems (not shown) and oil containment structures. (*Courtesy* Wausheka Electric Systems)

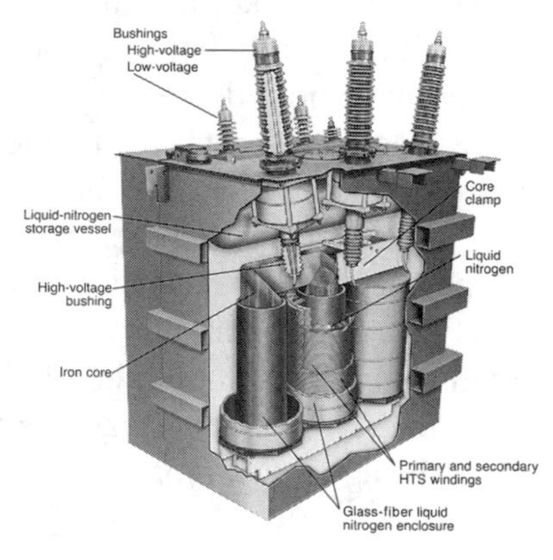

Figure 4-9B This artist's conception of a superconducting transformer resembles conceptually the one being developed by Intermagnetics General and Wausheka Electric. Note in particular the location of the liquid-nitrogen storage vessel, inside the top of the transformer box, and the enclosure containing liquid nitrogen around the primary and secondary superconducting windings. (Courtesy Wausheka Electric Systems)

Figure 4-9C The 630-kVA demonstration transformer based on high-temperature superconductor windings is presently under test by ABB, ASEA Brown Boveri Ltd., on the grid in Geneva, Switzerland. Designed to convert power from 18.7 kV to 420 V, its superconducting windings are made of powder-in-tube BSCCO-2223. (*Courtesy* Wausheka Electric Systems)

additional internal inspection. Connections to ground should receive special attention as a safety measure. Radiators should be inspected for dents, holes, rust, and discoloration.

3. Deenergizing makes possible closer inspection of the transformer both inside and outside the unit. Gaskets may be inspected closely, samples of oil taken, discoloration of terminals, core, coils, terminal boards, connections, tap-changer contacts may be observed at close range; loose connections, core laminations and bracing may be checked by actual touch. Scale and rust on internal tank, radiator, and piping surfaces may also be observed closely.
4. Ordinary mechanical maintenance includes: cleaning and repainting rusted, colored, or discolored areas; cleaning and tightening connections, bolts, lagging and internal supports; and replacing gaskets.
5. Electrical maintenance includes the checking of oil and coil and bushing insulation. Oil may be checked by comparing the color of a sample to standard colors (Fig. 4-2), which give an indication of emulsification. In a breakdown test, a sample is placed between two electrodes and a voltage applied until flashover or breakdown takes place, which indicates the insulating adequacy of the oil (Fig. 4-3). Oil may be filtered and replaced, core, coils, terminal boards should be first washed down with clean hot oil under pressure.
6. Coil and bushing insulation (as well as oil) can be checked by a "Megger," an instrument which measures resistance (Fig. 4-4); the indicated values are compared to standard acceptable values. The Megger is a portable dc generator which produces a voltage of about 500 volts; the small currents flowing through the insulation (coil, bushing, or oil) are calibrated to read resistance in ohms or megohms (Fig. 4-5).
7. When indications of moisture are encountered, the transformer may be dried out by heat applied externally through the circulation of hot clean air or oil, or both; or by internally generated heat obtained by short circuiting the windings and allowing current to flow through them; or a combination of both these methods (Fig. 4-7).
8. Electrical grounds provide low resistance paths to unbalanced, or neutral currents, which may be particularly heavy during fault conditions; this is especially true of large installations, such as substations, where such grounds help insure positive operation of protective relays. They should be checked for continuity and connections tightened. Their resistance may be measured with a Megger (Fig. 4-8).
9. Precautions on energized units include clearances, barriers, and signs. On deenergized units, leads should be disconnected with visible open breaks in the circuit, on both primary and secondary sides. If this cannot be done, visible grounds should be applied to both sides. Units should not be lifted or handled by bushings or attachments other than the lugs provided for that purpose. Care should be taken that nothing falls into the tank, or is left inside the tank, which may cause flashovers or unexplainable noises.

STUDY QUESTIONS

1. What are the causes of overheating of transformers, and how may they be discovered?
2. What is the limiting consideration in the safe operation of a transformer? What factors affect this limitation?
3. When should transformers be inspected and what external and internal inspections should be made?
4. Why should the condition of the oil in a transformer be checked?
5. How should the condition of the oil be checked in the field?
6. How may the condition of the oil or askarels in a transformer be restored? What special precautions should be taken?
7. What is a Megger and where is it used in the maintenance of transformers?
8. How may transformers be dried out and what precaution should be taken?
9. Why are electrical grounds associated with transformers of great importance and how may they be checked?
10. What precautions should be taken when working on deenergized transformers?

5

PRINCIPAL TRANSFORMER CONNECTIONS

A transformer may be connected in a number of different ways, depending upon the purpose of its installation. There are several ways of connecting a single transformer to the power supply, and also of connecting it to the load. Two or more transformers may be connected in different combinations to meet different requirements.

POLARITY

Unlike direct current, there are no fixed positive and negative poles in alternating current, and hence, transformers cannot have fixed positive and negative terminals.

The relative direction in which primary and secondary windings of a transformer are wound around the core determines the relative direction of the voltage across the windings. In a single-phase transformer, if the direction of the applied voltage at any instant is assumed, as from "a" to "b" in Fig. 5-1, the direction of the voltage across the secondary circuit will be either from "c" to "d" or "d" to "c," depending on the relative direction of the windings.

Since it is essential, if two or more transformers are to be connected together, to know the relative direction of the voltage of each transformer, certain conventions have been established for designating the "polarity" of a transformer.

The designation of polarity may be illustrated by Fig. 5-1. If one high-voltage lead is connected to the adjacent opposite low-voltage lead (say "a" to

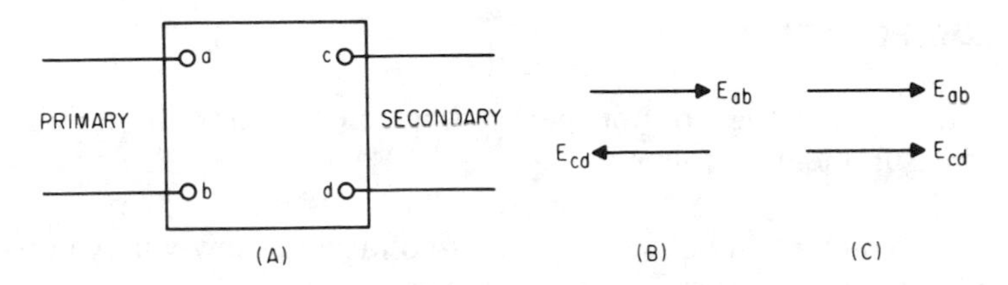

Figure 5-1 Polarity of a transformer. (A) Single-phase transformer. (B) Additive polarity. (C) Subtractive polarity.

"c") the voltage across the two remaining leads ("b" and "d") is either the sum or difference of the primary and secondary voltages, depending on the relative directions of the windings. If the voltage "b" to "d" is the sum, the transformer is said to have "additive" polarity; if it is the difference, the transformer is said to have "subtractive" polarity.

To indicate whether a single-phase transformer is of additive or subtractive polarity, the leads are marked as in Fig. 5-2.

If the windings of the high and low voltage coils are in opposite directions, the applied voltage and the induced voltage will have opposite directions and the transformer is said to have "subtractive" polarity. The H_1 terminal and the X_1 terminal will be on the *left,* when facing the low voltage bushings.

If the windings of the two coils are in the same direction, the applied voltage and the induced voltage will have the same direction and the transformer is said to have "additive" polarity. The X_1 will be found on the *right,* when facing the low voltage bushings.

To connect the secondaries of two (or more) transformers in parallel, similarly marked terminals are connected together irrespective of the polarity, provided similarly marked primary terminals have been connected together.

The polarity of distribution transformers rated below 200 kVA and less than 8600 volts, has been standardized (American Standards Association) as "additive"; above 200 kVA or above 8600 volts in any rating as "subtractive." The nameplate on the transformer usually identifies its polarity.

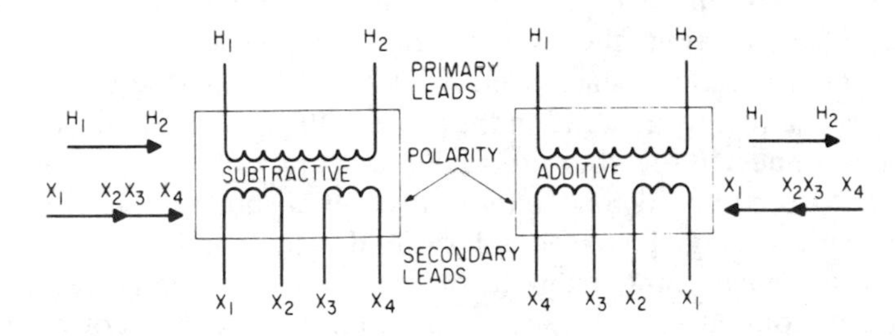

Figure 5-2 Polarity markings of a single-phase transformer.

TEST FOR POLARITY

If the polarity is unknown or not specified, it may be determined by a simple voltage measurement, as follows:

1. Place a connection between the high-voltage and low-voltage terminals on the right, when facing the low-voltage bushings.
2. Apply a low voltage, say 120 volts, to the two high-voltage terminals; measure this voltage with a voltmeter.
3. Measure the voltage from the left-hand, high-voltage terminal to the left-hand, low-voltage terminal.

If the latter voltage is lower than the voltage across the high-voltage terminals, the transformer has *subtractive* polarity. If it is higher, the transformer has *additive* polarity.

IMPORTANCE OF IMPEDANCE

In addition to the polarity consideration, transformers connected in parallel should have the same high-voltage rating, the same low-voltage rating, and very nearly the same impedance.

When transformers are connected in parallel, the largest current will flow in the one with the least impedance and that transformer may become overloaded. It is important, therefore, that the impedance of all transformers in parallel be close to the same, within a fraction of a percent.

SINGLE-PHASE TRANSFORMER CONNECTION

The simplest and by far the most widespread of all transformer connections used is that of one single-phase transformer.

A typical arrangement of the method of bringing the terminals or "leads" of the primary and secondary coils out through the tank of a single-phase distribution transformer is shown in Fig. 5-2. To provide for flexibility of connection, the primary and secondary coils are usually each arranged in two sections, each section of a coil having the same number of turns and consequently generating the same voltage. The two primary sections are usually connected together inside the tank and only two primary leads are brought out from one side of the tank through bushings, which insulate them from the tank casing. Four secondary leads may similarly be brought out through insulating bushings from the opposite side of the tank, two leads being brought out from each of the secondary coil. Transformers of older vintage have the four secondary leads brought out of the tank, with the two inner leads

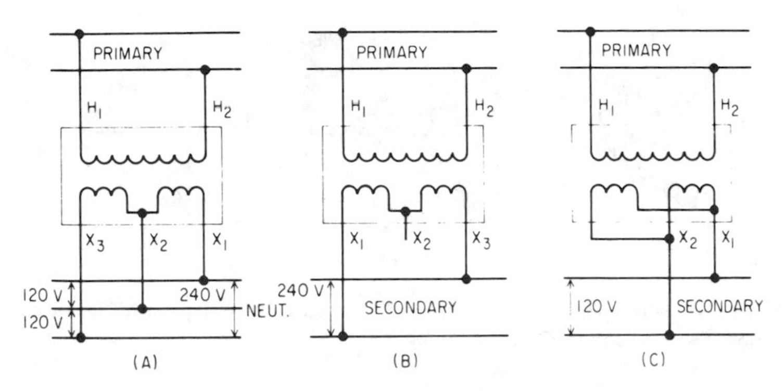

Figure 5-3 Single-phase transformers connected for: (A) 3 wire, 120/240 volt secondary. (B) 2 wire, 240 volt secondary. (C) 2 wire, 120 volt secondary.

from this secondary transposed before being brought out through the casing. In later units, these two transposed secondary leads are connected together inside the tank and only one common lead brought out through the casing. The center secondary bushing is generally called the "neutral bushing" and, in many cases, is a stud connected also to the tank casing providing a means also for grounding the transformer tank. Three methods of connecting such a transformer are illustrated in Fig. 5-3.

POLYPHASE SYSTEMS

There are two types of ac circuits: (1) single-phase circuits and (2) polyphase (multi-phase or more than one phase) circuits. In single-phase circuits, only one phase or set of voltages of sine-wave form is applied to the circuits and only one phase of sine-wave current flows in the circuits. In the polyphase circuits, two or more phases or sets of sine-wave voltages are applied to the different portions of the circuits, and a corresponding number of sine-wave currents flow in those portions of the circuits. Each different portion of the polyphase circuit is usually called a "phase" and these are usually lettered for identification, i.e., "A" phase, "B" phase, "C" phase. The voltages applied to the separate phases of the circuits are correspondingly referred to as the "A-phase voltage," the "B-phase voltage," etc. The phase currents of the different portions are also correspondingly identified as the "A-phase current," the "B-phase current," etc.

The voltages for polyphase systems are supplied from polyphase generators, each phase of voltage being generated in a separate winding. A three-phase generator has three such separate windings. These three separate coils may be arranged for connection in different ways to form the three-phase system.

Two commonly used methods of connecting the coils of three-phase

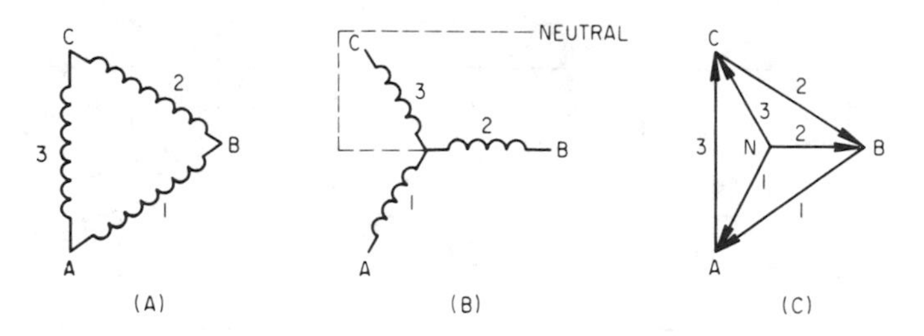

Figure 5-4 Three-phase connections and voltage relationships. (A) Delta connection. (B) Wye connection. (C) Voltage vectors.

generators to supply a three-phase system are shown in Fig. 5-4, A and B. One method (Fig. 5-4A) employs the "delta" (from the Greek letter Δ) connection; the other (Fig. 5-4B) the "star" or "wye" (Y) connection. The voltage relationships for these connections are represented by straight arrow lines as illustrated in Fig. 5-4C.

It will be noted that while the voltages between the terminals A, B, and C are the same in both the delta and wye systems, the voltages across the windings 1, 2, and 3 in the two systems are not only different in magnitude, but their directions do not coincide. This is important in connecting transformers and may be the cause of difficulty when such transformers may be connected together.

For example, two transformers may be connected to give the same secondary voltage, but these two secondary systems may have a voltage difference, or displacement, between them of 30° (Fig. 5-5). Hence, these two secondary systems cannot be connected together with safety.

These "angular" relationships between voltages of delta and wye systems are of paramount importance and must always be considered when making three-phase (or polyphase) transformer connections.

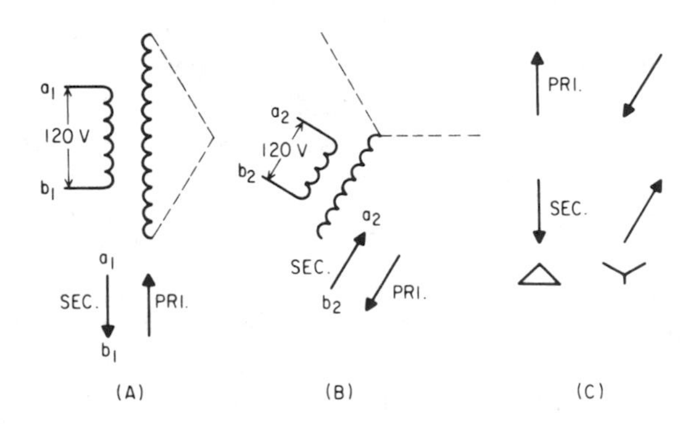

Figure 5-5 Delta and Wye systems-voltage angular displacements. (A) Δ system. (B) Y system. (C) Comparison of systems.

TWO-PHASE TRANSFORMER CONNECTIONS

Although two-phase systems are being replaced by three-phase systems, they are being included here because many still exist. Two-phase transformation of power is usually made with two single-phase transformers connected as shown in Fig. 5-6. Transformers illustrated are additive polarity.

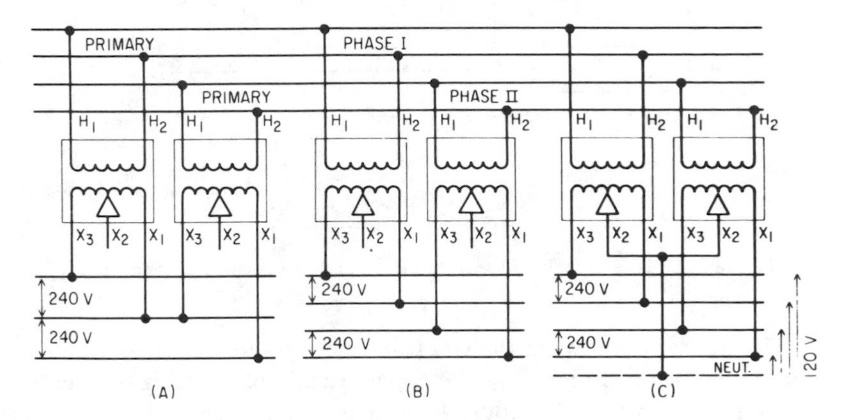

Figure 5-6 Two-phase transformer connections for: (A) 3 wire, 240 volts secondary. (B) 4 wire, 240 volt secondary. (C) 5 wire, 120/240 volt secondary.

THREE-PHASE TRANSFORMER CONNECTIONS

Three-phase transformations can be accomplished by means of three single-phase transformers or by means of a three-phase transformer. The methods of connecting the windings for three-phase transformation are the same whether the three windings of one three-phase transformer or the three windings of three separate single-phase transformers are used. The most common connections used are the delta and the star or wye described above, or combination of these. Additive polarity transformers are shown in Fig. 5-7 (on page 76); these usually represent distribution transformers while station, large, or industrial transformers are usually of subtractive polarity. Note the difference in voltage ratings of the transformers connected to the delta and wye primary supply mains.

DELTA-DELTA CONNECTION

In completing the delta connections, precautions should be taken that the third transformer is connected so that its voltage will be proper in relation to the other two. Despite the polarity identification on the transformer, it sometimes occurs that they may be in error. Hence, instead of a voltage of

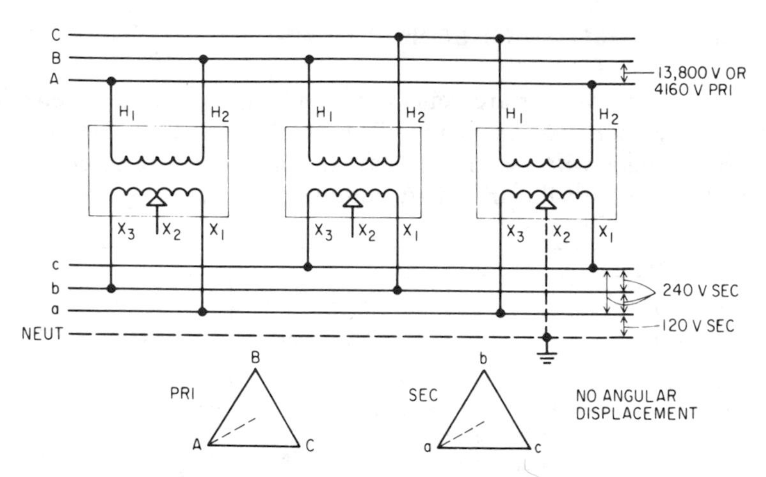

Figure 5-7 Three, single-phase transformers, additive polarity, connected Δ-Δ.

approximately zero between the two transformers closing the delta, a voltage of double that on one phase may be experienced, and closing the delta under these conditions would be equivalent to a short circuit. This condition is indicated in Fig. 5-8. Before the two transformers are connected to close the delta, therefore, a reading between them should be taken by a voltmeter (or appropriate device if on the high voltage side) or a fuse of relatively low capacity (but great enough to withstand exciting current) installed temporarily.

When three transformers are operated in a close delta bank, the impedance of the three units should be very closely the same. Otherwise, the difference in voltages caused by unequal voltage drops will cause currents to circulate in the delta creating unnecessary heating of the transformers. The same condition may arise if the voltage ratio of the three transformers is not the same.

The delta-delta connection is generally used to supply three-phase power loads. Comparatively small single-phase lighting requirements at approximately half the power supply voltage may be supplied from one of the transformers in the delta by establishing a neutral or ground connection at the midpoint of the secondary of one transformer, as shown in Fig. 5-7.

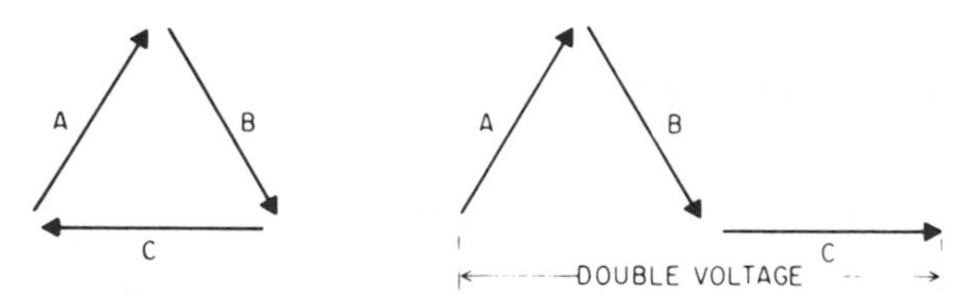

Figure 5-8 Correct (A) and incorrect (B) methods of making a delta connection.

OPEN-DELTA CONNECTION

Three-phase transformation may also be accomplished by using two similar transformers in open delta (Fig. 5-9). Because of the manner in which the current of the three phases divides among the two transformers as compared to that when three transformers are used, each transformer can only be loaded to 86.6 percent of its rating.

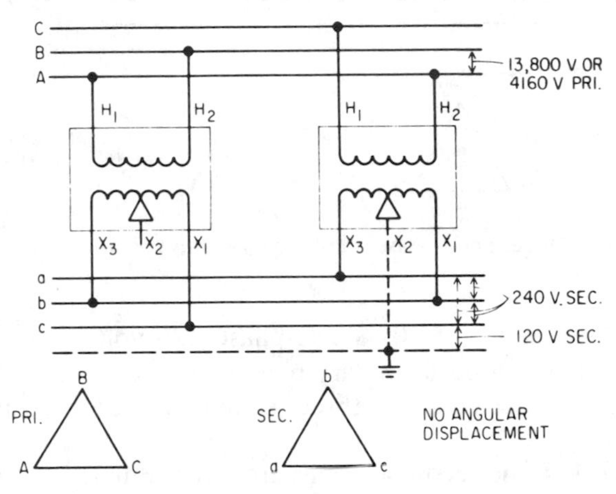

Figure 5-9 Two, single-phase transformers additive polarity, connected open Δ.

In the open delta connection, the impedances of the two transformers need not be the same, although it is usually preferable that they be so. Should it become necessary to close the delta with a third transformer, then the three must have practically identical impedance.

The open delta connection is often used as a temporary expedient, either in an emergency should one transformer fail, or pending a contemplated increase in load.

The regulation of an open delta bank is not as good as a closed delta bank. The voltage drop across the open delta is greater than across each of the separate transformers.

Similarly, as in a closed delta bank, the midpoint of the secondary of one of the transformers can be used to provide a supply to small lighting or other single-phase requirement.

DELTA-WYE CONNECTION

Frequent use is made of this type connection in supplying combined three-phase and single-phase loads, where the single-phase loads may be comparatively large. The two halves of the secondary winding are reconnected in

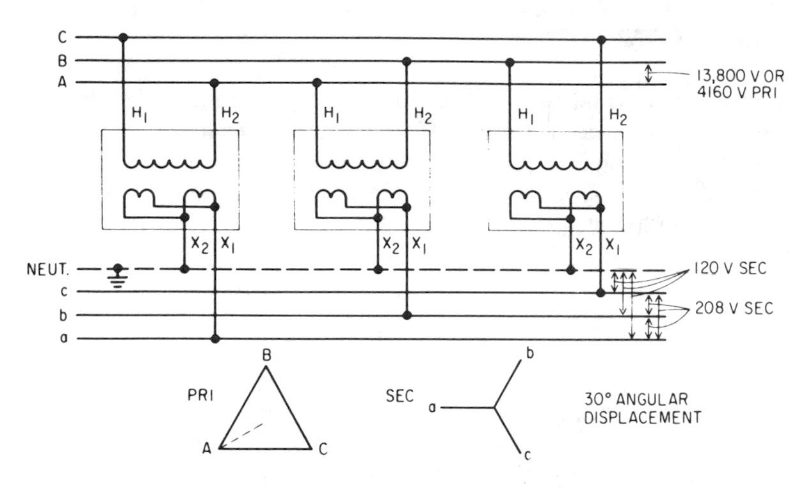

Figure 5-10 Three, single-phase transformers additive polarity, connected Δ-Y.

parallel as shown in Fig. 5-10. Single-phase 120 volt loads are connected between each of the three secondary phases and neutral, while three-phase power loads are connected to the three secondary phases receiving its supply at 208 volts.

In some cases, the secondary coils are not paralleled but brought out as in other connections to supply a 240/416 volt, three-phase power supply, with the three common connections not connected to a separate neutral or ground. Small single-phase 120 volt requirements may be supplied by bringing out the midpoint of one of the transformers, establishing it as a separate neutral, as in other cases mentioned above.

In large power transformers, the delta-wye connection is often used for step-up transformers. The wye high-voltage connection provides a neutral for grounding the high-voltage circuit; it also results in a lower voltage per transformer than would be the case for a delta connection, with a smaller number of high voltage turns and lessened insulation requirements. The angular displacement between incoming and outgoing voltages should be considered before paralleling the stepped-up voltage line with other lines.

WYE-DELTA CONNECTION

This connection is often used to supply large three-phase power loads at 240 volts from a wye primary system. As in other types of connections, small single-phase 120 volt requirements may be supplied by bringing out the midpoint of the secondary winding of one of the transformers, as shown in Fig. 5-11.

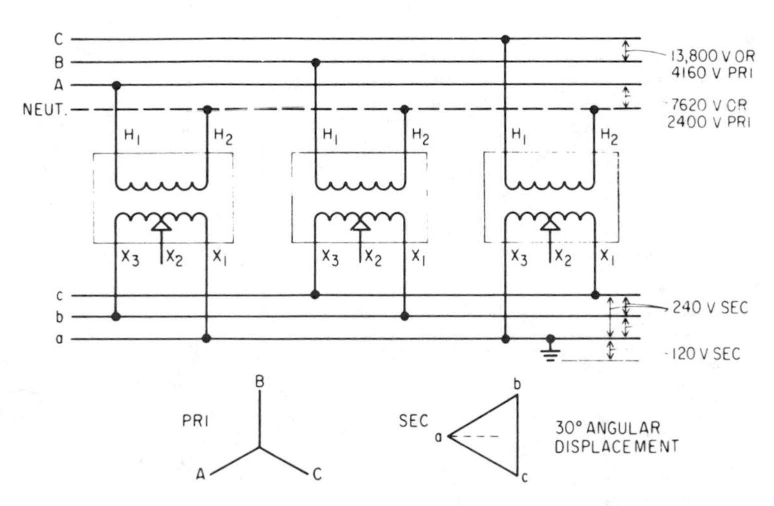

Figure 5-11 Three, single-phase transformers additive polarity, connected Y-Δ.

The impedance of the three transformers need not be the same in this type connection.

By reconnecting the primary windings from a delta connection to this wye connection, the incoming primary or transmission voltage may be raised (increasing the capacity of the circuit) without the need of replacing the transformers. It is often used in this fashion.

The common connection, on the wye high-voltage side, may be or may not be grounded and carried as a fourth primary conductor to the source of power.

OPEN WYE-OPEN DELTA CONNECTION

This is similar to the open delta connection, with the primary of each transformer connected between the common or neutral point and one of the three primary wires (Fig. 5-12 on page 80). The secondaries of the two transformers are connected to the secondary mains in the same fashion as for the delta connection, except that the third transformer is not used. The secondaries are connected in open delta and only 86.6 percent of the rated capacity of the two transformers can be utilized.

Like the open delta connection, this connection may be used in emergency or for expediency awaiting increases in load to install the third unit. It is also often used when relatively small three-phase power loads require supply in which case the "power" transformer, the one without the midpoint of the secondary brought out, may be of very small capacity as compared to the other transformer.

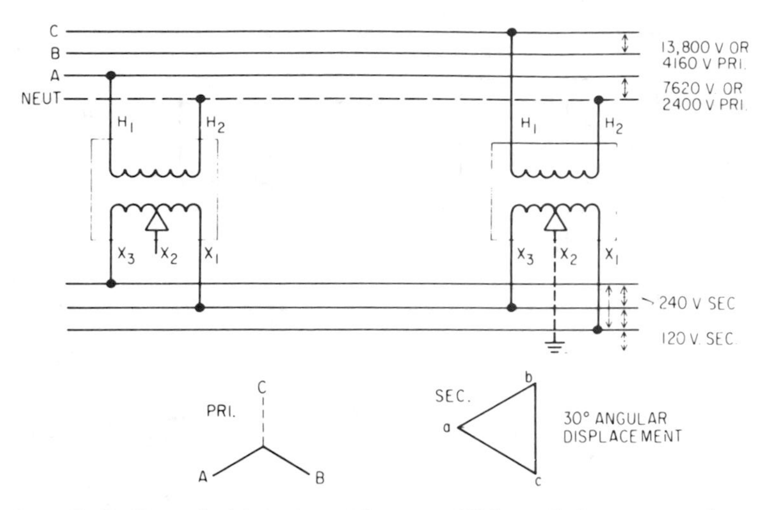

Figure 5-12 Two, single-phase transformers additive polarity, connected open Y-open Δ.

WYE-WYE CONNECTION

This connection is ordinarily used to supply large single-phase loads together with three-phase loads. Often the neutral of the primary and the neutral of the secondary are connected together electrically, usually in the form of a single conductor, or common neutral. The two halves of the secondary windings are connected in parallel, as shown in Fig. 5-13 providing a source of single-phase 120 volt supply between each of the phases and the neutral and a source of three-phase 208 volt power supply between the three-phase conductors.

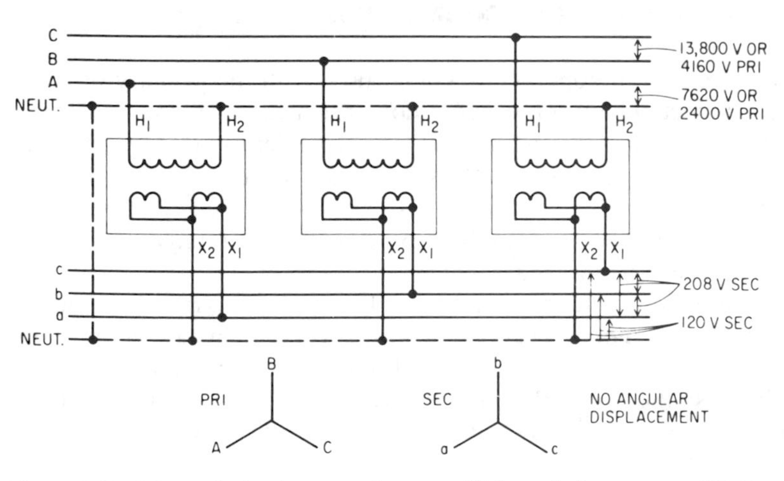

Figure 5-13 Three, single-phase transformers additive polarity, connected Y-Y.

The impedance of the transformers need not be exactly the same, but it is desirable so as to hold down current in the neutral resulting from unbalances due to unequal voltage drops.

SINGLE-BUSHING TRANSFORMER

In the wye-wye connection, transformers having only one high-voltage or primary bushing are often used in a three-phase bank. With this type of transformer, the single, high-voltage bushing is connected to the power line. The special connection on the outside of the tank takes the place of the second high-voltage bushing and must be connected among the three transformers and to the neutral wire or ground. The distribution transformers have a factory-built connection installed between the neutral low-voltage bushing and the tank. While the elimination of one high-voltage bushing restricts this type transformer to use only in the wye primary connection, it does make for a less expensive transformer, and simpler installation.

HARMONICS: THIRD WINDING

When a voltage is transformed from the primary to the secondary, certain side reactions take place. The sine wave of the primary voltage becomes distorted because of the uneven molecular movements within the iron core, producing distorted sine-wave magnetic fields which influence the shape of the sine waves in both the primary and secondary. This distorted wave can be considered to be made up of the original sine wave and a series of harmonices, or waves, whose frequency may be 3 times the 60 cycles per second, 5 times the 60 cycles per second, etc., which are referred to as the third harmonic, fifth harmonic, etc., of the original sine wave. These harmonics or fluctuations in the sine wave can either flow to ground when the transformer is provided with a ground, or when transformers are connected in delta, these fluctuations circulate around the delta producing a little heat, but resulting in a sine wave being formed in the secondary. This is a simplified explanation of the phenomenon.

In large, high-voltage station-type power transformers connected to a wye transmission circuit without a neutral back to the source, these harmonics are particularly bothersome. To overcome this, each of the transformers is provided with an extra small-capacity auxiliary winding. The three auxiliary windings are connected in delta as shown in Fig. 5-14 on page 82 (even though the main primary and secondary windings are connected in wye), and this provides a circuit in which the triple-frequency (and other harmonics) components of the exciting current can flow. A sine wave applied voltage on the high side will then produce a more pure sine wave voltage on the low side.

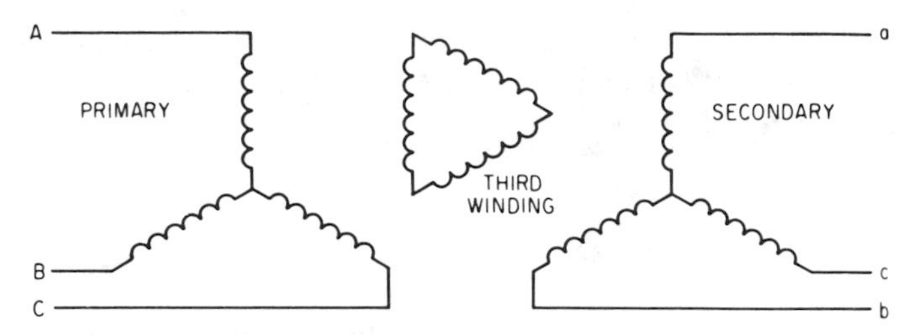

Figure 5-14 Connection of third winding in each of three single-phase transformers connected Y-Y.

POLARITY OF THREE-PHASE TRANSFORMERS

The order of bringing out and marking leads in a three-phase transformer is shown in Fig. 5-15.

The determination of the polarity of three-phase transformers is not as simple as for single-phase transformers; considerations of phase rotation, marking of leads, and types of internal connection are also involved.

Consider first delta-delta or Y-Y connections. For these two connections, the corresponding line voltages may have either a 0° or 180° displacement. With 0° displacement, if two corresponding leads are connected together (say H_1 and X_1), then H_1 H_2 is in phase with X_1 Y_2, and H_2 H_3 with X_2 X_3, etc. With 180° displacement, on the other hand. H_1 H_2 will be in phase opposition to X_1 X_2, etc. The vector diagrams for 0° and 180° displacement phase are shown in Fig. 5-16.

In the case of Y-delta connection, when two corresponding leads of high and low side are connected together, the other corresponding voltages (H_1 H_2 and X_1 X_2, H_2 H_3 and X_2 X_3, H_3 H_1, and X_3 X_1, may be 30° leading or lagging from each other. The vector diagrams for Y-delta or delta-Y connections, for this condition, are shown in Fig. 5-17. Figure 5-18 shows the vector diagram for a 30° lag of wye side to a delta side. It will be noted that in Fig. 5-17, if X_1 is placed on H_1 the voltages X_1 X_2, X_2 X_3, and X_3 X_1 will lead H_1 H_2, H_2 H_3, and H_3 H_1 by 30° for a delta-Y connection and lag by 30° for the Y-delta connection. The reverse is true for Fig. 5-18.

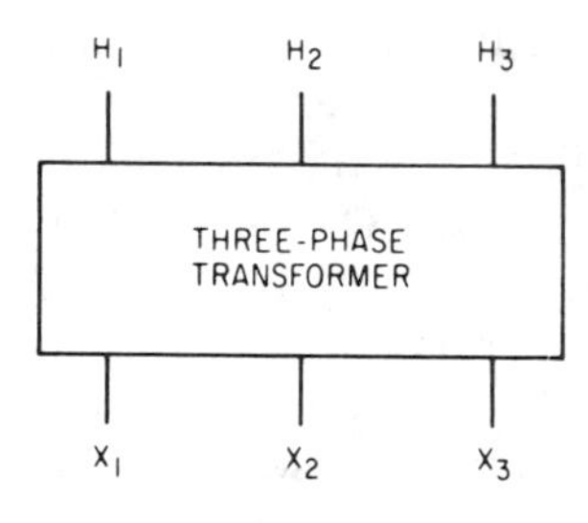

Figure 5-15 Arrangement of leads on a three-phase transformer.

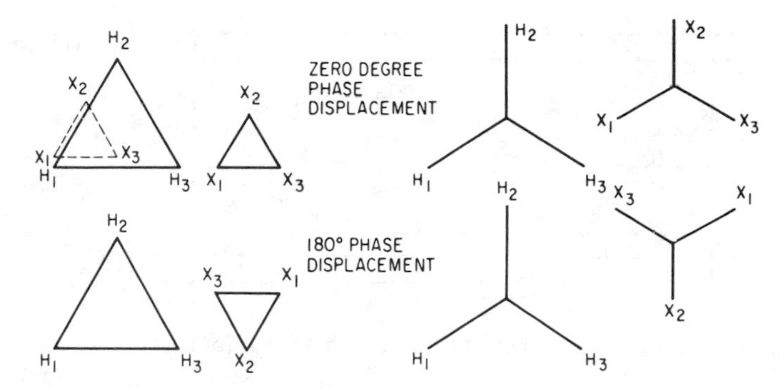

Figure 5-16 Polarity of three-phase transformer, showing phase displacement. (A) Zero degree phase displacement. (B) 180° phase displacement.

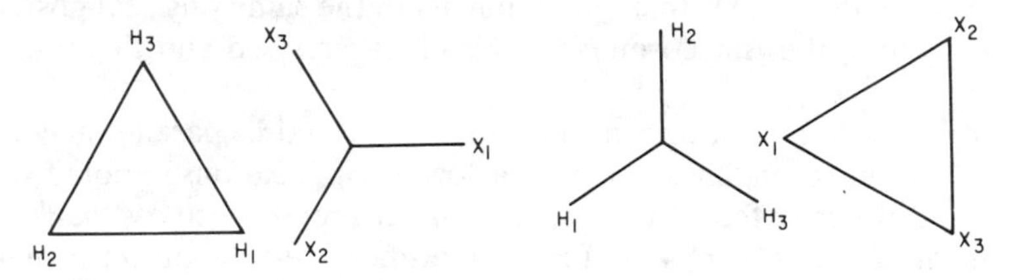

Figure 5-17 30° lead of wye side to delta side.

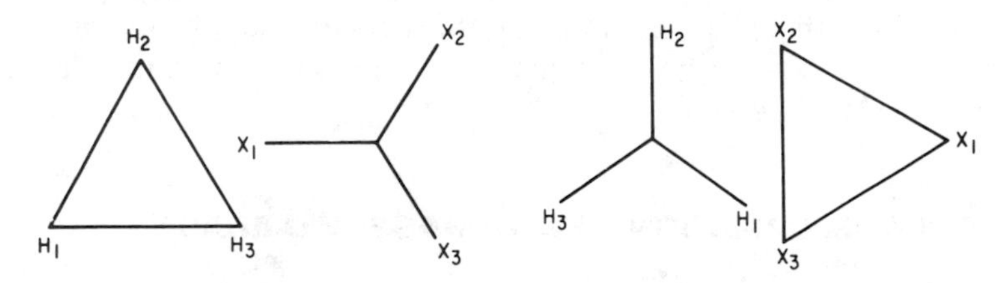

Figure 5-18 30° lag of wye side to delta side.

OTHER CONNECTIONS

Some transformer connections used less frequently or for special purposes are described below. (Refer to Chap. 1, Fig. 1-17.)

PARALLEL CONNECTION: SINGLE-PHASE

The two transformers connected in parallel (Fig. 5-19 on page 84) should preferably have the same characteristics. They should not only have the same voltage ratings and ratio of transformation, but preferably should also have the same impedance and the same polarity. If the impedances are not fairly close together, the load will not divide equally between them and one unit will reach its limit before the other thus reducing the capacity available from the other unit.

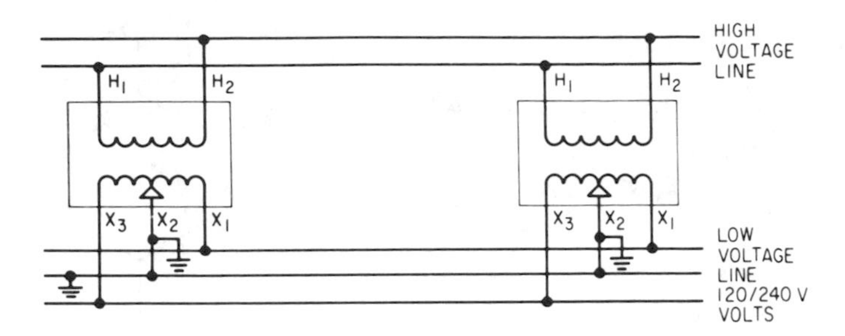

Figure 5-19 Parallel connection of two single-phase transformers to a 3-wire circuit.

In the occasion where it may be necessary to connect transformers with different polarities, the corresponding numbered terminals should be connected together rather than the terminals in the same physical position. If this is not done, the transformers maybe short-circuited and the units damaged.

On occasion, the transformers to be connected in parallel may not be located alongside each other, but may be located at locations remote from and out of sight of each other. Such connections are referred to as "banking" of the transformer secondaries, and are sometimes used to correct or improve local load and voltage conditions. The same considerations and precautions should be observed as for those located together. An added precaution is necessary, that the phase relation of the primary supply in the locations remote from each other be the same, otherwise short-circuits damaging the units will result.

PHASE TRANSFORMATION: THREE-PHASE/TWO-PHASE

Often, it may be desirable and economical to supply a two-phase load from a three-phase source, or vice versa. There are several ways this may be done, but two of particular commercial importance are the Scott connection and the wye to two-phase connection.

SCOTT CONNECTION

In this connection (Fig. 5-20), single-phase transformers having 50 percent and 86.6 percent taps on the three-phase side are required. One transformer, called the MAIN transformer is connected between two of the three-phase wires; the other, called the TEASER is connected between the third phase wire and its 86.6 percent tap is connected to the 50 percent tap of the MAIN transformer.

The two-phase side is shown as a four-wire circuit, but may also be three-wire or five-wire. Conductors a-2 and b-1 can be connected together or

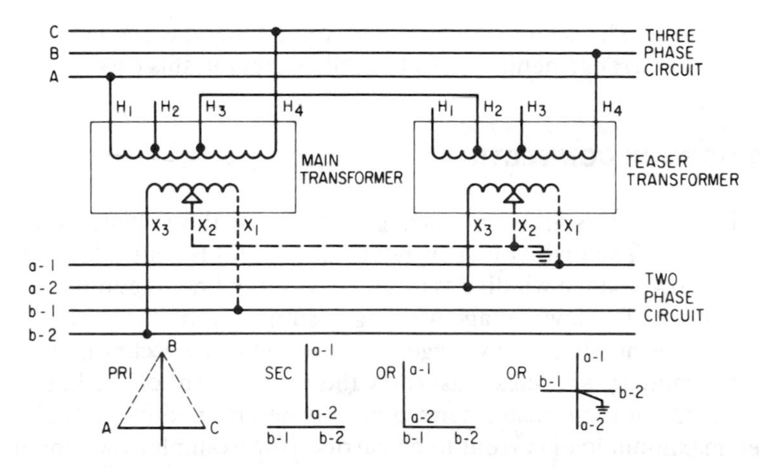

Figure 5-20 Scott connection of two single-phase transformers, additive polarity, for converting three-phase to two-phase.

combined into one conductor (dotted line) giving a three-wire, two-phase circuit. If a five-wire, two-phase circuit is desired, a fifth conductor can be connected to the midpoints of the secondaries (and usually grounded) as shown by the dotted lines in the diagram.

WYE TO TWO-PHASE CONNECTION

This connection (Fig. 5-21) employs standard transformers, but of different voltage ratings. Each transformer will have a rating equal to one-half of the

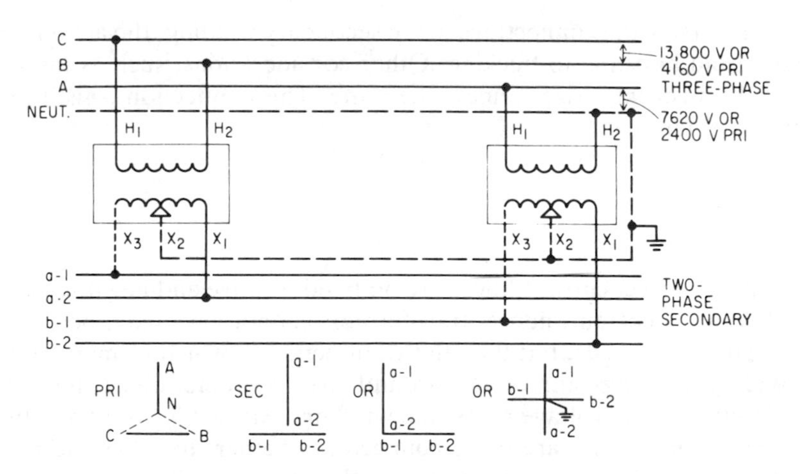

Figure 5-21 Connection of two single-phase transformers of different voltage rating for converting three-phase to two-phase.

two-phase load. The same observations concerning the secondary three-, four-, five-wire arrangements made above also apply in this case.

BOOST OR BUCK CONNECTION

Standard transformers are sometimes used to boost the voltage of a long or overloaded line. The high-voltage winding (Fig. 5-22) is connected across the line and the low-voltage winding is in series with the line to add its voltage to the line voltage. The low-voltage winding is subjected both to the voltage of the high-voltage circuit and to voltage stresses which may occur on that circuit. Since the secondary windings must carry the load current of the high-voltage circuit, the size of the transformer required is determined by multiplying the expected maximum load current to be carried (for example, 100 amperes) by the voltage of the secondary (for example, 240 volts), and the result expressed in volt-amperes (24,000 or 24 kVA; the nearest standard size would be 25 kVA).

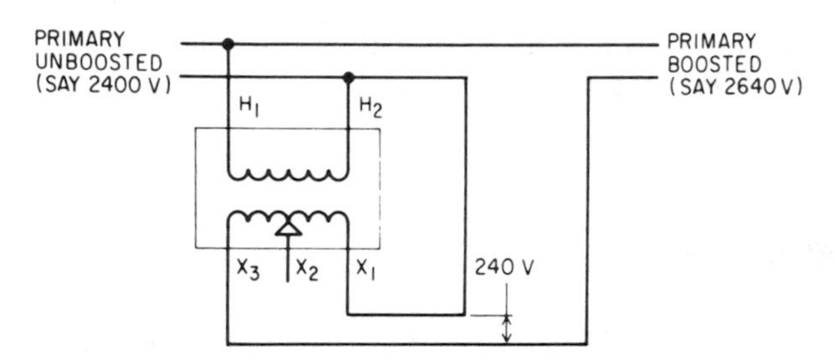

Figure 5-22 Standard single-phase transformer, additive polarity, used as a booster.

By reversing the connections to the secondary winding, the action can be reversed from boosting to bucking. Other considerations, such as insulation stress and transformer size, remain the same. This connection constitutes an autotransformer.

REVIEW

1. Unlike a direct current, there are no fixed positive and negative poles in alternating current; hence, transformers cannot have fixed positive and negative terminals, but the relative direction in which primary and secondary windings are wound around the core determines the relative direction of the voltage across the windings. Since it is essential if two or more transformers are to be connected together, to know the relative direction of the voltage transformed, their "polarity," or direction of the voltages, needs to be established (Fig. 5-1).

2. The two windings of a transformer may be wound in two ways. If the windings of the primary and secondary are in opposite direction, the applied voltage and the induced voltage will have opposite direction, and the transformer is said to have "subtractive" polarity. If the windings of the two coils are in the same direction, the applied and induced voltages will have the same direction, and the transformer is said to have "additive" polarity (Fig. 5-2).
3. In addition to polarity, transformers connected in parallel should have the same primary and secondary voltage ratings and should have nearly the same impedance.
4. Transformers may be connected to supply single-phase loads, with secondary windings connected to supply a two- or three-wire output (Fig. 5-3). They may also be connected in "banks" to supply two-phase or three-phase outputs. Two-phase supply may have three-, four-, or five-wire output. Three-phase supply may have three-, four-, or six-wire outputs.
5. Three-phase loads may be supplied by two or three transformers connected in delta (Δ) or wye (Y), or combinations of both, to accommodate the desired input and output voltages (Figs. 5-7 through 5-14).
6. When three transformers are operated in a closed delta bank, the impedance of the three units should be very closely the same. Otherwise, the difference in voltages caused by unequal voltage drops will cause currents to circulate in the delta, causing unnecessary heat in the transformers. Care should be taken in closing the delta, for, if one of the transformers is connected so that its voltage is in improper relation to the others, a dangerous voltage of approximately twice line to line voltage may result.
7. Open delta or open wye banks of two transformers are sometimes used for economy when the third transformer may not be immediately required, or in emergency, when the third unit is unavailable for any reason.
8. For economy and space reasons, and for ease in installing, three-phase transformers are used, which may comprise two or three single-phase units connected internally to supply the desired three-phase output. Hence, when such units are connected in parallel, it is again necessary to establish "polarity," but, in addition to relative voltage directions, their relative angular displacement from each other is also necessary (Figs. 5-15, 5-16, 5-17, and 5-18).
9. Transformers may also be connected to change three-phase to two-phase supply, or vice versa, (Figs. 5-20 and 5-21). Also for boosting or bucking supply voltage for higher or lower output voltages (Fig. 5-22).

STUDY QUESTIONS

1. What is meant by the polarity of a transformer and how is it usually indicated?
2. How may the polarity of a transformer be determined?
3. What factors should be taken into consideration when connecting transformers in parallel and why?
4. What is the most widespread distribution connection used, and how may the secondary coils be connected?
5. What is the difference between a single-phase and a polyphase system?
6. How may transformers be connected in a three-phase system and what are the relationships between their primary and secondary voltages?
7. What precautions should be taken in making a delta connection?
8. What are some advantages of the delta connection?
9. What are some of the advantages of a wye system?
10. What precautions should be taken in connecting three-phase transformers in parallel?

6

OTHER TRANSFORMER TYPES

AUTOTRANSFORMERS

Autotransformers may be viewed as a particular type of transformer, whose characteristics make it more advantageous to use than two-winding transformers under certain conditions.

When the ratio of transformation desired is low, usually not greater than about 5 to 1, and the electrical isolation feature of the two-winding transformer is not essential, use is made of the autotransformer.

The autotransformer has only one winding, a portion of which serves both as primary and secondary. In this type transformer, only a portion of the electrical energy is transformed and the remainder flows conductively through its windings. (In a two-winding transformer, all the energy is transformed.) The autotransformer, therefore, having to provide only for that portion of the energy to be transformed can be of smaller size than the equivalent two-winding transformer; this results in lower cost, not only of the unit, but also in its installation. Also, since it handles only a portion of the transformed energy, the losses caused by currents flowing through it are lessened resulting in greater efficiency and better voltage regulation.

A schematic diagram of an autotransformer is shown in Fig. 6-1 (on page 90).

If a source of voltage E_P is applied across all the turns between a and c, this part of the coil will serve as a primary winding. Some of the turns, between b and c, will also serve as a secondary winding giving the voltage E_S. This arrangement uses the transformer for step-down purposes, reversing E_S and E_S will allow the transformer to be used equally well as a step-up transformer.

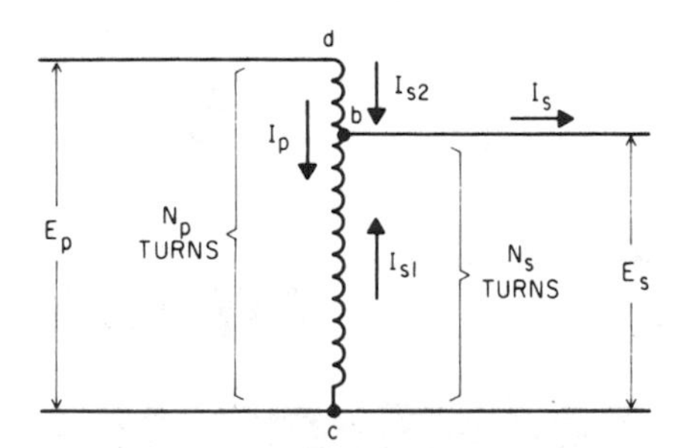

Figure 6-1 An autotransformer.

In this transformer, the same ratio of transformation holds as in the two-winding transformer:

$$\frac{N_P}{N_S} = \frac{E_P}{E_S} = \frac{I_S}{I_P}$$

Where the primary current I_P flows in the direction shown by the (closed head) arrow in the figure, the secondary current I_S will flow in the opposite direction. Hence, in the portion of the winding between b and c, the current is only the difference between I_P and I_{S1}. When the ratio of transformation is small, the difference between E_P and E_S is small, as is the difference between I_P and $I_{S1.}$ so that the portion of the winding between b and c, which carries difference of these currents, can be made of smaller conductor since it will carry only a small current. The portion of the winding between a and b, however carries the sum of the currents I_P (the exciting current) and I_{S2} which constitutes the relatively large untransformed current, and must be made of a conductor sufficiently large to carry this sum.

The percent of the volt amperes transformed is the same as the percent of the voltage transformed (using the high voltage as a base). For example, if the autotransformer boost the voltage 10 percent, it actually transforms only 10 percent of the volt-amperes (or kVA) supplied to the load. And since the size of the unit depends on the volt-amperes (or kVA) transformed, the size of the autotransformer need be only 10 percent of the load—if the load to be supplied was 50 kVA, the size of the autotransformer would be only 5 kVA.

Under these circumstances, it is obvious that the autotransformer is more economical than the two-winding transformer. However, as the ratio of transformation increases, the advantages decrease and generally disappear at a ratio of approximately 5 to 1. The need for a large part of the coil to be made of heavy conductor necessarily having the same heavy high-voltage insulation since both portions are connected electrically adds considerably to the cost.

One disadvantage of the autotransformer is that, because of the metallic connection between the high and low-voltage circuits, each circuit is affected by electric disturbances originating in the other. For example, a ground on either circuit is a ground on both circuits, and a ground on the high-voltage side may subject the low-voltage side (and the connected loads) to the high voltage. The low-voltage portion of the autotransformer is therefore insulated to withstand this high voltage. But, while this may protect the transformer, it does not protect the connected load appliances.

Another disadvantage of the autotransformer stems from its comparatively lower impedance which causes short-circuit currents of greater magnitudes to flow through it during fault conditions. hence, the autotransformer must be built very ruggedly to withstand the greater mechanical stresses produced, or external impedances should be connected in the circuit to limit the magnitude of these short-circuit currents, or a combination of both.

AUTOTRANSFORMER CONNECTIONS

Autotransformers may be used on both single-phase and polyphase circuits, including phase transformation. The single-phase transformation has already been discussed and illustrated in Fig. 6-1. Two-phase transformation from two-phase to two-phase circuits are merely two single-phase applications, one for each of the two phases.

Another connection (Fig. 6-2) known as the T-connection, and somewhat similar to the Scott connection for two-winding transformers, may be used for both three-phase transformations and for transformations from three-phase to two-phase, or vice-versa.

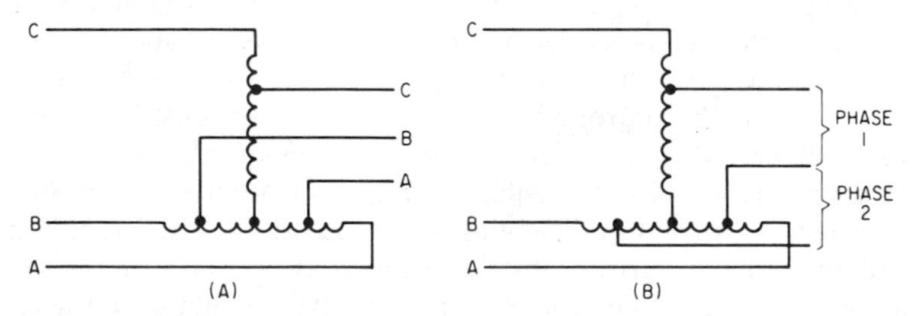

Figure 6-2 T-connection of auto-transformers. (A) Three-phase to three-phase, three wire. (B) Three-phase to two phase, three wire.

VOLTAGE REGULATORS

A voltage regulator is used to maintain the voltage of a circuit, and is essentially a special application of an autotransformer. The voltage in the portion of the winding which is added or subtracted in an autotransformer is made variable, so the outgoing voltage may thus be kept approximately at the rated value. These are two types of voltage regulators commonly used, and the principle of operation of both is the same as for the autotransformer.

One type of voltage regulator, called an induction-voltage regulator (Fig. 6-3 on page 92) has a primary which is used as a high-voltage winding

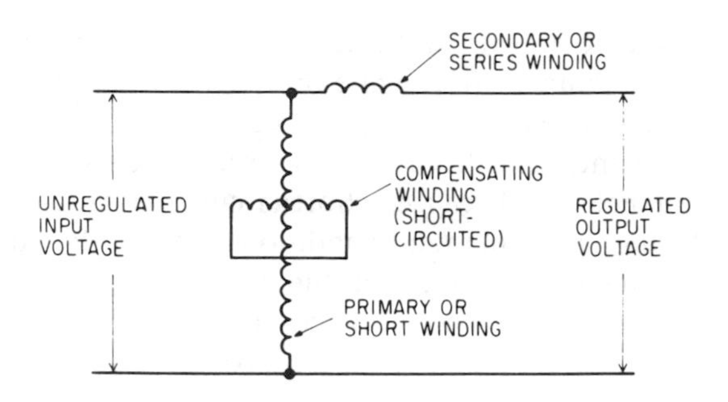

Figure 6-3 An induction regulator.

connected across (or in shunt with) the circuit to be regulated. The secondary or low-voltage winding is connected in series with the circuit. The primary (high-voltage) winding is mounted so that it can rotate on the axis of the secondary winding. The voltage induced in the secondary or series winding depends on the position of the primary winding. The high-voltage winding thus can be placed so that the voltage induced in the series winding will add or subtract from the input line voltage. When the load current is small, the primary will be rotated in one direction, thus lowering or bucking the line voltage. When the load current is larger, the primary will be rotated in the opposite direction, thus boosting it. The reactance of the secondary can cause a large voltage drop during the rotation of the primary. To avoid this drop (or cancel it out), a third coil is added and mounted on the movable core at a right angle to the primary winding and short-circuited on itself. The turning of the primary coil is usually controlled by means of a voltage-sensitive relay (called a contact-making voltmeter) connected in the output side of the circuit.

A second type voltage regulator, (Fig. 6-4) often called a Step-Type Voltage Regulator or a Tap Changing Under Load (TCUL) transformer, is more often found in electrical substations and is associated with the supply of large amounts of power. The connection of this type regulator in the circuit is the same as that of the induction-voltage regulator. The changes in voltage are accomplished by varying the ratio of transformation by changing the number of turns in the primary by means of taps connected as shown schematically in Fig. 6-5. Other arrangements accomplishing the same purpose include a separate winding (with the taps) which is capable of being reversed in connection so that the voltage within that separate winding can be added to or subtracted from that in the remainder of the primary winding.

The application of another small autotransformer capable of being connected across the taps makes the operation of tap changing under load successful. The function of this autotransformer is to aid in preventing the transformer from being disconnected from the line while stepping from tap to tap on the transformer winding. Figure 6-6 (on page 94) is a schematic diagram of how the tap-changing equipment operates under load. The selector

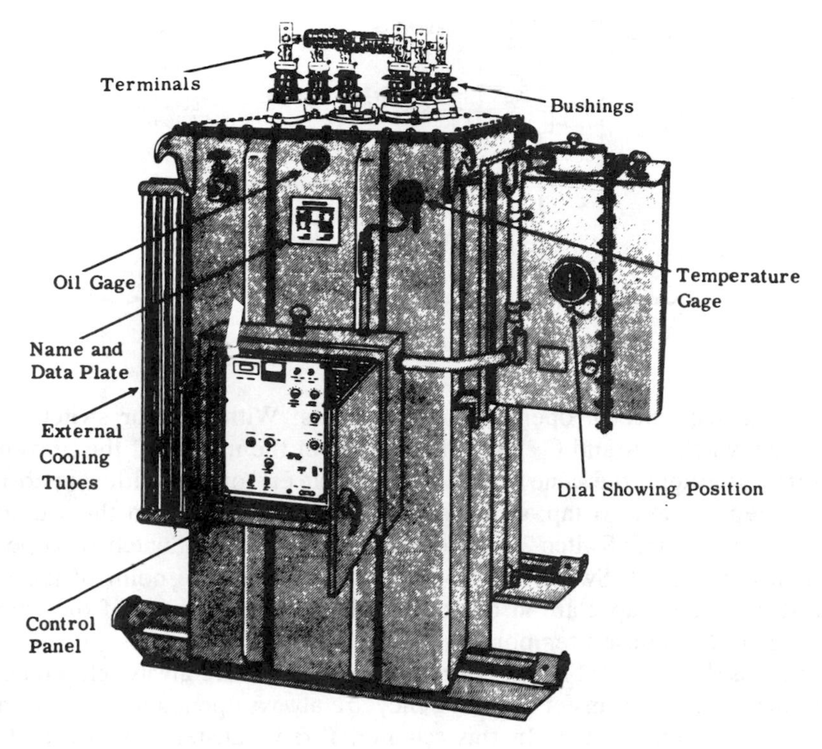

Figure 6-4 Step-type voltage regulator.

switches, numbered 1 through 9, are never used to make or break the circuit; generally, they are installed in the same enclosure as the transformer itself. The transfer switches, lettered A, B, and C, serve to make and break the circuit under oil and are installed in a separate compartment, so that they may be maintained, and the oil changed when it becomes contaminated from the switching arcs which may occur. The preventive autotransformer is connected as shown, and its midpoint connected to the main portion of the primary winding.

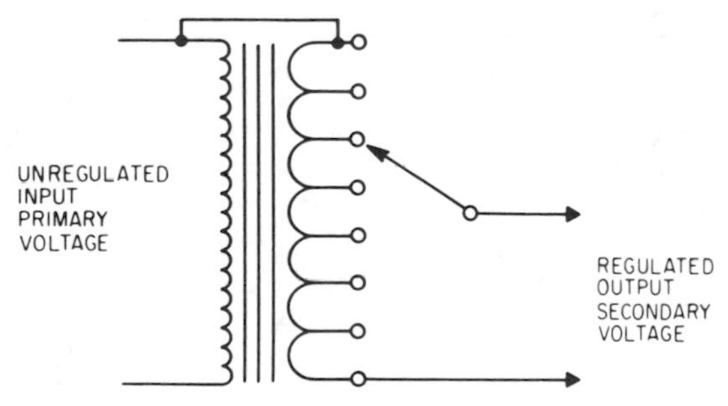

Figure 6-5 Top changing under load step-type voltage regulator (TCUL Transformer).

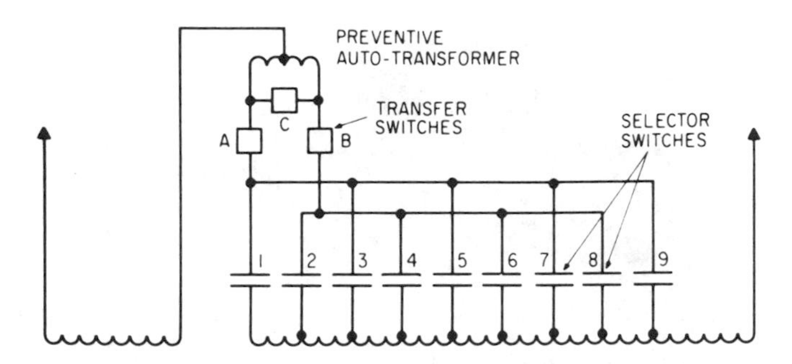

Figure 6-6 Operation of tap-changer under load.

The sequence of operation is as follows: With selector switch 1 and transfer switches A and C closed the voltage at the middle of the preventive autotransformer is the same as the voltage at the end of the winding at position 1. To step to the next tap, switches are operated as shown in the sequence: Switch C is opened; Switch 2 is closed; Switch B is closed; Switch A is opened; Switch C is closed; Switch 1 is opened. Now the middle point of the autotransformer and tap 2 are at the same voltage, and the part of the winding from 1 to 2 is unused temporarily. The sequence of switches for other tap changes is shown in Table 6-1. The selector switches are always closed before the corresponding transfer switches; they are always opened after the transfer switch has been opened. In this scheme, three transfer switches and one selector switch for each tap are required. Other schemes and modifications are designed to serve the same purpose.

When transfer switch C is closed, both ends of the preventive autotransformer are at the same voltage and one-half the current flows through each half. Since these currents flow around the core in opposite directions, only a small voltage drop occurs from the reactance of the circuit under these

TABLE 6-1 Sequence of Operation of Switches

Position	1		2		3		4		5		6		7		8		9
Transfer Switch A	X	X		X	X	X		X	X	X		X	X	X		X	X
Transfer Switch B		X	X	X		X	X	X		X	X	X		X	X	X	
Transfer Switch C	X		X		X		X		X		X		X		X		X
Selector Switch No. 1	X	X															
Selector Switch No. 2		X	X	X													
Selector Switch No. 3				X	X	X											
Selector Switch No. 4						X	X	X									
Selector Switch No. 5								X	X	X							
Selector Switch No. 6										X	X	X					
Selector Switch No. 7												X	X	X			
Selector Switch No. 8														X	X	X	
Selector Switch No. 9																X	X

circumstances. When transfer switch C is open and either transfer switch A or B is open at the same time during the short time of operation, the whole current flows through half of the preventive autotransformer winding. To prevent an undesirable voltage dip from occurring during this period, the preventive autotransformer is built with an air-gap in its core. This prevents the core from handling too great a magnetic field which would cause excessive voltage drop in the coil.

The operation of the tap changers may be made by hand or controlled automatically. For larger tap changers, the control mechanism is mounted outside the transformer case. A motor mechanism, actuated by relays, opens and closes both the selector switches located inside the transformer tank and the transfer switches located in a separate compartment. The mechanism is so arranged that the switches operate in their proper sequence. A time delay device, used in connection with the control relay prevents its operation during voltage dips of short duration.

Such TCUL transformers may be used to regulate voltage over a plus 10 percent to minus 10 percent range, and may have thirty-six or more steps incorporated in the tap changing mechanism.

INSTRUMENT TRANSFORMERS

When values of voltage or current are large, or when it is desired to insulate the meter, relay or other control device from the circuit in which the electrical values are to be measured, an instrument transformer is used.

Instrument transformers (Fig. 6-7) are of two kinds, voltage or potential transformers, and current transformers. Both work on the same principle of the two-winding transformers. They differ from power or distribution

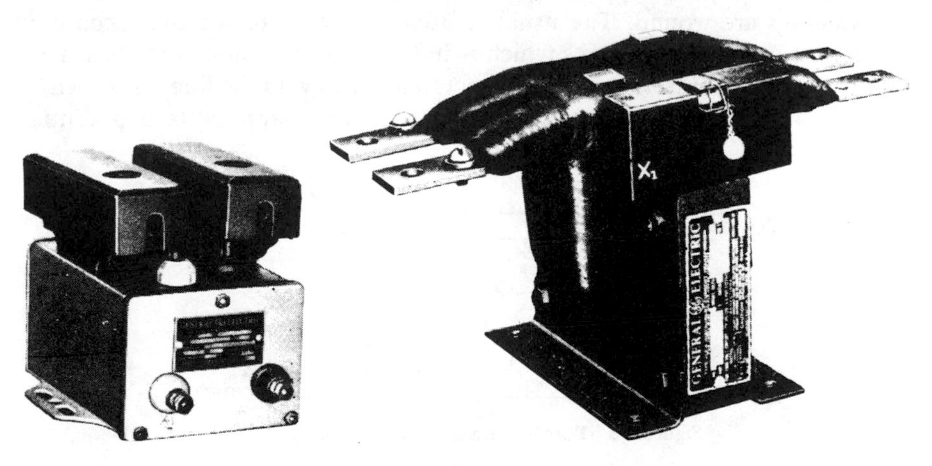

Figure 6-7 Typical instrument transformers.

transformers in that they are of comparatively small capacity and are designed to maintain a much higher degree of accuracy in their ratios of transformation than is necessary in other forms of transformers. To differentiate from other transformers, the load carried by these instrument transformers is referred to as their "burden."

In using instrument transformers with various meters, relays or controls, it is necessary to know the relative direction of the current flowing in the leads at any particular instant. Hence, it is necessary to know the relative polarity of the primary and secondary windings. This is indicated by marking one terminal of each winding with a mark, usually a colored dot, but the marking may be something else; sometimes the letter H_1 is used to mark the high voltage winding and X_1 to mark the low voltage winding. The marked leads are of the same polarity when current flows toward the transformer in the marked primary lead, but current will flow away from the transformer in the marked secondary lead, and vice versa.

Potential transformers (PT) have a small output and have high accuracy; that is, they have small regulation or voltage drop; the windings are also compensated to give its exact ratio at some specified rated burden. Compensation is an adjustment in the turns ratio to compensate for the voltage drop at the specified burden. They are usually designed for a secondary voltage of 150 or 300 volts. Potential transformers are connected in parallel or across the line as shown in Fig. 6-8. They are used with voltmeters, wattmeters, watt-hour meters, relays, and other controls. Several instruments may be connected to the same transformer if their sum does not exceed the burden for which the transformer is designed.

Current transformers (CT) also have a small output and high accuracy. The primary winding may consist of one or few turns and must be insulated to sustain the line voltage; in some instances, the insulated line conductor itself may serve as the primary, passing through the core around which the secondary windings are wound. The usual standard design calls for the secondary current rating to be 5 amperes which is full scale for the ammeter associated with it. Current transformers are connected in series with the line, as shown in Fig. 6-9. They are used in conjunction with the same apparatus as potential

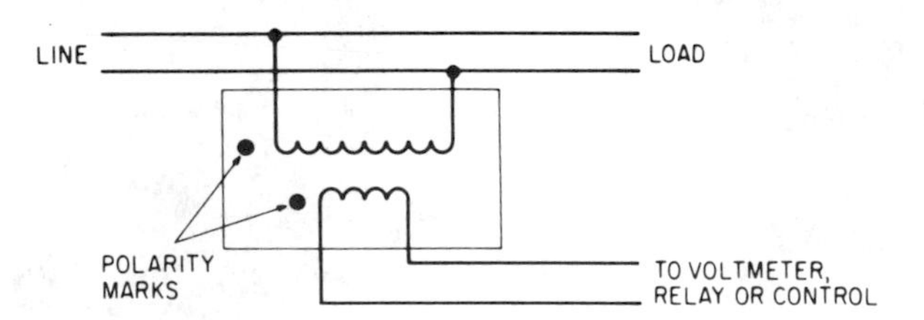

Figure 6-8 Potential transformer connection.

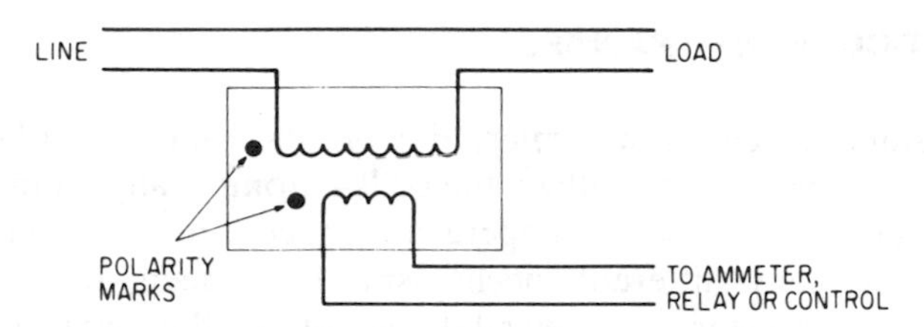

Figure 6-9 Current transformer connection.

transformers, and can have several items connected to it so long as its burden, rated in amperes, is not exceeded.

Because of the high turn ratio usually met with in a current transformer, it is possible to obtain an unsafe high voltage across the secondary terminals of the current transformer. The regulation of these transformers, however, is designed so that the voltage drops very rapidly as soon as the ammeter or other burden is connected to it. For safety reasons, therefore, the terminals of the current transformer should always be short-circuited when instruments are not connected to it; the instruments should first be connected before the short-circuit strap is removed.

Both potential and current transformers may be of the dry-type or oil-filled. They may be equipped with solid porcelain or oil-filled high-voltage bushings, through which the high-voltage leads are brought out of their case. Windings are insulated from each other in a similar manner as other transformers. Their construction must also be sufficiently rugged to carry short duration fault currents safely.

STREET LIGHTING TRANSFORMERS

Street lighting may be served from multiple circuits or from series circuits. In the first, the lamps are connected across the two conductors of the circuit, much as the rungs of a ladder connect to the runners on both sides. This is the type of circuit that commonly exists in houses, offices and factories, supplied from two-winding distribution transformers previously described. In the second or series circuit, the lamps are connected in succession, like beads on a string, so that whatever current passes through one passes through all of them. This type circuit cannot be supplied from the usual distribution transformer, but requires a transformer which will produce a constant current, but with a voltage at the outgoing terminals which will vary in magnitude proportionately to the load, or number of lamps, it is called upon to supply. Such a transformer is known as a "constant-current transformer," and is usually designed to supply a current of 6.6 amperes, the values designed to give optimum operation of the lamps.

CONSTANT-CURRENT TRANSFORMER

In the constant-current transformer, the primary and secondary coils are movable with respect to each other, the coils automatically taking the correct position to deliver the rated 6.6 ampere current continuously to the circuit.

Although many different mechanisms have been developed for the constant-current transformer, they all do essentially the same thing. The secondary coil moves up and down to vary the magnetic field (Fig. 6-10). The primary coil is usually fixed and the secondary floats. The latter is balanced by a weight which is proportional to the repulsion force of the rated current. If more than the rated secondary current flows, the repulsion becomes greater than the weight and pushes the secondary farther away from the primary. This automatically produces a greater leakage flux, which in turn automatically lowers the voltage, thus automatically lowering the current to its rated value again.

Hence, if part of the load is short-circuited or removed, the coils will move apart and the secondary current brought back to its normal value. Constant-current transformers are usually designed so that, with normal voltage on the primary, the secondary may be short-circuited without causing quite full separation of the coils.

Constant-current transformers differ in several important respects from the usual power and distribution transformers. Their load is a series of lamps

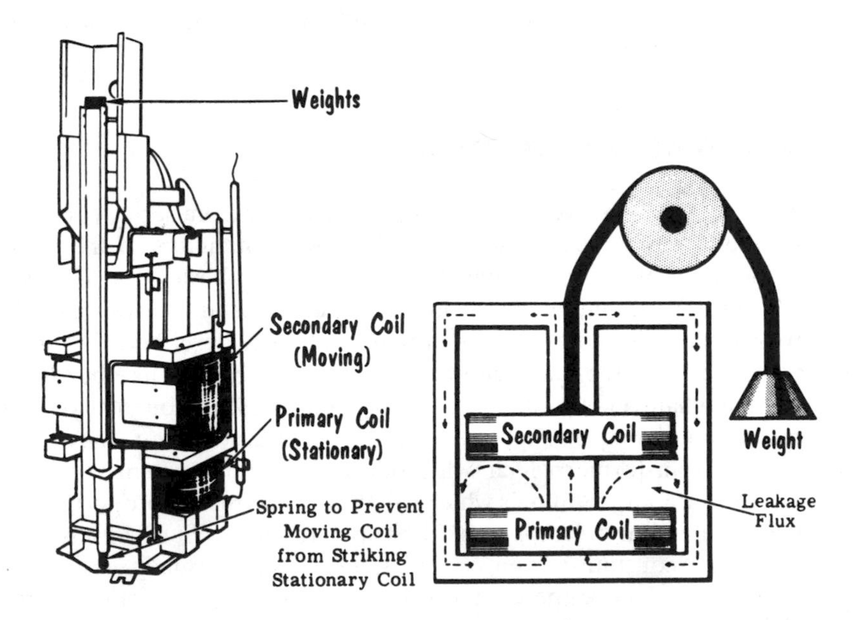

Figure 6-10 Constant current transformer.

in which the current is constant; increased load, therefore, means increased resistance in the secondary circuit and consequently increased voltage delivered by the secondary. It is impossible to overload a constant-current transformer. As load is added (secondary resistance increased), the coils come closer and closer together, increasing the output voltage. Maximum voltage will be delivered when the moving coil has reached it limit of travel with respect to the stationary coil. If still more load (resistance) is added, no more voltage is available and the secondary current must decrease.

Power and distribution transformers reach their maximum temperature at maximum load, but constant-current transformers reach their maximum temperature at zero load, when the secondary is short-circuited. Since the currents in the windings of these transformers are the same for all conditions of load, the copper conductor losses are the same, and since the applied voltage is the same, the loss in the iron core is somewhat less at full separation than at minimum separation. But the magnetic field also cuts the mechanical parts which help support the transformer and flows over a longer path of the iron core. As the coils separate, these effects become more pronounced so that maximum temperature will result where maximum coil separation is reached.

Constant-current transformers may have taps connected in their primary windings and sometimes also in their secondary windings. They are usually constructed for station operation, in which case the coils and mechanism may be exposed for better ventilation, and the whole enclosed in a cage for safety. They are also installed in tanks for both overhead and underground installations, in which case the leads are brought through the case by means of bushings, similar to other distribution transformers. They are rated in kVA, the usual standard sizes being 15 kVA and 25 kVA.

Although obsolete, installations will exist for some time into the future.

"IL" AND "SL" TRANSFORMERS

As the voltage at the socket of a series lamp may be high, it is sometimes desirable to feed the lamp through a transformer which will reduce the voltage to safe values. This is particularly applicable to metal street standards. This transformer, small in size, is known as an "isolating" or "individual lamp series transformer"—usually referred to as an IL transformer (Fig. 6-11 on page 100). The ratio of the transformer in this case is 1:1. Sometimes, these transformers are used to supply larger 15 or 20 ampere lamps from a 6.6 ampere system, in which case the ratios are different.

Another type of "series transformer" works on the same principle, but may feed several lamps connected in series, this is commonly referred to as an SL transformer (Fig. 6-11).

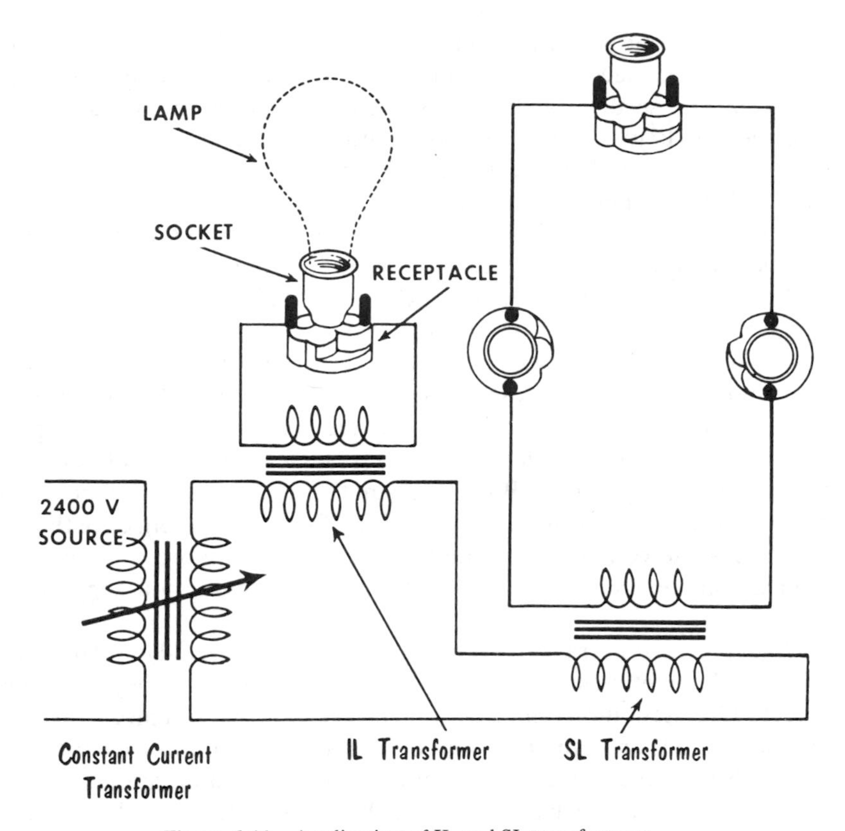

Figure 6-11 Application of IL and SL transformers.

OTHER CUSTOMER-TYPE TRANSFORMERS

Where voltages of a value other than that of the supply source are required or desired for safety and operating reasons, transformers or autotransformers are employed. Some examples, by no means complete, are given below.

SIGN LIGHTING

Signs made of luminous gas tubes require higher voltages than the usual normal secondary distribution supply and are essentially constant-current devices. The impedance of the tube increases as the length increases and is higher for a tube of small diameter than for one of larger diameter. Hence, the transformer chosen should deliver the proper current and voltage for the tube. For small signs, an autotransformer may be used. For larger signs, where the applied voltage may be several thousand volts in magnitude, a two-winding transformer is used with the midpoint of the high-voltage secondary

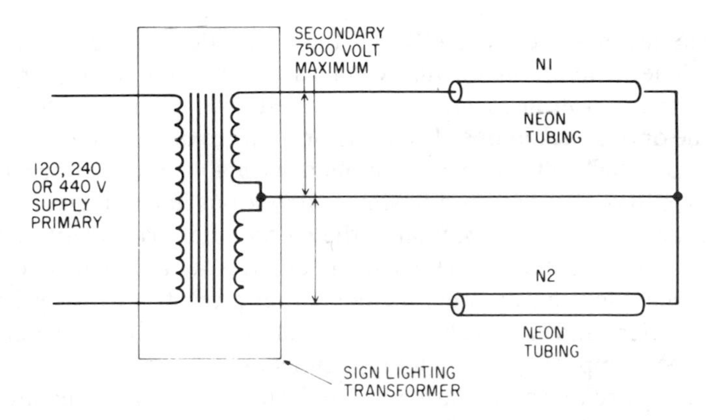

Figure 6-12 Sign lighting, two winding,transformer with midpoint of high voltage secondary winding grounded.

grounded, connected as shown in Fig. 6-12. This enables twice the length of tube to be served using a voltage measured from ground of only half the required value. For safety reasons, the voltage from these transformers is usually limited to a maximum of 15,000 volts. Where the tubing for the sign may be too long to be adequately supplied at this voltage, it is sectionalized into lengths which may be supplied safely. As an additional safety feature, the transformer may be placed physically on or near the sign near the middle of the total length of tubing to be served, to keep the lengths of cable operating at high voltage as short as possible. The transformers may be oil filled for outdoor use and air cooled enclosed types for indoor use; these latter have their high-voltage terminals usually sealed so that they cannot be readily accessible and may be so arranged that the primary circuit is automatically deenergized when the enclosure is opened.

FLUORESCENT LIGHTING

Fluorescent lamps operate somewhat similarly as the gas tubes described above. They are fixed dimensions and the voltage applied is raised from the usual normal 120 volts secondary supply by a "ballast" which is incorporated in one end of the fixture in which the fluorescent tube is installed (the "ballast" is essentially a small autotransformer designed for the specific lamp size). Standard ballasts are designed for ranges of 110–125 volts, 199–216 volts, and 220–250 volts.

MERCURY, SODIUM AND HALOGEN LAMPS

Mercury lamps are high-intensity lights and sizes run from 400 watts to 3000 watts. They operate on the principle of vaporizing a pool of mercury into gas, and the current carrying vapor (ionized gas) gives off a greenish light. For the sizes indicated, they are more efficient than incandescent lights and give off less heat. They operate at starting voltages of 600 volts or higher and have small dry-type, autotransformers incorporated with each unit; the unit may contain one or two such lamps. The transformers are designed for operation at 120, 240, and 480 volts, and are supplied with four-line voltage taps of 5 percent each to permit proper adjustment to various line voltages.

Sodium lamps are very similar to the mercury lamps, except that sodium is used in place of mercury, and the light given off is yellowish instead of green.

Halogen lamps gasify other elements that give off a softer yellow light, are more efficient and are quicker to reach full brilliance than the mercury and sodium vapor lamps. Operating temperatures are very high.

For lamps requiring ballasts or transformers, line voltages higher than maximum limit of the range will shorten the life of the lamp and transformer; below the limit will cause uncertain starting, shorter lamp life, and reduce illumination.

MOTOR STARTING

Practically all ac motors, except those of fractional horsepower, are started at reduced voltage to reduce their starting current. For the usual motor of several horsepower, an autotransformer is used together with switches or breakers to accomplish this purpose. Figure 6-13 shows two methods of connecting these in the motor circuit. Note that the transformer is usually connected to the neutral terminal so as to hold down the voltage that its insulation must withstand.

The autotransformers used are "under-sized" for economy. Since the starting period is usually short, and the rating of the transformer limited by the permissible temperature rise, the amount of current they are designed to handle will be larger than that for continuous operation. Should the starting duty recur at more frequent of very close intervals, it has to be taken into

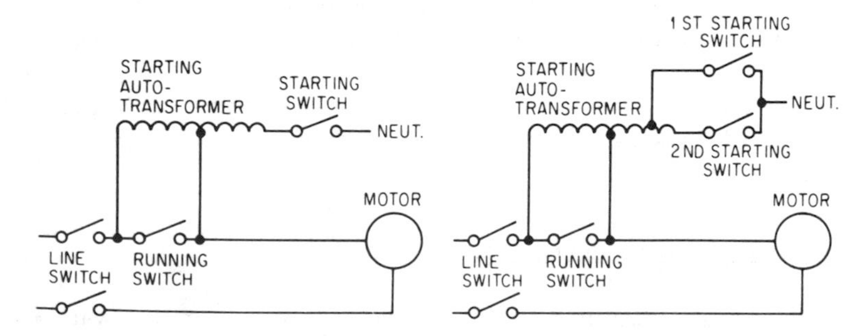

Figure 6-13 Motor-starting arrangement using auto-transformers. (A) One-step connection. (B) Two-step connection.

consideration in the choice of transformer size and design. Starters are usually designed so that they cannot rest in the starting position, or in any position which will make inoperative the protective devices in the circuit.

X-RAY EQUIPMENT

X-ray equipment usually requires dc voltages from 10,000 volts to 500,000 volts or higher for operation. The voltage wave may be a sharp peaked one and the duration of the current flow only a fraction of a second. Current values may range from milliampere (1/1000 ampere) to several amperes. The high voltage is obtained by means of a transformer, usually operating at 240 volts primary, and is converted from alternating current to direct current by means of vacuum tubes or rectifiers. The transformer may be oil filled or air cooled, and is usually situated as near the x-ray tube as possible to reduce the length of high-voltage wiring required. Should the transformer fail, the secondary may come in contact with the primary voltage of only 240 volts; hazards from the primary voltage are low and hence elaborate construction or precaution is not usually required. In addition to medical uses, x-ray equipment is also employed in the examination of fabrics, metals, plastics, fruits, and other purposes.

WELDING TRANSFORMERS

Welding transformers are usually single-phase and operate with primary voltages of 240 or 480 volts. Secondary voltages are low as may be expected from the high current requirements. The load created by the arc is highly inductive and, if the transformer should be disconnected on the primary side (by safety devices or for other reasons), the arc across the switch or breaker contacts would tend to persist. Switches used to control equipment of this type should have ample interrupting capacity.

SMALL INTERMITTENT-USE EQUIPMENT

Small transformers for such uses as bell ringing, annunciator systems, indicating lights, and other intermittent uses are usually of the dry type and operate from a 120 volt source for their primary and up to 15 volts for their secondary voltage. Their input is usually not over 50 watts. They are intentionally designed to have poor regulation, that is, the secondary voltage falls very rapidly as the load is increased, the current flowing increases only slightly, thereby decreasing their probability of failure. They may be protected with a small fuse (5 ampere) on the 120 volt side.

IGNITION COILS

The ignition coil in the ignition system of an internal combustion engine is essentially a two-winding transformer having a primary of a few turns operating usually at 6 or 12 volts with a secondary of many turns of fine wire, from 40 to 100 times the number in the primary (Fig. 6-14). An iron core serves to strengthen the magnetic field. The primary is connected to a battery with a mechanism in the circuit to open and close the circuit periodically. The dc from the battery builds up the magnetic field; as the circuit is open, the field collapses and cuts the secondary, thereby inducing a voltage in the secondary coil. The secondary circuit is completed through a spark gap at the end of the sparkplug in the cylinder; the induced voltage is high enough to cause a spark to jump across the spark gap and ignite the gasoline and air mixture in the cylinder.

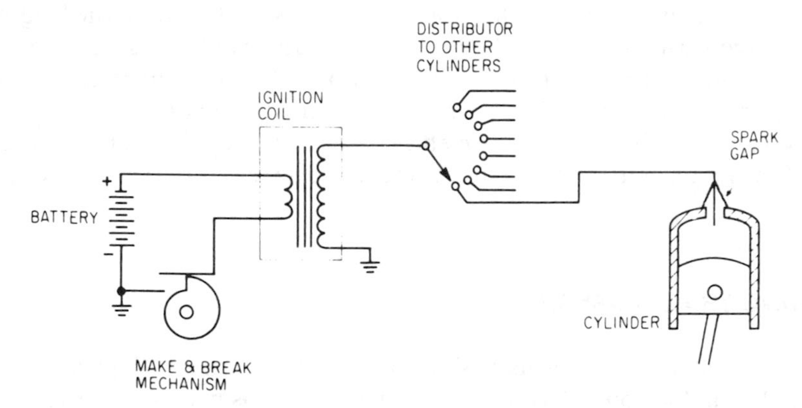

Figure 6-14 Ignition system for internal combustion engine.

It is obviously impractical to describe all types of transformers and their application. Examples of representative types have been included above. For other special types or applications, references should be made to more complete technical texts, manufacturers' data, and instruction manuals.

REVIEW

1. Autotransformers have only one winding, the entire winding being used as either primary or secondary, while a portion of it serves as secondary or primary (Fig. 6-1). It is used where ratios of transformation are low, usually less than about 5 to 1, and where the electrical isolation between the primary and secondary circuit is not essential.
2. In the autotransformer, only a portion of the energy is transformed, and the remainder flows conductively through its winding; hence, it can be

smaller and have less losses than a corresponding two-winding transformer. Its relationship between voltage ratio, current ratio, and turn ratio is the same as for two-winding transformers.

3. Autotransformer connections may be the same as for two-winding transformers. They may be connected in either delta or wye, but may not be connected in combination of these. They may also be used for three-phase to two-phase transformation (Fig. 6-2).
4. Voltage regulators are a particular application of the autotransformer. The voltage in the portion of the winding which is added or subtracted is made variable, as the outgoing voltage may be kept at approximately rated value. In the induction regulator (Fig. 6-3), this portion of the coil can be rotated with reference to the major coil, so that the voltage induced in it can vary as the coils deviate from parallel positions. In a second type, the portion of the coil whose voltage is to be added or subtracted is equipped with taps; selecting the proper tap adds or subtracts more or less voltage to or from the incoming voltage so that the output voltage may be kept at approximately the rated value (Figs. 6-5 and 6-6).
5. Transformers are also used for metering and relaying purposes, when only a small portion of the original values need be used (Fig. 6-7). These are known as potential transformers (PT) for measuring voltage; and current transformers (CT) for measuring a current. PT's are connected in parallel with the main supply circuit (Fig. 6-8), while CT's are connected in series (Fig. 6-9); relative polarity of primary and secondary windings of these transformers is also necessary. The capacity of these transformers is usually very small; and be of dry or oil filled types.
6. PT's have small output and high accuracy, they have small regulation or voltage drop, the windings are also compensated to give its exact ratio at some specified rated burden. CT's also have small output and high accuracy. The primary may be one or a few turns and insulated for line voltage, the insulated conductor itself may serve as the primary passing through the core around which the secondary windings are wound. Because of the high turn ratio usually met in CT's, it is possible to obtain an unsafe high voltage across the secondary terminals. The regulation is so designed that the voltage drops very rapidly as the burden is connected to it. For safety reasons, the terminals of the CT's should be short-circuited when instruments are not connected to them; instruments should first be connected *before* the short-circuit strap is removed.
7. Series street lighting employs types of transformers including a constant-current transformer (Fig. 6-10), where the primary and secondary coils move away or closer to each other on the iron core causing the secondary voltage to change, but delivering an essentially constant value of current. Other two-winding transformers are used to supply an individual light

(IL) or several lights (SL) from such series circuits (Fig. 6-11), either for safety in having the lights electrically isolated from the main circuit, or when the lamps operate at different current ratings than the main circuit.

8. Other transformers and autotransformers having characteristics special to their application include those for sign lighting (Fig. 6-12), fluorescent, mercury, sodium, and halogen vapor lamps, for motor starters, (Fig. 6-13), for x-ray equipment, for welding, for small intermittent use as in bells and annunciators. Ignition coils for use with internal combustion engines are also essentially two-winding transformers (Fig. 6-14).

STUDY QUESTIONS

1. What is an autotransformer and where is it used?
2. How does the operation of an autotransformer differ from that of the two-winding transformer?
3. What are some advantages and disadvantages of an autotransformer?
4. How are autotransformers connected in polyphase circuits?
5. What is an induction voltage regulator and where is it used?
6. What is a step-type regulator? What is it also called?
7. What are instrument transformers and where are they used?
8. How are current transformers (CT) and potential transformers (PT) connected in a circuit?
9. What precautions should be taken with current transformers and potential transformers?
10. What is a constant-current transformer and where is it used?

7

TESTING AND TROUBLESHOOTING

Tests are made on a transformer and its accessories for a variety of reasons: during manufacture, to check the condition of the components making up the assembly; at delivery, to ascertain that the prescribed standards are met; when in service, to determine its operating condition and maintenance requirements; as part of the maintenance process; under contingencies, to determine the sources and kinds of trouble; and, after repairs, to ascertain that the unit will meet the demands that will be made on it.

The tests described are somewhat basic. Their application may vary with the individual transformer under test. Further, these tests may be changed, to accommodate a particular type, class, or vintage of transformer, and to conform with equipment available, local conditions, and other factors. Specifically, the method of cooling may have important impact on the tests chosen and any variations made in them. Natural air cooled, forced air cooling, oil cooling, forced oil cooling, askarel cooling, water cooling, and combinations of these all present special problems and the test methods chosen should take these, among other factors, into account.

LOW-VOLTAGE TESTS

These tests can be conveniently divided into two classes: low and high-voltage tests. The following are low-voltage tests.

RESISTANCE TESTS

These tests are made to measure the resistance of each winding to calculate power (I^2R) loss, the resistance component of the voltage drop under load, to determine the temperature rise under load, and to ascertain that the internal connections are correct. A dc source should be used, together with ammeters and voltmeters of proper ranges. Temperature measurements of the windings should be taken while the resistance measurements are made—these may be thermometers or thermocouples. As a precaution, to avoid possible damage from voltages induced by the dc current causing the magnetic field to build up and collapse on being switched on and off, the winding whose resistance is not being measured should be temporarily short-circuited (Fig. 7-1). The re-

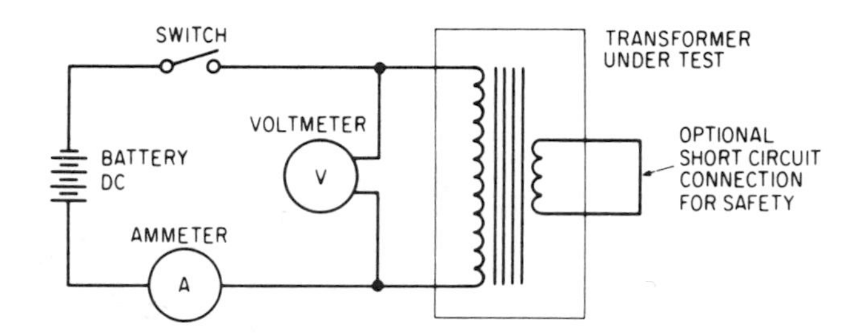

Figure 7-1 Connections for measuring transformer winding resistance.

sistance of each winding is obtained from the Ohm's Law derived equation: $R = E/I$ or voltage value (volts) divided by the current flowing (amperes).

POLARITY TESTS

These tests are made to determine how the coils of a transformer are wound in relation to each other, so that the "direction" of the secondary voltage may be known when connecting transformers in parallel or in polyphase banks. In general, the leads are marked on the high side with the letters H_1, H_2, etc., reading from right to left; on the low side with the letters X_1, X_2, etc., reading from left to right for subtractive polarity and from right to left for additive polarity.

Single-phase Transformers

A single polarity check may be made by connecting a low-voltage ac to the high side of the transformer and having the H and X windings connected in series, one H lead being connected to the adjacent X lead (Fig. 7-2). If the

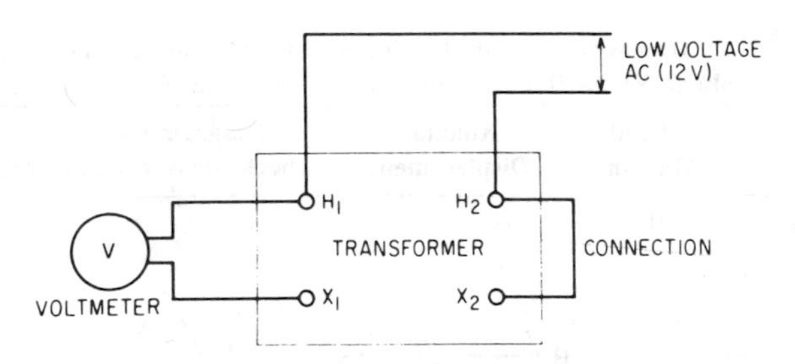

Figure 7-2 Connection for testing polarity of single-phase, two-winding transformer (subtractive polarity shown).

voltage across the open leads is greater than that applied across the H or high side winding, the polarity is "additive," if less, the polarity is "subtractive." If there are additional windings, the third is marked with the letter Y and the fourth with Z, the letter H for the highest-voltage winding, X for the next highest, etc.

Three-phase Transformers

Polarity for three-phase transformers is somewhat more complex. In addition to the direction of the winding, there may exist an angular displacement between the secondary voltages, depending on how the high voltage windings are connected to the three-phase primary; these will be described below.

To simplify the connection of three-phase transformers, the H_1 lead is brought out on the right-hand side facing the high-voltage side of the transformer case. The X_1 lead is brought out on the left-hand side facing the low-voltage side. The remaining H and X leads are brought out numbered as shown in Table 7-1. Winding leads marked Y and Z, if any, are brought out and numbered in the same manner as the X leads.

A single-phase check of the three-phase transformers can be made by testing each phase, one at a time. Referring to one of the connections in Table 7-1 (e.g., the delta-delta connection), a low ac voltage may be applied between H_1 and H_2 and the voltage measured between H_2 and X_2 should be equal to the difference of the voltages measured between H_1 and H_2 and between X_1 and X_2. Similarly, a low ac voltage may be applied between H_1 and H_3 and the voltage measured between H_3 and X_3 should be equal to the difference of the voltages measured between H_1 and H_3 and between X_1 and X_3. Again, applying the low ac voltage between H_2 and H_3, the voltage measured between H_2 and X_2 should be equal to the difference of the voltage measured between H_2 and H_3 and between X_2 and X_3.

TABLE 7-1 Transformer Lead Markings—IEEE Classification of Polarities Showing Voltage-Vector Diagrams for Three-Phase Transformer Connections

Groups	Lead Markings	Angular Displacement	Diagram for Check Measurement	Check Measurements
NO. 1: ANGULAR DISPLACEMENT ZERO DEGREES	H_1 H_2 H_3 / X_1 X_2 X_3	H_2, H_1, H_3; X_2, X_1, X_3 — DELTA-DELTA CONNECTION	H_2, X_2, H_1, X_1, H_3	CONNECT H_1 TO X_1 MEASURE H_2X_2; H_3X_3; H_1H_2; H_2X_3;
	H_0 H_1 H_2 H_3 / X_0 X_1 X_2 X_3	H_2, H_0, H_1, H_3; X_2, X_0, X_1, X_3 — WYE-WYE CONNECTION	H_2, X_2, X_1, H_1, X_3, H_3	VOLTAGE RELATIONS: (1) $H_2X_3 = H_3X_2$ (2) $H_2X_2 < H_1H_2$ (3) $H_2X_2 < H_2X_3$
NO. 2: ANGULAR DISPLACEMENT 180 DEGREES	H_1 H_2 H_3 / X_1 X_2 X_3	H_2, H_1, H_3; X_3, X_1, X_2 — DELTA-DELTA CONNECTION	H_2, X_3 X_1, H_1, H_3, X_2	CONNECT H_1 TO X_1 MEASURE H_2X_2; H_3X_3; H_1H_2; H_1H_3;
	H_0 H_1 H_2 H_3 / X_0 X_1 X_2 X_3	H_2, H_1, H_3; X_3, X_1, X_2 — WYE-WYE CONNECTION	H_2, H_1, X_3, X_1, H_3, X_2	VOLTAGE RELATIONS: (1) $H_2X_2 = H_3X_3$ (2) $H_1H_2 < H_2X_2$ (3) $H_1H_3 < H_3X_3$

RESISTANCE AND POLARITY TESTS

These two tests may be made at the same time using the same dc source for the polarity test. Referring to Fig. 7-1, the dc battery is connected in place of the ac source across one winding (normally the high-voltage one) and the voltmeter placed across its terminals to obtain a positive deflection. The two voltmeter leads are then transferred directly across the transformer to the low-voltage terminals, without crossing the leads. The current in the first winding is interrupted, inducing a momentary voltage in the second. If a negative "kick" is observed on the voltmeter, the polarity of the transformer is subtractive; if a positive "kick," the polarity is additive.

PHASE DISPLACEMENT TEST

As mentioned earlier for three-phase transformers, it is necessary to know the polarity of each set of phase windings, as well as the angular displacement that

TABLE 7-1 (cont.) Transformer Lead Markings—IEEE Classification of Polarities Showing Voltage-Vector Diagrams for Three-Phase Transformer Connections

Groups	Lead Markings	Angular Displacement	Diagram for Check Measurement	Check Measurements
NO. 3: ANGULAR DISPLACEMENT 30 DEGREES	H_1 H_2 H_3 X_0 X_1 X_2 X_3	H_2 H_1 H_3 X_2 X_1 H_0 X_3 DELTA-WYE CONNECTION	H_2 X_2 H_1 H_3 X_1 X_3	CONNECT H_1 TO X_1 MEASURE H_2X_2; H_3X_3; H_1H_3; H_2X_2; H_2X_3
	H_0 H_1 H_2 H_3 X_1 X_2 X_3	H_2 H_0 H_1 H_3 X_2 X_1 X_3 WYE-DELTA CONNECTION	H_2 H_1 X_2 H_3 X_1 X_3	VOLTAGE RELATIONS: (1) $H_3X_2 = H_3X_3$ (2) $H_3X_2 < H_1H_3$ (3) $H_2X_2 < H_2X_3$ (4) $H_2X_2 < H_1H_3$

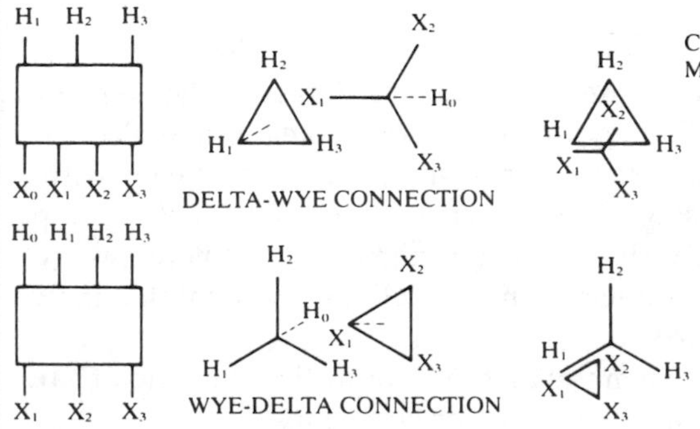

EXAMPLE OF A THREE-PHASE TRANSFORMER WITH TAPS

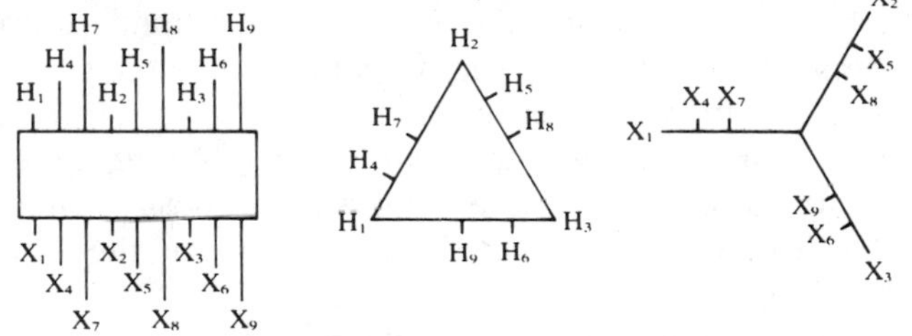

H_7 H_8 H_9 H_4 H_5 H_6 H_1 H_2 H_3 / X_1 X_2 X_3 X_4 X_5 X_6 X_7 X_8 X_9

H_2 H_5 H_8 H_7 H_4 H_1 H_9 H_6 H_3

X_2 X_5 X_8 X_4 X_7 X_1 X_9 X_6 X_3

may exist between corresponding secondary windings before such units can be connected in parallel or in banks.

The typical three-phase transformer has a nameplate which includes a diagram (called a voltage-vector diagram) showing the relations between the voltages of the three phases in accordance with one of the groups shown in Table 7-1. The angular displacement between the H and X voltages (the high-voltage and low-voltage winding) is the angle between lines passing from the neutral point of the voltage-vector diagram through H_1 and X_1 respectively.

Leads are marked so that the sequence of the H and X voltages (usually referred to as the "phase rotation") in the order H_1, H_2, H_3 and X_1, X_2, X_3 will be the same. That is, if a three-phase motor could be connected across the terminals H_1, H_2, and H_3 (let the imagination forget the difference in voltage magnitudes for time being) it would rotate in the same direction as when connected across the terminals X_1, X_2, and X_3.

To check for phase displacement, one of the H leads is connected to one of the X leads, as shown in Table 7-1. The primary of the transformer is connected to a low-voltage, three-phase source (preferably much lower than the rated secondary voltage of the transformer—a precautionary safety measure) and the voltage measured between leads as shown; the voltage relations should be in agreement with the comparisons indicated.

RATIO TESTS

The ratio of a transformer is the ratio of voltage of the high-voltage winding to the low-voltage winding for two-winding transformers. When there are more than two windings, there are several ratios, all taken with respect to the high-voltage winding. The various voltages obtainable on a particular transformer are usually shown on the nameplate. Where two (or more) transformers are to be connected in parallel, it is essential that the voltage ratios of each should be correct.

There are two general methods for testing the ratio of a transformer: one for small transformers, and one for large ones.

For smaller transformers, used for general distribution purposes, the transformer to be tested is connected in parallel with a standard transformer; between the terminals of the secondaries in parallel, a voltmeter is connected as shown in Fig. 7-3. If the voltmeter reads zero, the two transformers have the same ratio and the same polarity; if the voltmeter reads double the secondary voltage, the polarities of the two transformers are opposite. Should the two transformers be alike in polarity but different in ratio, the voltmeter will indicate the difference.

For larger transformers, including station-type transformers, the test is made by connecting voltmeters to both the primary and secondary sides (usually employing potential transformers). The voltage on the high or primary side is held at a particular value, while the voltmeter on the secondary side is read. The voltmeters are interchanged to compensate for meter errors, and the test repeated. The average of the two is used to calculate the ratio. An ammeter is sometimes connected in the circuit to read the exciting current,

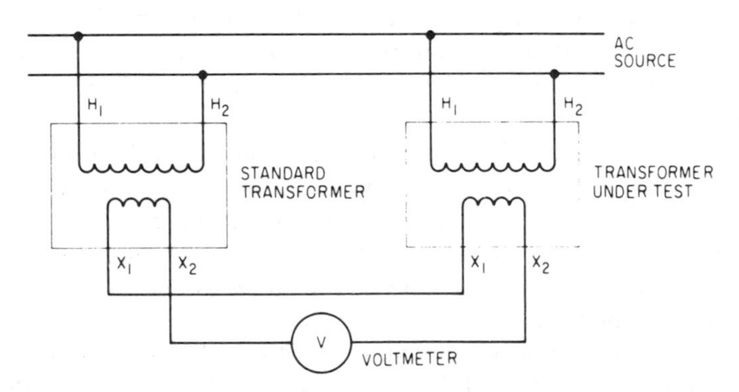

Figure 7-3 Connections for ratio (and polarity) tests using standard transformer.

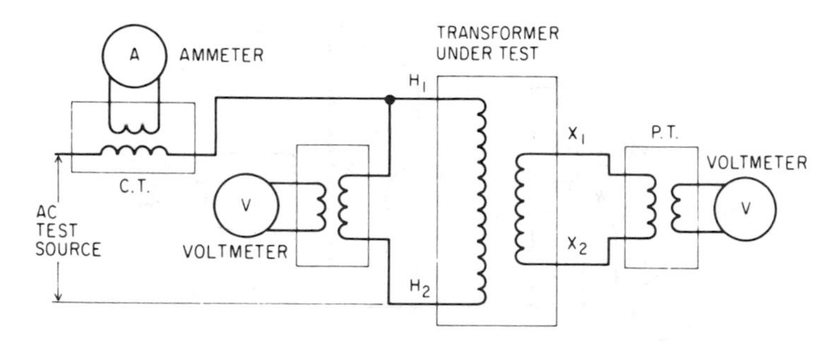

Figure 7-4 Connections for ratio test using two voltmeters.

which should be kept to approximate design values to eliminate errors from excessive voltage drop in the primary.

For three-phase transformers where three-phase power supply is used for testing, it is important that the voltmeter readings be taken on corresponding windings. For power transformers the ratio usually must be within plus or minus 1 percent. Figure 7-4 shows the test circuit diagram.

For induction voltage regulators, the ratio may be tested by applying a voltage on the primary and reading the voltage across the secondary with the armature in the maximum boost position. For tap changing under load transformers, the same procedure is applied with the unit set for maximum boost position; tap position voltages may also be measured. For three-phase regulators, measurements may be taken for maximum lower positions in addition to the maximum boost position.

IMPEDANCE TESTS

The impedance of a transformer is important since, to a large extent, it determines the regulation of the transformer. It is made up of the internal resistance and reactance and will create a voltage drop caused by current flowing in the windings. Impedance losses (measured in watts) are a combination of heat (I^2R) losses in the conductor, eddy-current loss, and losses in other parts of the transformer caused by stray currents. Impedance tests show whether the coil and core arrangements produce satisfactory voltage drops and losses in the transformer. They also determine whether transformers may be satisfactorily paralleled since the manner in which such transformers divide the load depends on their impedance voltages; that is, the one having the higher impedance will take the smaller part of the load, and vice versa.

The impedance of a transformer is measured by short-circuiting one winding (usually the secondary) and with instruments connected (Fig. 7-5). An ac voltage is impressed on the primary winding and adjusted until the transformer is carrying rated current, as shown by the ammeter. When full-load current flows in the primary, full-load current will also flow in the short-

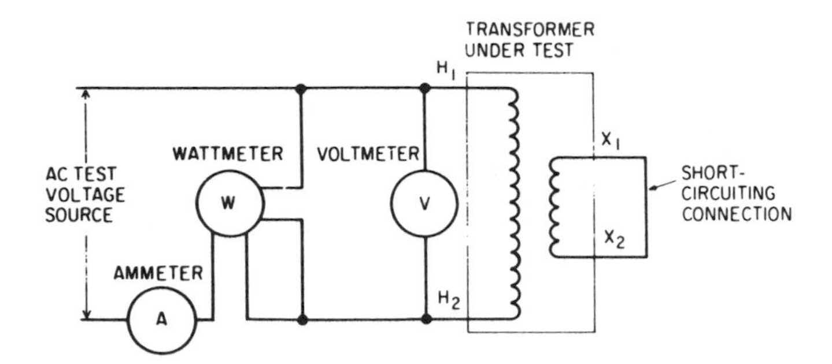

Figure 7-5 Connections for impedance test.

circuited secondary. The transformer should be allowed to heat to approximately its normal full load operating temperature (of about 75°C). Simultaneous readings of the three meters are taken. The temperature reading should also be taken so that the so-called copper losses (heat losses in the winding) may be adjusted to the standard temperature of 75°C. The value of the ac test voltage applied to create full-load current is the impedance voltage, and the percent impedance of the transformer may be determined as follows:

$$\text{percent impedance} = \frac{\text{measured impedance voltage}}{\text{rated voltage of the primary (windings)}} \times 100$$

From the test data, the total equivalent resistance, the reactance and impedance, in ohms, may be determined and the data may also be used for calculating the efficiency of the transformer. The value of impedance volts, usually expressed as percent impedance, is usually placed on the nameplate.

Similar procedures are employed in measuring the impedance of a three-phase transformer. Three-phase source voltage and three sets of instruments may be used, although the power loss may be measured by two voltmeters, rather than three, properly connected to read total three-phase power.

The same procedures are used when obtaining the impedance of an induction voltage regulator. The impedance is measured on the regulator primary with the secondary in the maximum boost position. The secondary is short-circuited on itself, and the ratio current is held on the primary. Both volts and watts are read.

TRANSFORMER LOSSES

For practical purposes, the major transformer losses, the core loss of a transformer and the copper (or winding) losses may be determined separately by

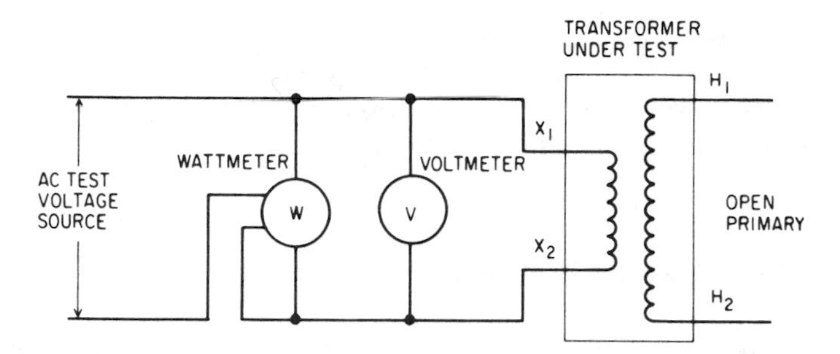

Figure 7-6 Connection for determining core loss.

comparatively simple tests, and the total losses of the transformer found by adding them together.

Core Loss

These may be determined by a connection similar to that for the impedance test (Fig. 7-6). Care should be taken with the primary terminal leads as they are left open-circuited and a high voltage may be produced across them. The rated low voltage is applied across the secondary winding and the wattmeter reading will represent the core loss.

Copper Loss

The transformer under test is connected again as for the impedance test (Fig. 7-5). The low voltage winding is short-circuited, and the voltage applied to the primary or high voltage winding is adjusted so that full-load (primary) current is indicated by the ammeter. The wattmeter reading will represent the full load copper loss.

CONTINUITY TEST FOR TCUL TRANSFORMERS

This test is to ascertain that the circuit remains closed throughout the whole range of tap changes in a Tap Changing Under Load transformer. The connections for this test are similar to those for impedance tests; in fact, this test is often made just before the impedance test is run.

With the low-voltage winding short-circuited, and full-load current in both windings, the tap changing mechanism is operated through its full range and the ammeter observed throughout to make sure the current is not interrupted, which would be an indication of an open circuit at the tap position where such interruption is noted. The current is recorded for each tap position.

Whether the taps are on the high voltage or low-voltage position does not make any difference. If they are on the high-voltage side and the circuit is open, the current will drop to only the exciting current in the primary produced by the (impedance) voltage being applied.

TEMPERATURE TEST OR HEAT RUN

This test is made to determine the temperature rise of the transformer under rated load. Thermometers or thermocouples are installed to measure ambient temperatures, oil surface temperatures in the transformer tank and at some point (perhaps 2 or 3 inches) below the oil surface where maximum temperatures may be expected. Readings on these are taken before the test is started to obtain base temperatures on which to determine the rise at rated load. These are also read periodically during the test, and continued until temperatures do not vary more than a few degrees over a period of hours, and hence may be considered as stabilized.

LOADING TRANSFORMERS UNDER TEST

There are three ways of loading the transformer under test: the direct method; the opposition method; and the compromise short-circuit method.

The Direct Method

Used for very small transformers, which may be energized in the normal manner and loaded by means of a lamp bank or other resistance load. For most transformers, this is not practical.

The Opposition Method

The unit under test is connected to a duplicate transformer as shown in Fig. 7-7. The two transformers are connected in parallel on both the high voltage and low voltage sides. Full rated voltage is applied to the low voltage or secondary side, and a voltage from an external source is inserted in the high voltage or primaries in parallel as shown, just large enough to circulate rated current through the primaries of the two transformers. The two transformers are thus both carrying their normal currents in both their primary and secondary windings. The loading transformer should be interposed between the test source and the primaries of the transformers under test to avoid imposing a high voltage and possible insulation breakdown on equipment constituting a test source.

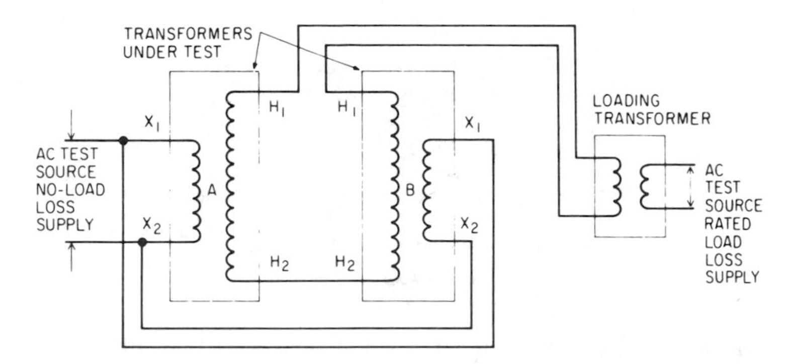

Figure 7-7 Connections of "opposition method" for temperature test.

Compromise Short-Circuit Method

One winding, usually the secondary, of the transformer under test is short-circuited, similar to the connection for the impedance test, shown in Fig. 7-5. A current is circulated in the other winding which will give an impedance loss equal to the sum of the core loss and the normal impedance loss. (See also Impedance Tests and Transformer Losses previously discussed.)

With each method, it is necessary to obtain the winding resistances both when the transformer is cold and after the oil heating is stabilized. The hot transformer is temporarily disconnected while the resistance test is made. (See Resistance Tests at beginning of chapter.) The final temperature of the windings is calculated from the formula:

$$\text{Final Temperature} = \frac{\text{Hot Resistance}}{\text{Cold Resistance}} \times (234.5 + \text{Starting Temperature}) - 234.5$$

The temperature *rise* of the windings is found by subtracting the final ambient temperature from the final winding temperature, found above. The oil temperature *rise* is found by subtracting the final ambient temperature from the final oil temperature.

Connections for testing three single-phase transformers or for a three-phase transformer when the windings can be connected in delta is shown in Fig. 7-8.

HIGH-VOLTAGE TEST

Those tests in which high voltage may be applied are usually associated with the insulation of transformers.

The object of this test is to show that the insulation in the transformer winding is adequate. A higher than rated voltage is applied between the

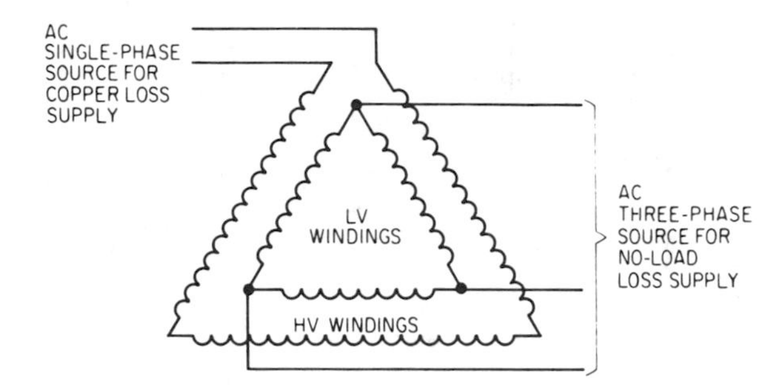

Figure 7-8 Connections for temperature test of three, single-phase transformers or three-phase delta transformer.

winding under test and the other windings and ground. The terminals of the winding under test are connected together and to one terminal of the testing equipment. The terminals of the other windings are connected together and to the core and tank and to the other terminal of the testing equipment; this second series of connections is usually connected to ground as a safety precaution. The connection is shown in Fig. 7-9. The voltage applied is raised gradually to a predetermined value, held at that value for a predetermined time, usually 1 minute, and then reduced to a low value before the circuit is opened. The test voltage is usually about twice the rated voltage of the winding. Repeated testing may damage the winding; hence, the original acceptance test may be done at this voltage, but succeeding tests for maintenance may be held to 60–70 percent of the original test.

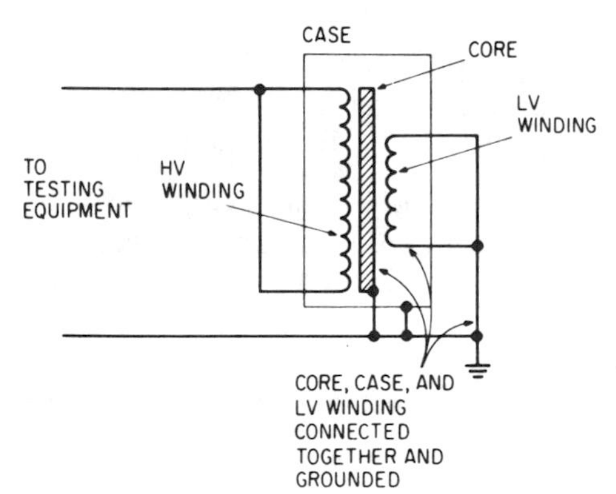

Figure 7-9 Connection for high voltage test.

IMPULSE TEST

This test is made to prove that the transformer insulation will withstand voltage surges which may be caused by lightning or switching; this includes insulation to ground, insulation between turns and windings, and the flashover

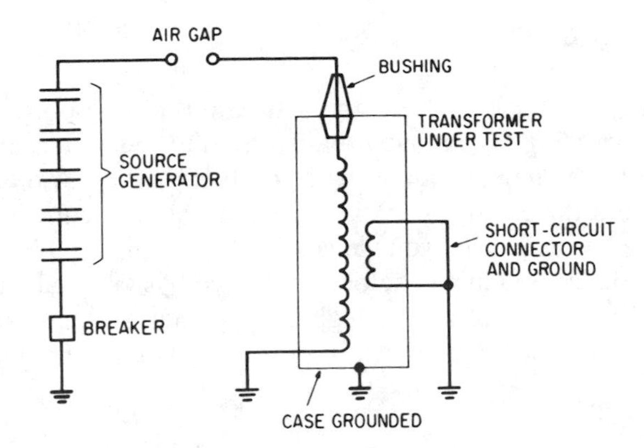

Figure 7-10 Connection for impulse test.

value of the associated bushings. A high-voltage wave of standard values, and approximating a lightning surge, is imposed on the unit to be tested. The basic circuit is indicated in Fig. 7-10. The surge generator usually consists of a number of capacitors connected so that they be charged in parallel from a relatively low-voltage source and discharged in series to give a high voltage across the test piece. A "standard" impulse wave is illustrated in Fig. 7-11.

The "standard" wave reaches its peak voltage value in 1½ microseconds and reduces to half the voltage value in 40 microseconds. The value of the voltage applied depends on the rating of the insulation, and may vary from 5 to 30 times the voltage rating of the insulation. Since rather elaborate and costly equipment is needed, impulse tests are usually performed only on large station-type power transformers, and rarely made after they leave the factory.

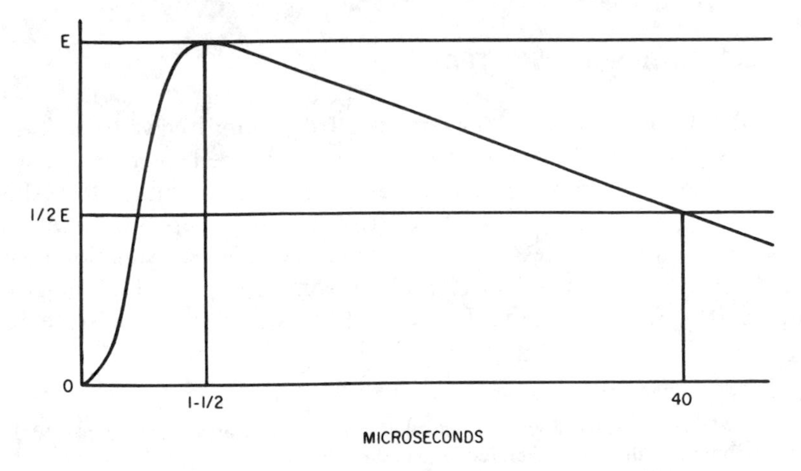

Figure 7-11 Shape of standard 1½ x 40 microsecond wave.

INDUCED-VOLTAGE TEST

The purpose of this test is to check the insulation between turns, layers, and sections of the winding. A voltage usually equal to twice the rated voltage is applied to the low-voltage windings with the high-voltage winding left open-circuited. To limit the exciting current, voltages having frequencies above the rated (60 cycle) are used. The voltage is raised gradually and held for about 1 minute and reduced gradually before the circuit is opened. The time the voltage is held is decreased as the higher frequency voltages are used. The diagram for this connection is shown in Fig. 7-12. The sphere gaps connected

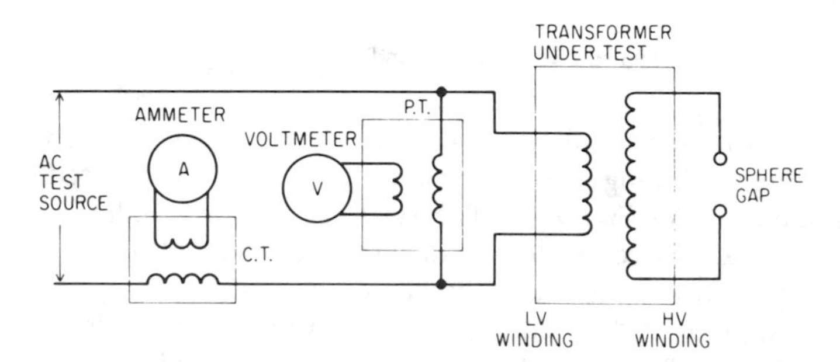

Figure 7-12 Connection for induced voltage test.

across the open-circuited winding act as a safety precaution for high-voltage tests over 50 kV. For windings designed for operation with one end solidly grounded, this test is the only low-frequency insulation test that can be made.

INSULATION-RESISTANCE TEST

This is the often mentioned Megger* test. It is meant to give some indication of the condition of the insulation, and is often used in maintenance procedures. The connections are the same as for the high-voltage test, shown in Fig. 7-9. The insulation resistance of a transformer depends largely on the temperature and cleanliness and dryness of the windings. Insulation resistance should be at least 1 megohm (1 million ohms) for each 1000 volts of test voltage. If it falls below this figure, presence of dirt or moisture may be indicated.

* A Megger is a small dc generator, turned by hand or motor, which is connected through a meter (mounted on the case) calibrated to read directly the megohms in the circuit connected to the terminals. See Chapter 4.

LIQUID INSULATION TESTS

This usually calls for a cup of oil or askarel in which two insulated terminals are set, usually 0.1 inch apart, and across which a voltage of 20,000 volts or more is applied. Good oil or askarel should not break down at voltages under this value. (The process is described in Chap. 4, under Maintenance.)

BUSHING POWER-FACTOR TEST

In addition to the resistance and high-voltage tests, the measurement of power factor of a bushing is used to determine the condition of the insulating materials in a bushing. The bushing may or may not be removed from the transformer, but it must be disconnected from other apparatus and the connecting bus. The power factor is obtained by reading volts, amperes, and watts in the circuit (Fig. 7-13). The instruments and instrument transformers are sometimes contained in a package, with only a power factor indicating scale and voltage control device mounted on the package. Power factor values are in the range of 5 to 10 percent. It is not the single reading that necessarily determines the condition of the bushing insulation. As it deteriorates, more current will leak through the insulation and the power factor, therefore, becomes increasingly greater, (approaching 100 percent). Power-factor readings are taken periodically, as part of maintenance programs. These readings are compared, not with readings of other bushings, but with previous readings on the same bushing. To what extent power factor increases may be acceptable is a matter of judgement. "High" values may depend on the importance of the

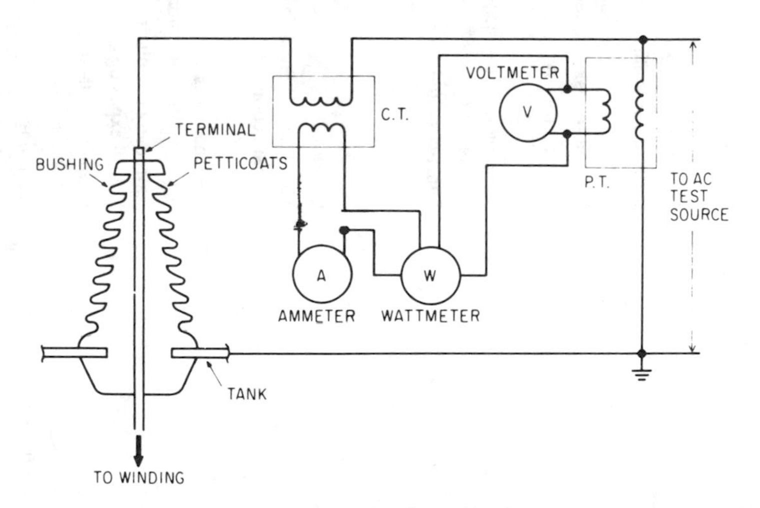

Figure 7-13 Connection for power factor test.

TABLE 7-2 Transformer Troubleshooting

Condition	Possible Cause	Suggested Remedy
Hot transformer	High ambient temperature	Improve ventilation or relocate unit to cooler location
	Overload	Reduce load; reduce amperes by improving power factor with capacitors; check for circulating currents for paralleled transformers—different ratios or impedances; check for open phase in delta bank
	High voltage	Change circuit voltage, taps.
	Insufficient cooling	If other than naturally cooled, check fans, pumps, valves, and other units in cooling systems
	Low oil (or askarel) level	Fill to proper level
	Dirty oil (or askarel)	Filter to remove dirt, sludge, or carbon; clean core, coils, and terminal board with clean, filtered oil or askarel
	Winding failure—incipient fault	See "No voltage—unsteady voltage" below
	Short-circuited core	Test for exciting current and no-load loss; if high, inspect core, remove and repair; check core bolts, clamps and tighten; check insulation between laminations; if welded together, return to factory for repair or replacement
Noisy transformer	Overload	See "Hot Transformer" above
	Metal part ungrounded, loose connection	Determine part and reason; check clamps, core other parts normally grounded for loose or broken connections, bolt fallen out, nuts vibrated off, bolts, etc; tighten loose clamps, bolts, nuts; replace missing ones
	External parts and accessories in resonant vibration	Tighten items as above; in some cases, loosen to relieve pressure causing resonance and install shims
	Extremely low oil or askarel level exposing live parts	Check cause (leak, etc.), repair or replace damaged part, refill to proper level
	Incipient fault—core or winding	See above under "Hot Transformer"
Bushing flashover	Lightning	Check lightning protection; arresters, connections, lugs, grounds
	Dirty bushings	Clean porcelains; check source of dirt
Broken bushing or parts	Strains on terminal connections	Flexible connections inserted between terminals and cables or busses to remove strain from bushing
No voltage—unsteady voltage	Winding failure—lightning; overload; short-circuit from	Check winding, remove foreign object or damaged material; repair or replace parts of insulation materials, if possible, otherwise return to

	foreign object or low strength dielectric (oil or askarel)	factory, replace oil or askarel; check relief diaphragms and seals for entry of moisture
Leaks	Cracks; holes; loose bolts or poor threading; imperfect welds, improper assembly; poor gasket installation	Repair damaged parts, or replace, if possible, otherwise return to factory; check and tighten bolts, clamps on cover, hand-holes, bushings, etc. Replace gaskets, do not attempt to salvage them
Rust and paint deterioration	Weather, polluted or salt atmosphere; overloads	Remove rust and deteriorated paint; clean surfaces; repaint with proper paints and sufficient coatings
Low dielectric strength of oil or askarel	Dirt; moisture; sludge or carbon	See "Hot Transformers"; "No voltage—unsteady voltage"; "Leaks" above
	LINE TROUBLES AFFECTING TRANSFORMERS	
	Overload Hot neutral line	Too small neutral conductor; replace; severe unbalance between phases; rebalance and equalize loads
	One leg of wye bank open	Check associated fuse; if blown, remove cause and replace; check for open circuit in winding of transformer in bank
Voltages unbalanced	Open neutral unbalanced loads	Check neutral connections; see "Hot neutral line" above
Voltages high and unbalanced	Open neutral on wye bank ground in winding of one transformer in wye	Check neutral connections and load balance; check values of voltages between phases and phase to ground voltages; vector should indicate source of trouble
Flashover of bushing on delta-connected bank—fuse possibly blown	Ground on one leg of delta	Transformer winding not designed for full phase-to-phase voltage
No voltage—one phase of delta connected bank	Grounds on two legs of delta (delta collapse—loads "single phasing")	Remove grounds from at least one leg of delta source
Overloads on two transformers in delta bank	Open in third transformer of bank; operating in open delta	Check fuses on supply to their bank; check winding of third transformer for continuity
Low voltage on two phases of delta	Open in one phase of delta supply; two transformers now connected across one same phase	Check fuse on supply; check supply circuit back to source for open circuit

installation and other factors. Values will vary with temperatures, and excessive readings should be compensated for any difference in temperatures at times of test.

TRANSFORMER TROUBLES

Troubles which may be indicated by transformers acting abnormally may be caused by electrical or mechanical failures within the units themselves and the parts associated with them or may be caused by line troubles which may affect the transformer's installation, or combinations of these. It is impractical to attempt to list all possible troubles, situations, and sources of trouble. A few of the more prevalent ones are listed in Table 7-2. These are meant to be representative; other troubles not listed in the table may be found, their causes determined and proper remedies applied by extending the logic and the methods outlined in the tabulation. Those which have been selected for inclusion have been derived from handbooks, manufacturers' maintenance manuals, technical papers, and the author's experiences over many years. To this list, the worker may add those of particular individual importance based on the particular circumstances.

REVIEW

1. Tests are made on transformers to ascertain that prescribed standards are met; to determine operating and maintenance requirements, as part of maintenance; to determine kinds and sources of trouble; and to ascertain that repairs have been effective. Tests can be made with the application of low voltage or high voltage.
2. Low-voltage tests include: resistance tests (Fig. 7-1) to obtain copper losses and check internal connections; polarity tests (Fig. 7-2) to check for relative direction of windings on single-phase transformers (for three-phase transformers see Table 7-1); ratio tests (Figs. 7-3 and 7-4) to check transformation ratio; impedance tests (Fig. 7-5) to determine regulation of the transformer; transformer loss tests (Fig. 7-6) to determine core losses and copper losses; continuity test for TCUL transformers to ascertain that the circuit remains closed while taps are changing; and temperature test (Fig. 7-7) or heat run to determine temperature rise of the unit under load (for three single-phase or three-phase transformer, with winding connected in delta) (Fig. 7-8).
3. High-voltage tests include: high-voltage test (Fig. 7-9) to check adequacy of insulation; impulse test (Figs. 7-10 and 7-11) to prove that insulation will withstand voltage surges caused by lightning or switching; induced

voltage test (Fig. 7-12) to check insulation between turns; insulation-resistance test, using Megger, to check condition of insulation; liquid insulation test to check dielectric strength of oils or askarels; bushing power-factor test (Fig. 7-13) to determine condition of bushing insulation.

4. Transformer troubles from mechanical or electrical failures may be observed, the cause ascertained and remedial measure applied. Similarly, transformers may be affected by line troubles; the causes and remedies may be ascertained and applied (Table 7-2).

STUDY QUESTIONS

1. When and why are transformers tested?
2. What types of low-voltage tests are made on a transformer and why?
3. What types of high-voltage tests are made on a transformer and why?
4. What voltage source is used in a Megger test for testing transformers?
5. What are the possible causes for a hot transformer?
6. What are the possible causes of a noisy transformer?
7. What are the possible causes of a bushing flashover?
8. What are the possible causes of a winding failure?
9. How may winding failure be detected?
10. What are the possible causes of low dielectric strength of oil?

Part II

ELECTRICAL POWER EQUIPMENT

8

CIRCUIT BREAKERS: DESIGN AND CONSTRUCTION

FUNCTION

Switching operations are a necessary part of any system of electrical circuits, primarily those used for the opening and closing of circuits, a function usually called *circuit breaking*. Circuit-breaking equipment varies from very simple devices for opening low-voltage circuits that carry small currents to large and very elaborate devices for interrupting high-voltage circuits that transport large currents and great amounts of energy. The term *circuit breaker,* however, is usually applied to heavy-duty devices capable of interrupting comparatively large electric currents safely.

Circuit breakers are designed not only to carry and interrupt the normal load currents flowing in circuits, but also, and more important, to interrupt any abnormally high current that might flow under fault conditions such as a short circuit. Although the same device is used both to close and open a circuit, the requirements associated with its opening under abnormal conditions are what determine many of its principal construction features and materially affect its cost.

INTERRUPTING DUTY

When an electric current flows through a conductor, a magnetic field is produced about that conductor, the intensity of which depends upon the intensity of the current (Fig. 8-1A). When two or more such current-carrying conduc-

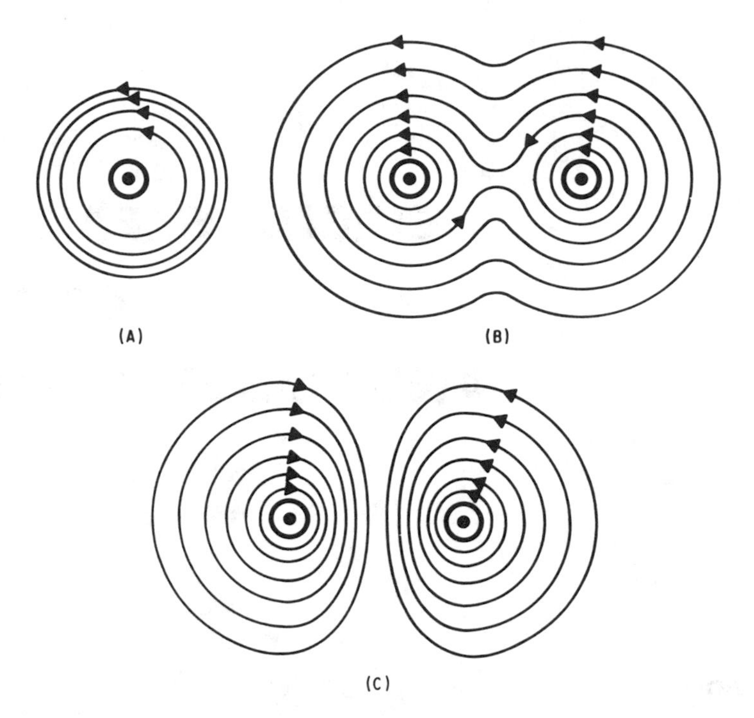

Figure 8-1 Magnetic fields: (A) Intensity of magnetic field around conductor depends on current flowing in conductor; (B) currents flowing in same direction tend to force conductors together; (C) currents flowing in opposite directions tend to force conductors apart.

tors lie adjacent to one another, a force of attraction (Fig. 8-1B) or repulsion (Fig. 8-1C) will exist between them because of the interaction of their magnetic fields. If the currents are extraordinarily high, the forces produced may be enormous.

The magnitude of the current flowing in a circuit under normal conditions depends upon the load and the apparatus or appliances connected to the circuit; it is referred to as the *normal,* or *load current.*

When, however, current flow is not restricted by the load connected to the circuit, because of the failure of one of the circuit components (including the conductors themselves), there is nothing to limit current flow except the resistance of the circuit conductors from the source to the point of failure and the resistance of the fault itself (Fig. 8-2). The current is thus apt to be extremely large and its effects very destructive, making it imperative to interrupt the circuit as quickly as possible to hold damage to a minimum.

Consider a *dead short circuit,* that is, one with no resistance to limit the current flow, applied on conductors that are close to a generator or other source of electric power. Assume also that the voltage generated is 5000 V and

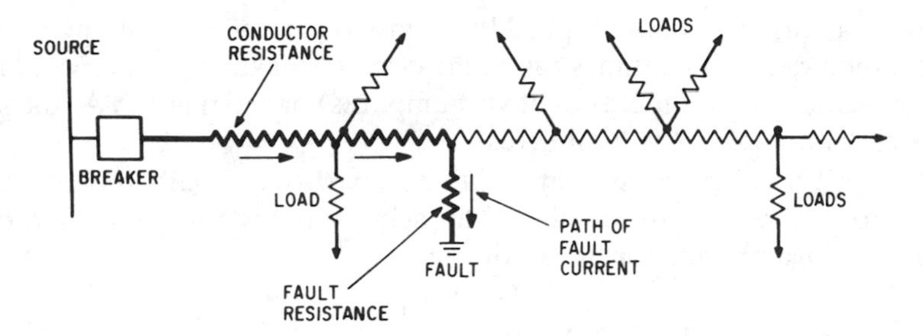

Figure 8-2 Schematic showing resistance to a fault current in a circuit.

that the resistance of the short length of conductors to the point of fault is on the order of 0.1 ohm. Then, by Ohm's Law (current in amperes equals voltage divided by resistance) the current flowing will be:

$$\text{Current} = \frac{5000}{0.1} = 50{,}000 \text{ amperes}$$

If this current were to flow through a breaker, it is obvious that enormous forces would be set up between the current-carrying parts of the breaker, perhaps large enough to destroy it.

Moreover, the magnetic fields set up by the currents in the conductors will induce or create currents in the metallic members of the circuit-breaker structure itself. These, in turn, will set up magnetic fields of their own, which will interact with the other magnetic fields, adding to the destructive forces created by the fault currents.

Since dead short circuits can literally never be realized, the magnitude of the short-circuit current flowing through any breaker will depend on the resistance (in a dc circuit) or the impedance (in an ac circuit) of the system back to the generating source (or sources) as well as on the actual contact resistance and nature of the short circuit. Generally speaking, the further away from the generating source, the lower will be the short-circuit current.

RATING

For the reasons stated, circuit breakers must be constructed mechanically strong enough to withstand the forces set up by enormous short-circuit currents. Conditions at the time of a short circuit are not very stable, that is, the voltage at the point of fault may be higher (from lightning or switching surges) or lower (because of greater loss of voltage from the larger current flow). For this reason, the ability of a breaker to withstand short-circuit forces is expressed in volt-amperes, that is, as the product of the nominal circuit voltage and the short-circuit current for which the breaker is designed. Since 1 VA

(1 volt-ampere, or 1 volt multiplied by 1 ampere) is a very small quantity, the short-circuit rating, sometimes called the circuit breaker *duty,* is given in kVA (kilovolt-amperes, or thousands of volt-amperes) or often in MVA (megavolt-amperes, or millions of volt-amperes).

Circuit breakers must conform to the particular characteristics on the system to which they are applied. The specifications for a circuit breaker, therefore, must provide for the following:

1. Operating voltage of the circuit
2. Normal operating or maximum load current
3. Maximum abnormal or fault current that must be interrupted

The first characteristic determines the insulating requirements; the second, the requirements of the normal or load-carrying parts; and the third, the mechanical requirements of the breaker itself and of its supporting structure.

PRINCIPLE OF OPERATION

In interrupting a circuit, the circuit breaker actually makes a physical separation in the current-carrying or conducting element by inserting an insulating medium sufficient to prevent current from continuing to flow. In so doing, the persistence of an arc across the gap is prevented.

The circuit is usually opened by drawing out an arc between contacts until the arc can no longer support itself. The arc formed when the contacts of a circuit breaker move apart to interrupt a circuit is a conductor made up of ionized particles of the insulating material. With the advent of higher voltage power systems, the length of arc that would be necessary before the arc would extinguish itself in air would require breakers of such physical size as to be impractical. Hence, whenever voltages and currents are going to be large, other forms of insulation are used in place of air to extinguish the arc as quickly as possible. The magnetic fields set up and the heat generated will persist as long as current continues to flow through the contacts and the arc. Should this current be "fault current," the enormous magnetic fields and the great amount of heat will tend to cause damage in terms of their own duration.

The most commonly used insulation is oil. For higher voltages and larger capacities, the insulating medium may be a vacuum or an inert gas, such as sulphur hexafluoride (SF_6).

Other means are also employed to limit and extinguish the arc as quickly as possible as well as to reduce the magnitude and duration of the fault current, both of which are factors in the size, reliability, and cost of circuit breakers.

TYPES OF BREAKERS CATEGORIZED BY INSULATION MEDIUM

Air Circuit Breaker

An air circuit breaker employs air as the interrupting insulation medium. Of all the insulating media mentioned, air is the most easily ionized and, hence, arcs formed in air tend to be severe and persistent.

The switching elements, for an air current breaker, therefore, may often consist of main and auxiliary contacts. The auxiliary contacts open before the main contacts do, and the arc is drawn on them, thereby avoiding severe pitting of the main contacts (Fig. 8-3). As the contacts separate, a puff of air is sometimes used to blow the arc into "arc chutes." Moreover, the current itself is made to flow through coils that set up magnetic fields that force the arc into these arc chutes. An electric arc is essentially an electric conductor with a surrounding magnetic field; hence it can be attracted or repulsed by another magnetic field.

The arc chutes are usually made of corrugated insulating material (such as porcelain or slate). Their function is to stretch out the arc until it can no longer sustain itself. In simple form, it is an enclosed channel in which the arc is dissipated and kept from striking surrounding objects.

Breakers of this type are usually employed in electric systems where fault

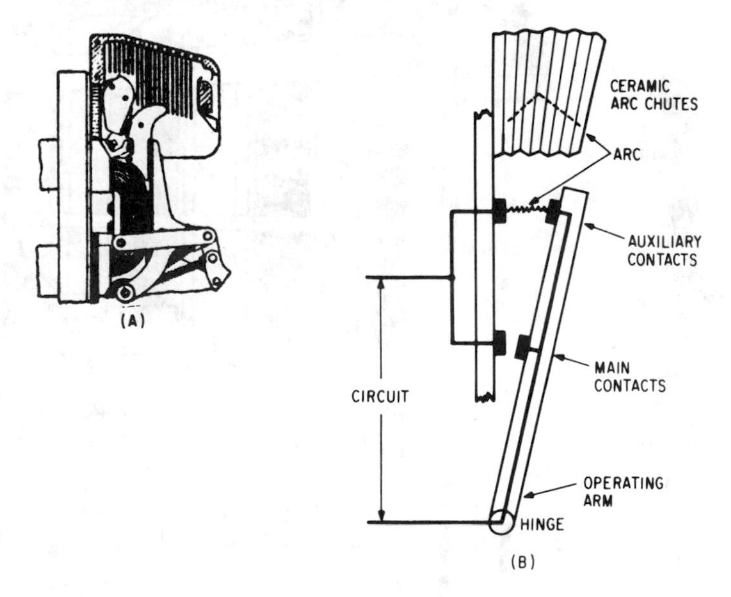

Figure 8-3 (A) Air circuit breaker, and (B) schematic showing main and auxiliary contacts and arcing arrangement.

currents are relatively small. Their advantages are simplicity of construction, low cost, and relatively low maintenance requirements.

As a means of extending the short-circuit duty of this type of breaker, a blast of air trained on the arc from a reservoir of compressed air enables a heavier arc to be extinguished.

Oil Circuit Breaker

Oil circuit breakers have their contacts immersed in insulating oil. They are used to open and close high-voltage circuits carrying relatively large currents in situations where air circuit breakers would be impractical because of the danger of the exposed arcs that might be formed (Fig. 8-4).

When the contacts are drawn apart, the oil covering them tends to quench the arc by its cooling effect and by the gases thereby generated, which tend to "blow out" the arc. At the instant the contacts part, the arc formed at

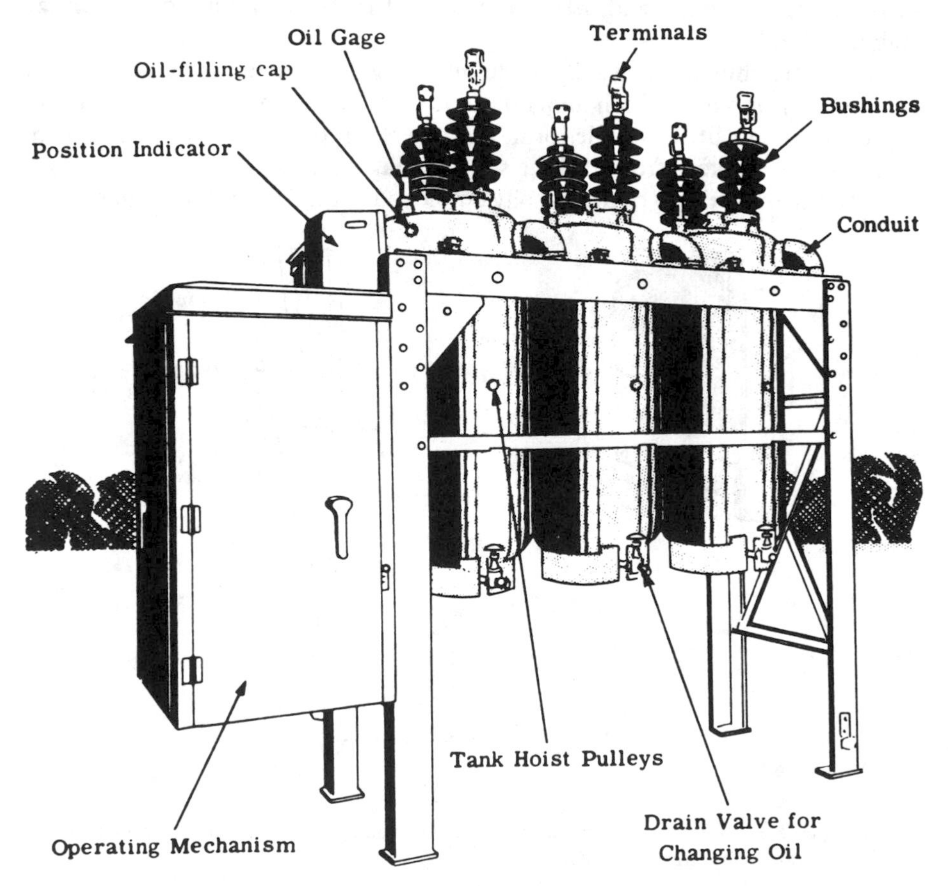

Figure 8-4 Oil circuit breaker.

each contact not only displaces the oil but decomposes it, creating gas and a carbon residue. If these carbon particles were to remain in place, as a conductor they would tend to sustain the arc formed. However, the violence of the gas and the resulting turbulence of the oil disperse these particles, and they eventually settle to the bottom of the tank. The rate of expansion and the volume of gas generated depend to a great extent on the energy to be interrupted, the speed of the separation of the contacts, the cooling effect, and the quantity of oil used as the insulating medium.

The use of oil as insulation also permits smaller distances between live parts and between live parts and ground (the structural and mechanical support parts of the circuit breaker) because of its better insulating qualities (higher dielectric strength) in comparison to air. The insulating oil normally used as a dielectric strength of around 30 kV per one-tenth of an inch (compared to a similar value of 1 kV for air). Oil is also an effective cooling medium. Askarels sometimes replace oil when required because of special fire hazards; refer to Chap. 2.

Vacuum Circuit Breaker

In this type of breaker, the contacts are drawn apart in a chamber from which air has been evacuated (Fig. 8-5). The electric arc is essentially an electric conductor made up of ionized air. Thus, if there is no air, theoretically the arc cannot form. In practice, however, a perfect vacuum is not likely to be obtained. The small residual amount of air that may exist permits only a small arc to be formed and one of only a very short duration. The same vacuum, however, will not dissipate the heat generated as readily as other insulating media. Although this type of breaker has certain advantages in terms of its size and simplicity, its interrupting ratings are not comparable to those of oil circuit breakers. Furthermore, maintenance procedures are more complex and, hence, more costly.

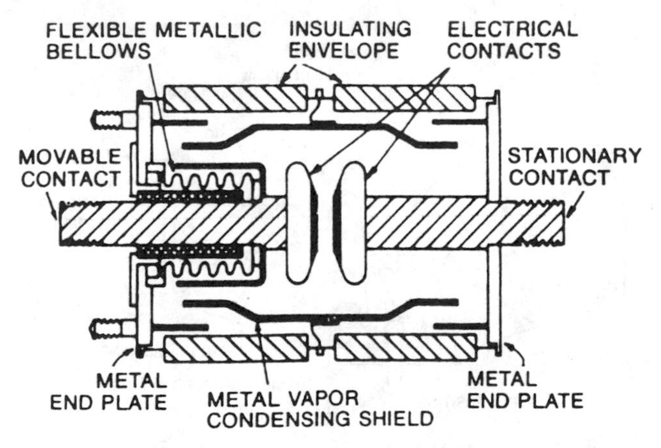

Figure 8-5 Vacuum circuit breaker.

Sulphur Hexafluoride (SF_6) Breaker

This type of breaker (Fig. 8-6) is similar to the vacuum type of breaker except that the vacuum is replaced by an inert, nontoxic, odorless gas—sulphur hexafluoride (SF_6). This gas, under pressures that may vary from 45 to 240 lb/sq in, extinguishes the arc so rapidly as almost to prevent its formation. It also has excellent heat-dissipating characteristics, and its dielectric strength is very much greater than that of oil.

The breakers are constructed in modules capable of operation at voltages from 34.5 kV at gas pressures of 45 psi to 362 kV at 240 psi. (Fig. 8-7). By

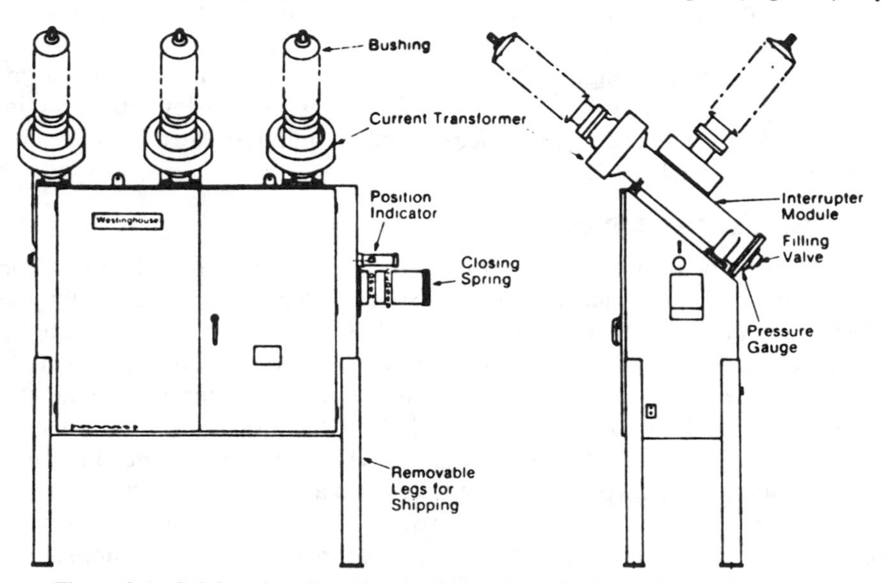

Figure 8-6 Sulphur hexafluoride circuit breaker, 34.5 to 69 kV. (*Courtesy*, Westinghouse Electric Co.)

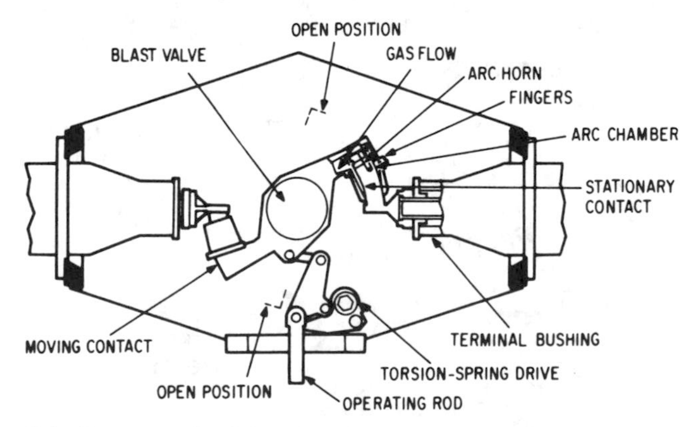

Figure 8-7 Features of sulphur hexafluoride interrupting module. (*Courtesy*, Westinghouse Electric Co.)

connecting two or three such modules in series, breakers capable of operating at 800 kV at 240 psi can be constructed, with two- to three-cycle interrupting time (Fig. 8-8). Installation and maintenance requirements of these breakers are less costly than for other types.

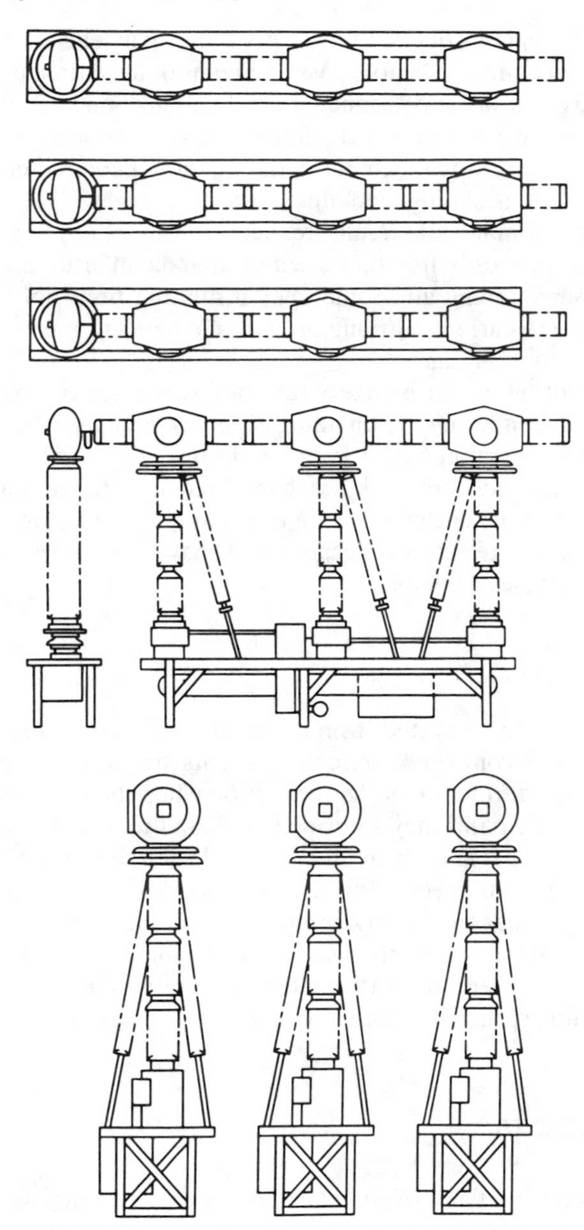

Figure 8-8 Modules in series in sulphur hexafluoride circuit breaker. (*Courtesy*, Westinghouse Electric Co.)

CONSTRUCTION

Tanks and Other Enclosures

The contact elements of circuit breakers are usually enclosed in metallic tanks, through which the conductors pass by means of insulating bushings; the conductors are usually made of copper rods. The tanks may contain a single contact element or two or more such elements in a common enclosure. Attached to the rods are stationary contacts that are either completely immersed or surrounded by the insulating medium.

The contacts of many air circuit breakers, however, are not enclosed in tanks or other receptacles. Often they are mounted on an insulating panel and exposed to the surrounding air. Sometimes insulating fireproof barriers are installed between the current-carrying breaker elements to prevent arcs from communicating between them.

In the case of oil circuit breakers, the tanks, are not completely filled, allowing some space in which the oil and gas generated may expand, particularly if a breaker is operating under fault conditions. The contacts of vacuum and sulphur hexafluoride breakers are contained in interrupting modules, which may or may not be enclosed within tanks or chambers (Fig. 8-8). These interrupting modules are interchangeable and may be removed for maintenance or other purposes.

Contacts

Movable contacts, operated from external mechanisms, bridge the stationary contacts and complete a circuit when its breaker is closed. These contacts are displaced from the bridging position when the breaker is opened, thus interrupting the circuit; they are usually held in the closed position by the action of a latch, lock, toggle, or similar device. When the circuit breaker has to interrupt a circuit, its mechanism usually receives a controlling impulse from a relay that is either electrically or manually operated or from a manually operated mechanical linkage. In this way, the restraining force is released, and the moving contacts recede from the stationary contacts under the action of gravity, accelerating springs, or both.

TYPES OF CONTACTS

There are four principal types of circuit breaker contacts: butt, wedge, brush, and bayonet.

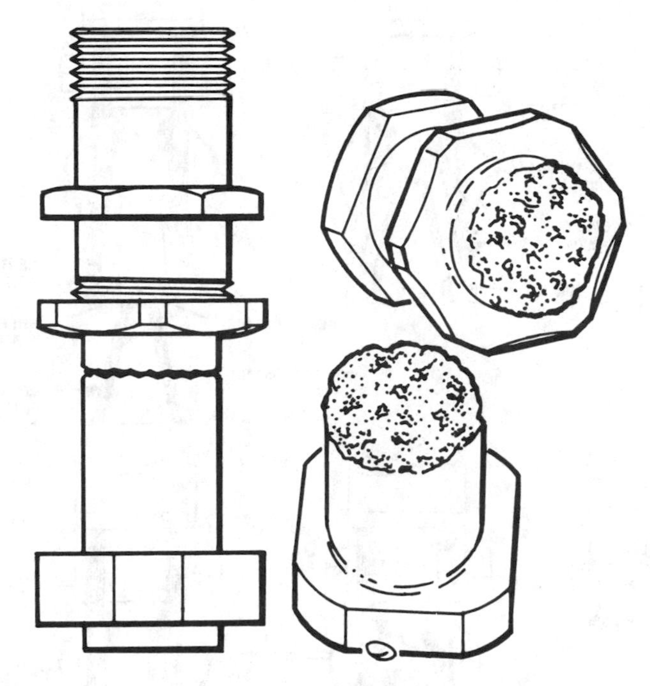

Figure 8-9 Butt contacts.

Butt Contacts

Butt contacts (Fig. 8-9) consist of two solid elements with flat or curved faces that butt together when a circuit is closed. (Silver facings will reduce the contact resistance and thus lead to less heating, pitting, and wear.) A spring held tight by a latch or other restraining device holds the breaker closed. The moving contact butts solidly against the stationary contact. When the breaker trips, the latch is released, and the spring throws the contacts apart rapidly.

Wedge Contacts

Wedge contacts (Fig. 8-10) are probably the most frequently used contacts for circuit breakers. The wedge is squeezed between spring jaws when the breaker is closed, providing a good area for contact pressure. Forcing the wedge between the spring jaws also produces a wiping motion that maintains a clean surface on both of them. Arcs are normally taken on the top of the blade, thus protecting the actual contact area from damage. The moving contacts are usually made of a hard copper alloy with renewable arcing tips. These tips must be of rather massive proportions in order to provide a high thermal capacity and thus minimize burning. The stationary contacts consist of

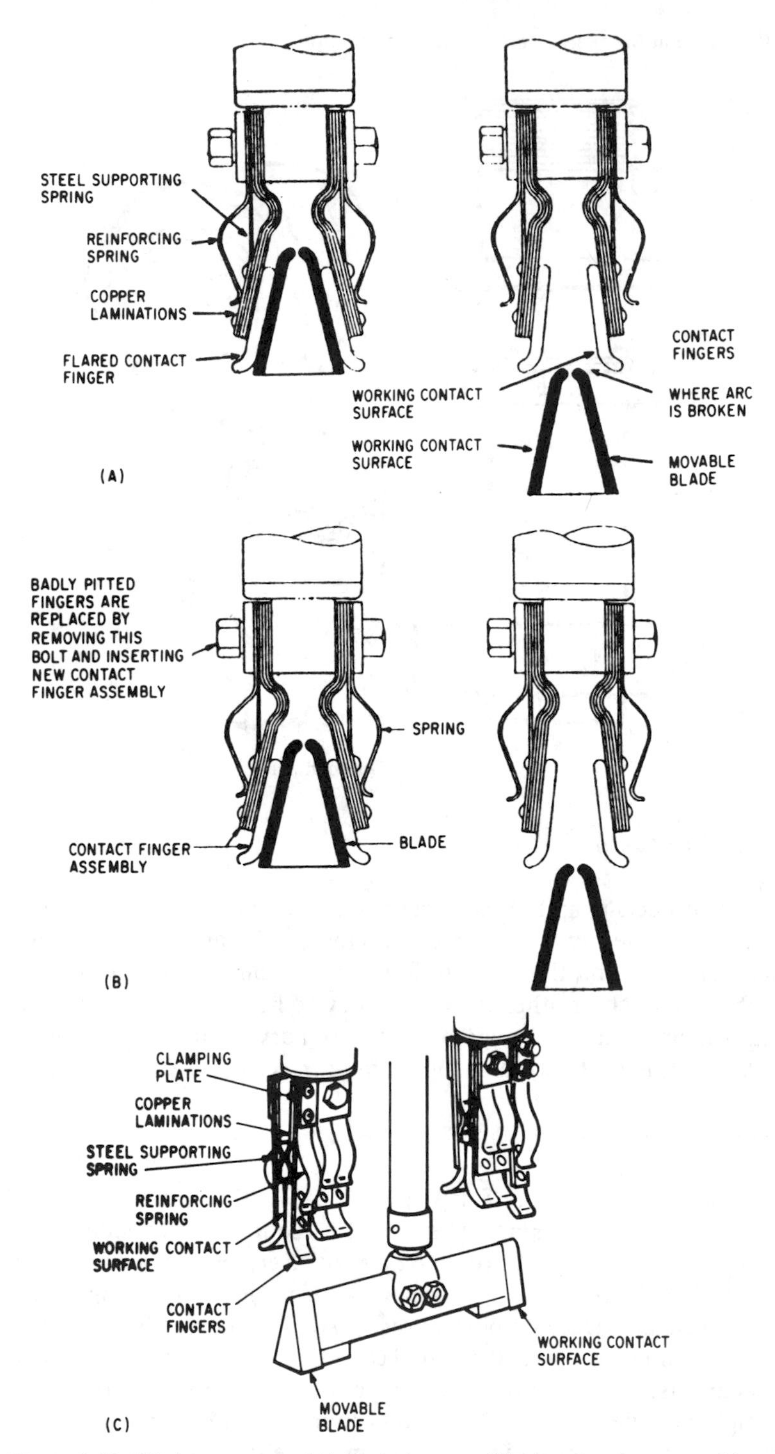

Figure 8-10 Wedge contacts: (A) Typical types, (B) closed and open positions, and (C) O.C.B. assembly.

fingers arranged in pairs so that they may surface well on both sides of the moving contact when the breaker is closed. Flat springs on the fingers permit the contacts to align themselves automatically on the wedge so that full current-carrying capacity may always be obtained. One or more pairs of fingers are used, depending upon the current capacity required. As a rule, the blade is also of large proportions to help carry off heat from the arcing tips.

Brush Contacts

Brush contacts (Fig. 8-11) are usually used to insure good contacts where heavier currents are involved. The stationary contact consists of a solid copper stud, and the moving contact is made up of laminations. A wiping action takes place when the breaker is closing to ensure a good clean surface on which the laminations may make good contact.

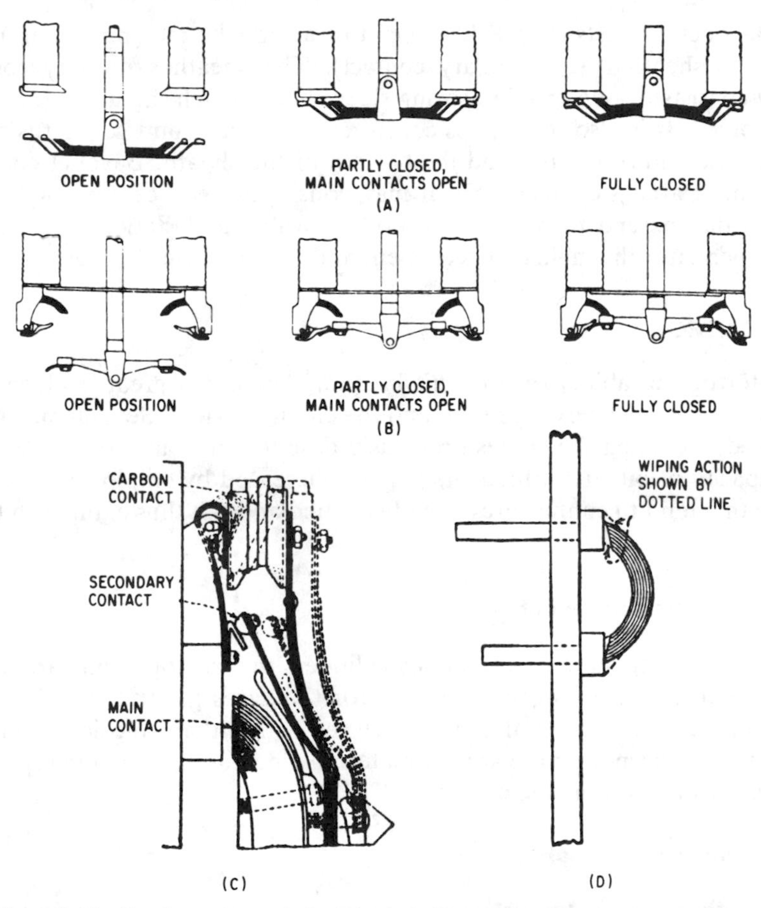

Figure 8-11 Brush contacts: (A), (B), and (C) several types, and (D) wiping action.

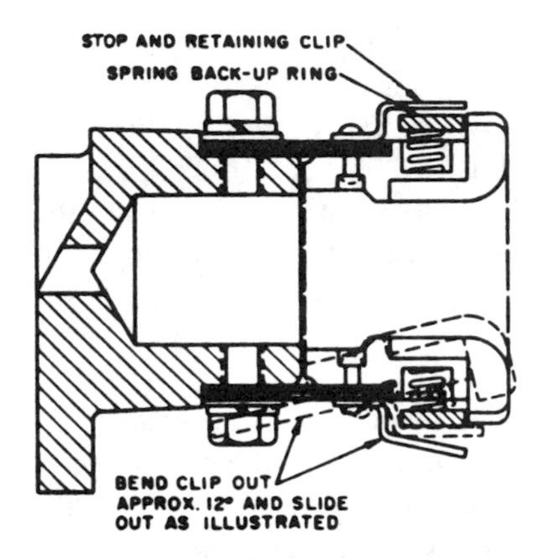

Figure 8-12 Bayonet contact.

Bayonet Contacts

Bayonet contacts (Fig. 8-12) consist of a rod (the moving contact) that fits into a sheath (the stationary contact). The sheath stationary contact usually consists of one or more spring sleeves into which the moving contact rod is forced. It is also sometimes constructed so that a butt contact is made between the end of the rod and the bottom of the sheath. Bayonet contacts usually have arcing contacts, which may consist of another auxiliary set of contacts that extend to one side of both movable and stationary contacts. These open after the main contacts open, and the arc forms between them.

ARC CONTROL

The interrupting ability of a circuit breaker depends a great deal on how quickly the arc can be extinguished and the circuit opened. Several means are employed, including the chutes previously described for air circuit breakers. The capacity of oil circuit breakers may be increased by employing auxiliary devices that tend to confine arcs already formed and can thus extinguish them more rapidly.

Explosion Chambers

Explosion chambers—essentially cylinders that surround any arc taking place—serve to create higher pressures for the gases generated (Fig. 8-13). These higher pressures, confined to a degree within the explosion chamber, not only tend to smother the arc but materially increase the opening speed of the breaker interrupting the circuit.

De-ion Arc Quencher

Another device, the de-ion arc quencher, confines, divides and extinguishes arcs much more rapidly when a circuit is broken. In this device, the

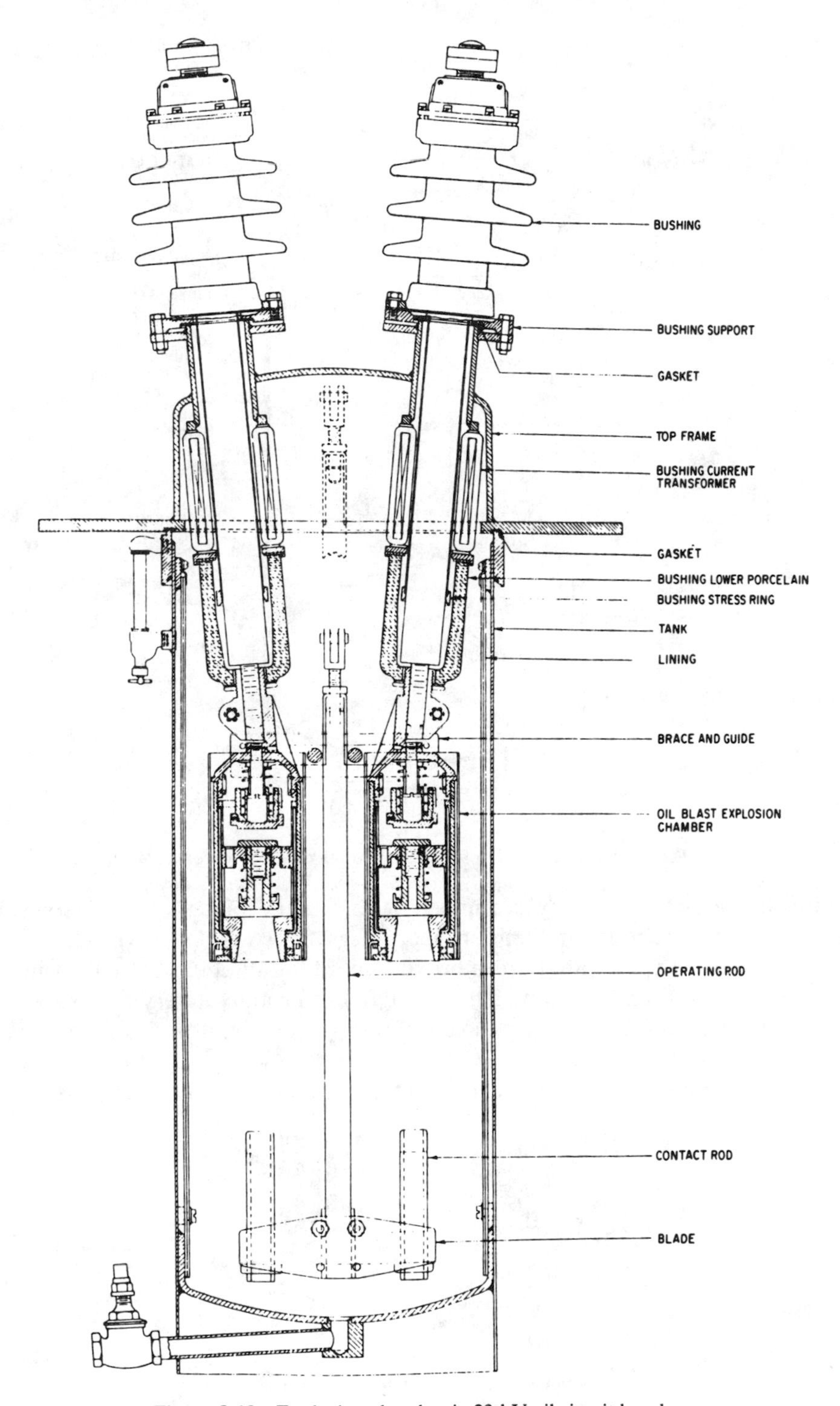

Figure 8-13 Explosion chamber in 23-kV oil circuit breaker.

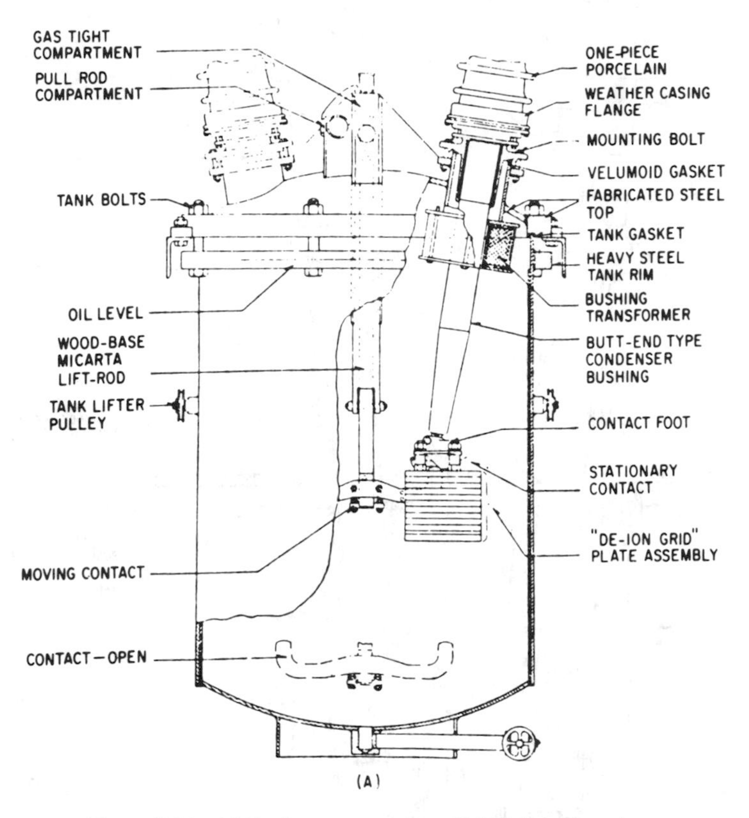

Figure 8-14 (a) De-ion arc quencher: 69-kV de-ion breaker.

contacts are surrounded by a number of thin metal plates or grids spaced about 1/16 inch or about 2 millimeters apart (Figs. 8-14A and 8-14B). When the contacts open, the resultant arc is broken up into a number of small 1/16-inch arcs, which are then caused by a magnetic field to rotate at very high speeds, thereby preventing overheating and burning of the metal; all arcing is confined to the grid area (Fig. 8-15).

In an ac circuit, when the zero point of the current cycle is reached (Fig.

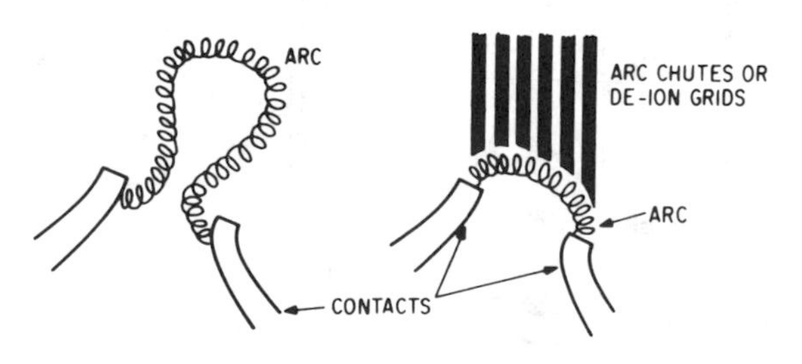

Figure 8-15 De-ion arc quenching.

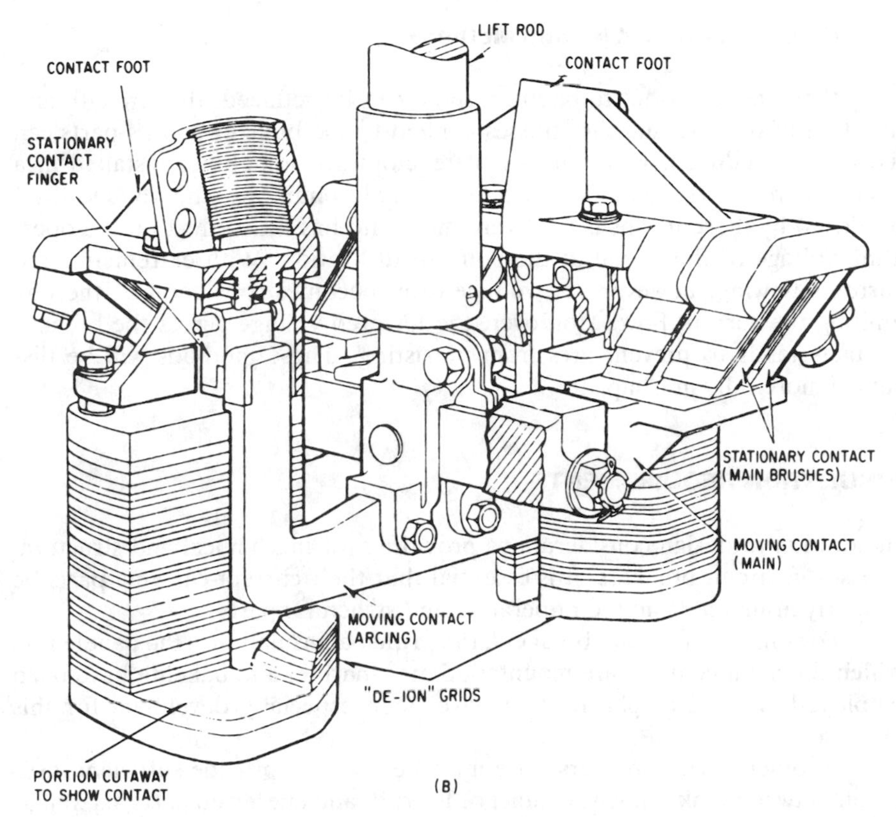

Figure 8-14 (con't) **(b)** De-ion arc quencher: details of contact assembly.

8-16), the insulation adjacent to and between the grids becomes instantly deionized, that is, it is no longer a conducting path. The voltage existing between the grids is not sufficient to re-establish the current flow, and the arc is extinguished, usually in about two or three cycles (as compared with anywhere from six to twenty cycles for other types).

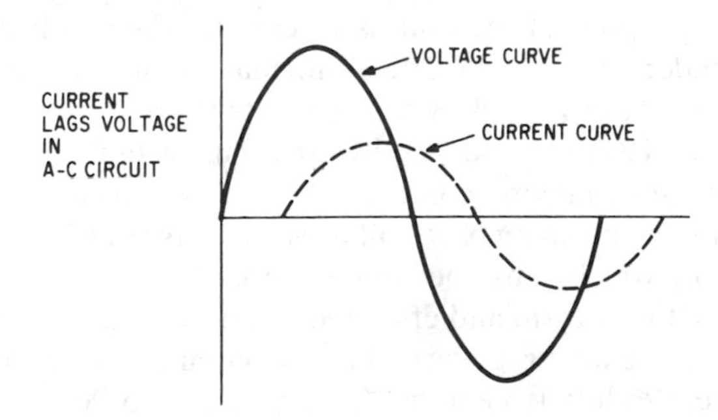

Figure 8-16 Points at which voltage is zero.

External Current-Limiting Methods

If the fault current to be interrupted can be reduced, the strength and duration of the arcs and the stresses applied to the breaker and its parts can likewise be reduced. Resistances and reactors are sometimes installed in a circuit to reduce current flow. Under normal conditions, the voltage drop produced by these means is relatively small and has little effect on the operating voltage of the circuit at the point of utilization. With extremely large currents flowing, however, the voltage drop becomes considerable, thereby limiting the current flow. Moreover, the lowered voltage across the breaker contacts tends to prevent arcs from persisting. These methods will be discussed more fully in Chap. 16.

INSULATION REQUIREMENTS

In addition to design considerations providing for mechanical and structural stresses in circuit breakers, it is essential that their current-carrying parts be properly insulated from their operating and structural parts.

For smaller air circuit breakers, the principal insulation is the panel upon which the main contacts are mounted. Slate, marble, and bakelite have been employed, as well as plastics that have been especially developed for this purpose.

In other circuit breakers, care must be taken to provide sufficient insulation between tank, moving contact or lift rod, and energized parts, as well as between the bushings that bring the energized conductors through the tank wall.

Tanks

Design of tanks for a circuit breaker must take into account insulation clearances, the depth in the case of oil, and the mounting of bushings. Although the main purpose of the tank is to contain the conductors, contacts, and insulation under both normal and abnormal circuit interruptions, it must also dissipate the heat produced by the currents flowing through the current-carrying parts as well as that produced by arcs. The distance between contacts, length of the stroke of moving contacts, insulation, mechanical struts, reinforcement, clearances between parts–all these factors contribute to the dimensions of the tank, which must be ample enough to provide for the large volumes of gas that may form and effectively reduce clearance distances.

The shape of the tank is governed by the circuit breaker service requirements. For moderate duty (to about 2500 amps and 15,000 V), all poles of a circuit breaker may be in one tank, and the tank may be rectangular in shape. Medium-capacity breakers (about 10,000 amps and 15,000 V) may be of

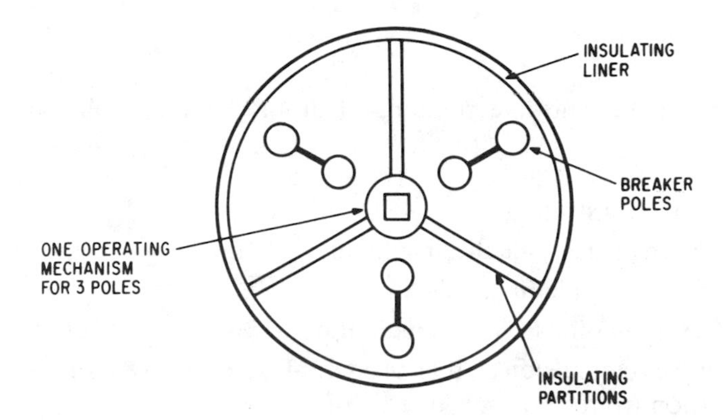

Figure 8-17 Top view of three-pole circuit breaker in one tank.

single-pole construction with elliptical tanks in order to resist the greater internal pressures. Breakers of larger capacity are usually single pole as well but employ circular or round tanks. Round tanks, however, may be used even for low-voltage, heavy-current breakers in which all (two or three) poles are placed in the same tank. Whenever all poles are mounted in one tank, insulating partitions (usually made of special insulating compounds and laminated for greater strength) may be installed between them. Tanks may also be fitted with liners of the same insulating material (Fig. 8-17).

Moving Contacts

Since moving contacts are usually designed to move upward to close, they are often mounted on a metal crossbar and operated by an insulated "lift rod" (Fig. 8-18). The rod may be of an insulation-impregnated, tough wood such as hickory; it may also be laminated for greater strength: For heavy duties, multiple rods may be used.

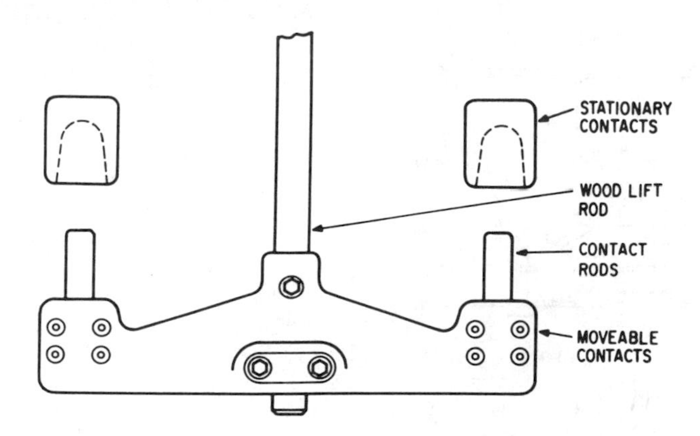

Figure 8-18 Lift rod assembly.

Oil

For effective circuit interruption, oil should have the following properties:

1. High dielectric strength
2. Freedom from acid, alkali, or sulphur
3. Low viscosity to aid in arc cooling
4. Low freezing point so as to remain fluid in outdoor temperatures
5. Resistance to emulsion so that accidental water will separate from it and any carbon from the arcs will settle out
6. A specific gravity that will not permit ice to rise through it and float.

Although the requirements for circuit breaker oil may be somewhat different from those for transformer oil, the same oil will satisfy both applications.

Porcelain Bushings

Porcelain bushings are solid porcelain cylinders that surround the conductors (Fig. 8-19). They are used for moderate voltage breakers (to about 20,000 V) of relatively small interrupting capacity where the repulsion effects on the bushings (caused by short circuits within the capacity of the breaker) are relatively limited and the solid porcelain will have sufficient mechanical strength. Smaller procelain bushings may have smooth sides, but larger ones, and those for use outdoors that will be exposed to the elements, especially rain, are corrugated or provided with "petticoats" to increase the creepage distances or leakage paths between the energized terminal and the tank.

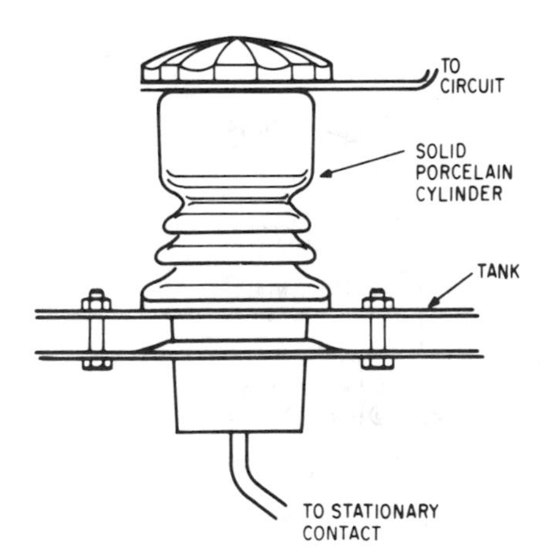

Figure 8-19 Solid porcelain bushing.

Oil-Filled Bushings

For breakers operating at greater voltages (to about 75,000 V), the conductor may be mounted inside porcelain cylindrical insulators filled with oil. The oil space may be divided by thin insulating cylinders of special materials, including plastics (for example, herkolite), with an oil gauge at the top to indicate the level of oil in the bushing. The insulation of these bushings is dependent on the oil, which has high insulating qualities and helps keep the bushing cool. These bushings are similar to those for transformers. See Fig. 2-18.

Condenser Bushings

For even higher voltages (usually over 75,000 V), conductors may be insulated with layers of oil-impregnated special paper, with metal foil inserted at several locations among the layers (Fig. 8-20). The layers of insulation at the

Refer to Appendix A for Polymer Insulation.

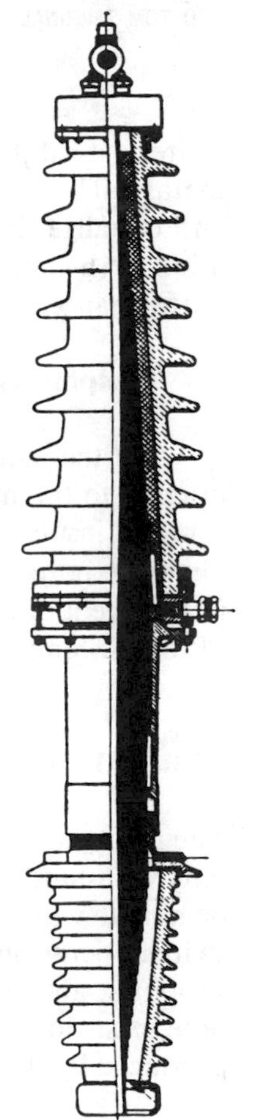

Figure 8-20 Condenser bushing for oil circuit breaker. (*Courtesy*, Westinghouse Electric Co.)

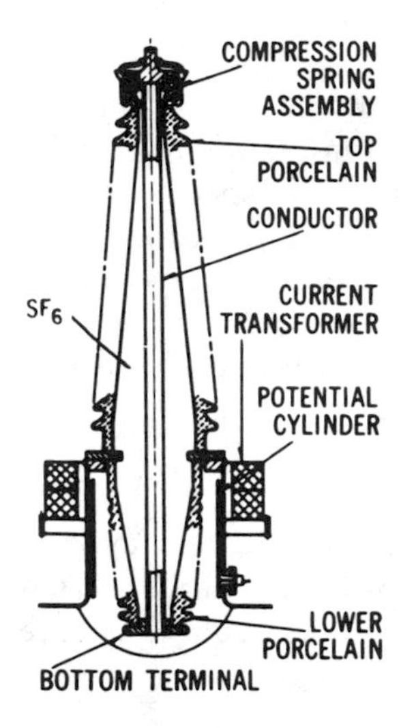

Figure 8-21 Sulphur-Hexafluoride filled bushing for sulphur hexafluoride circuit breaker. (*Courtesy*, Westinghouse Electric Co.)

center are subjected to greater electrostatic stresses than the outer layers. The metal foil between the layers forms a series of condensers that tend to even out and equalize the stresses among the layers. Such construction permits operation at higher voltages with less material, at lower cost, and with better performance.

Sulphur Hexafluoride Bushing

In the sulphur hexafluoride breaker, the hollow porcelain bushing is opened to the main breaker tank and filled with sulphur hexafluoride gas to serve as insulation (Fig. 8-21). There is no organic insulation to break down. Since the porcelain is under low pressure, it maintains the gas flowing outward in the event of a pin hole or crack, thereby preserving the insulating qualities of the bushing until repairs can be made.

INSULATION COORDINATION

The insulation associated with the several parts of the circuit breaker must not only withstand the normal operating voltages imposed on them, but must also be able to withstand the higher voltage surges resulting from lightning or from switch operations. The insulation from the energized parts to ground (tanks, supports, and the like), or from one pole to another, where more than one pole exists in the same tank, must be strong enough to withstand the voltages produced by the impulses at these points.

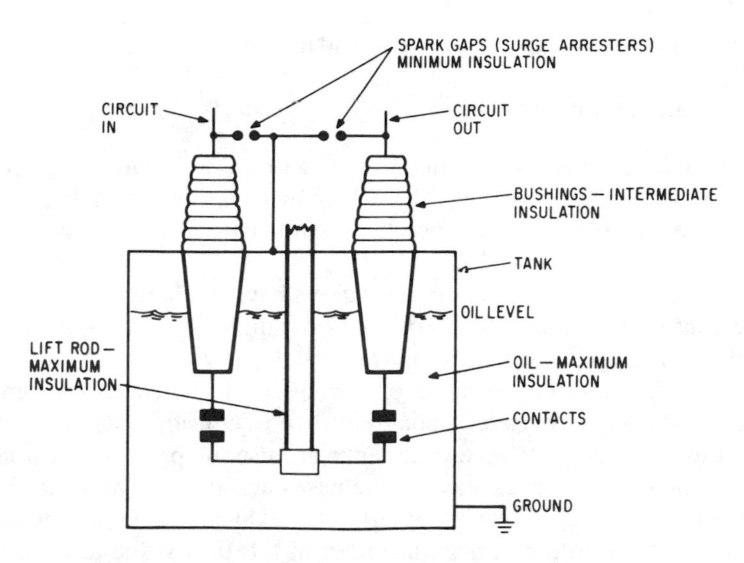

Figure 8-22 Schematic diagram of insulation coordination.

To provide adequate insulation without resorting to excessive and expensive insulating material, the insulation requirements of the several parts must be carefully coordinated. In general, this means that the internal parts are insulated as nearly equally as practical, but usually with more strength than the bushings (including any current transformer, potential transformer, or other accessory that may be mounted on the bushing within the breaker chamber, and which may be enclosed in individual compartments to afford protection against damage during heavy interruption duties). This approach ensures that any breakdowns that may occur will be outside the tank where the damage will be comparatively light, localized, and easier to repair.

Furthermore, since the bushing is usually protected by an air gap or lightning arrester whose insulating value under surge conditions is even lower than its own, any failure that occurs will be caused by flashover across the bushing and the tank rather than by puncture of the bushing insulation (Fig. 8-22). Finally, the insulation of the weakest point in the breaker should be weaker by a large enough margin than that of the principal equipment it is protecting. Such a coordinated arrangement of insulation tends to restrict the damage to main pieces of equipment.

Insulation coordination requires that the insulation of all components of a system should exceed a minimum level and that lightning or surge arresters or other protective devices operate below that insulation level. This is known as the *Basic Insulation Level* (BIL) and will be discussed further in Chap. 12.

TYPES OF INDOOR AND OUTDOOR BREAKERS

Station or Indoor Types

Small, automatically operated breakers are used indoors whenever very large currents are not involved. These may be either air or oil breakers suitable for manual or electrical operation, with a usual maximum voltage of 15,000 V.

Larger automatically operated breakers having a high interrupt capacity may be constructed as single-pole or multipole units. The multipole units may be made up of single-pole units or may have all poles supported on a common frame in a single tank. Because of the heavy short-circuit currents that breakers of this type are called upon to interrupt, considerable amounts of oil vapor or gas may be generated. The tanks are usually provided with mufflers to relieve the pressure and condense the gases and thus prevent escape of the oil (commonly called "throwing" of oil). Oil leakage is taken care of by means of drains (to safe holding areas) installed in the floors; the area containing the breakers may be encompassed by dams or curbs to form a basin that will contain any oil that flows from a breaker in trouble.

These larger breakers are electrically and fully automatically operated. They may be mounted on trucks or on structural steel frames.

Outdoor Types

Since outdoor oil circuit breakers are designed to withstand weather conditions of all kinds, their live parts usually have greater clearances and their physical dimensions are larger than those for indoor types (which are usually limited to breakers operating at lower voltages and less demanding interrupt duties). They are usually mounted on a self-supporting steel framework. The poles for each breaker are operated through a common shaft actuated by one operating mechanism. Outdoor breakers may also be installed in oil-containing concrete or earth encompassed basins.

Metal-Clad Switchgear

For smaller circuit breakers such as those found in industrial plants or in electric utility distribution substations, metal-clad switchgear installed outdoors (as well as indoors) will provide maximum safety to the operator. Since all high-voltage equipment and connections are enclosed in grounded metal compartments, danger from accidental contact is avoided. Access to low-voltage control wiring and secondary connection compartments is provided by hinged doors and panels that may be opened safely when the units are in service because steel barriers isolate these compartments from the high-voltage circuits (Fig. 8-23). Access to high-voltage areas—including current and potential transformers, busses, connections, and the breaker itself—is

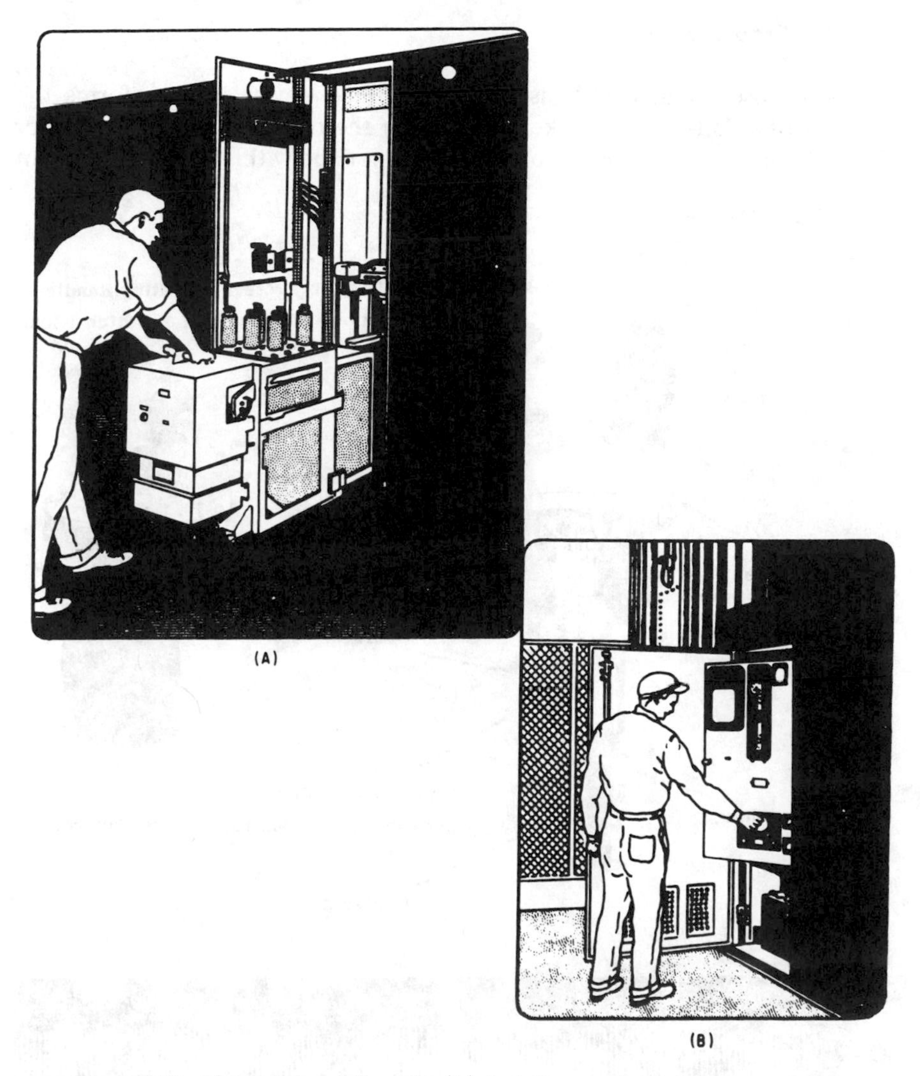

Figure 8-23 Metal-clad switch gear: (A) Installing a circuit breaker, and (B) adjusting a control.

provided by one or more steel covers or panels that cannot be removed while the circuit is energized, being prevented from doing so by a mechanical interlock. Similarly, an interlock prevents moving the circuit breaker unless it has been tripped. When the breaker is removed from its cubicle, it is automatically isolated from the circuit controlled and thus obtains the same protection afforded by air-break switches (see Chap. 10).

Heaters are installed in outdoor breakers, including metal-clad switchgear, to reduce condensation and prevent freezing of the mechanisms; some are permanently energized and others are thermostatically controlled.

Oil-Circuit Recloser

A recloser (Fig. 8-24A) is a self-contained device that performs the functions of an oil-circuit breaker except that the tripping and closing power and fault sensing and reclosing controls are contained within the unit. The unit

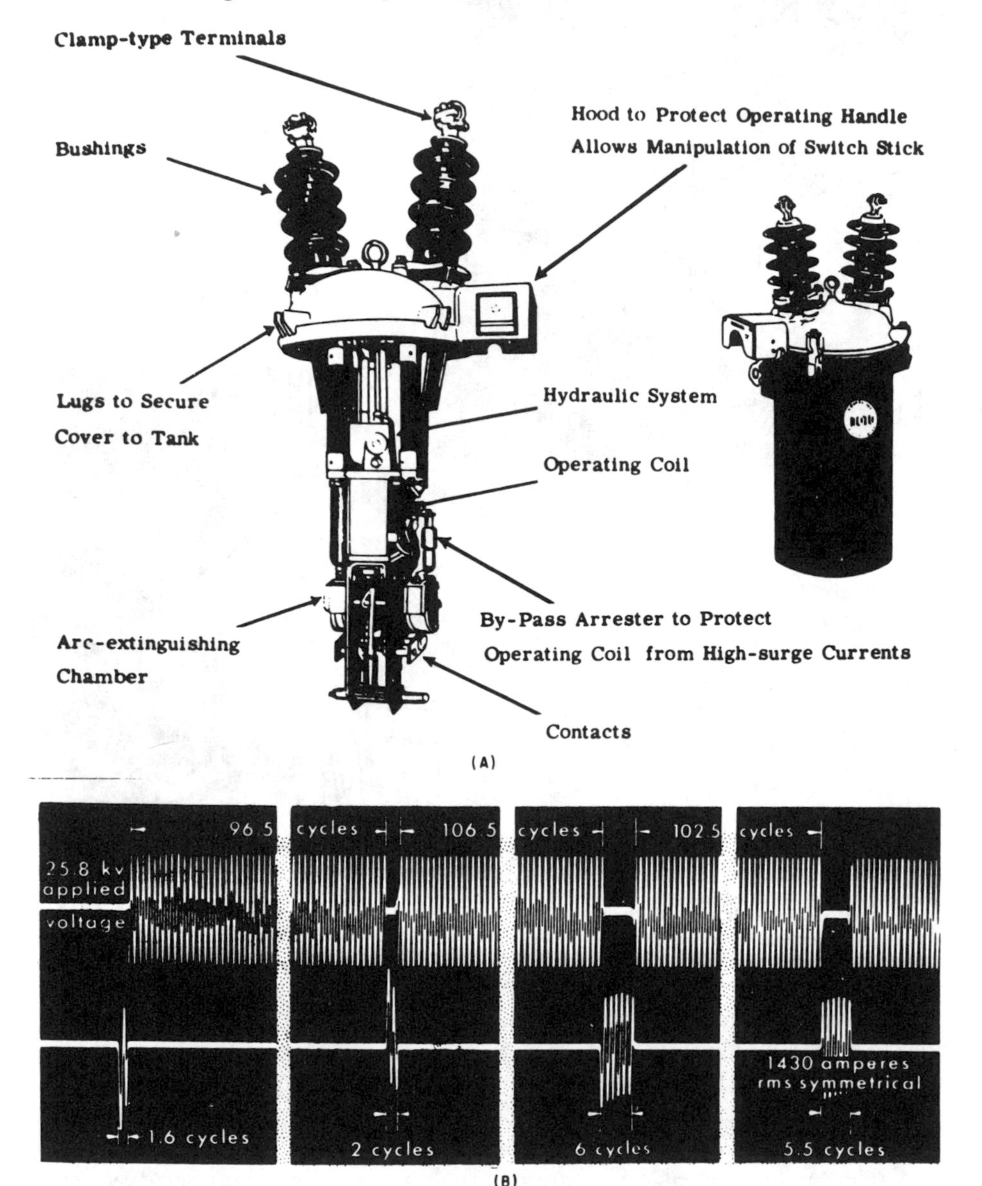

Figure 8-24 Single-phase oil-circuit recloser: (A) Construction details, and (B) oscillogram showing example of recloser operation.

gets its power from closing coils connected on the source side of the device; these closing coils also compress the springs used for tripping. Another coil in series with the closing coil trips a latch to open the breaker. A relay re-energizes the closing coils, closing the breaker automatically three more times before locking out.

The sequence of operation when a fault occurs is similar to that in any other breaker. Usually, the first trip occurs quickly after the fault is sensed; the second, third, and fourth may occur with time-delays inserted in the operation cycle (Fig. 8-24B). The unit may also be operated manually.

These units have a smaller interrupting capacity than regular circuit breakers and are usually installed at some critical or sensitive location in the circuit. The three reclosings serve to restore power to a circuit (or portion of a circuit) subject to a temporary failure such as might occur when a tree limb falls across power lines but drops to the ground immediately or when wires whip together in a severe wind.

GROUNDS

Tanks, metallic supporting structures, and other internal and external mechanical parts not associated with the current carrying or interrupting electrical function of the circuit breaker should be connected (or bonded) together and grounded to one or more grounds. Under fault or lightning conditions, surge voltages may develop in these parts that might prove unsafe for workmen or cause damage to the insulating materials in the breaker or to the metallic parts themselves. Ground connections from any associated lightning or surge arresters should also be interconnected with these ground connections to ensure proper functioning of the devices. The grounds should be checked periodically to make sure not only that the circuit connections are unbroken but also that their resistance is reasonably low.

REVIEW

1. Opening and closing of electric circuits is accomplished by equipment that varies from very simple devices for opening low-voltage circuits that carry small currents to large and very elaborate devices for interrupting high-voltage circuits that transport large currents and great amounts of energy. The term "circuit breaker," however, is usually applied to heavy duty devices capable of interrupting comparatively large electric currents safely.
2. Two or more adjacent conductors create magnetic fields that tend to attract or repel each other (Fig. 8-1). When the currents are extraordinarily high, such as during short circuit or fault conditions, mechanical

forces created may be so high as to damage or destroy the breaker. Hence, the unit must be built ruggedly enough to withstand these forces, the limits of which are included in the rating of the breakers and are referred to as the interrupting capacity or interrupting duty of the breaker.

3. Ratings of circuit breakers specify the operating voltage of the circuit in which they may be connected, the normal operating or maximum load current that will not result in heat values damaging to the breaker, and maximum abnormal or fault current that may be safely interrupted.
4. Circuit breakers are generally filled with oil, not only as an insulating medium, but also to aid in quenching arcs and in cooling the unit (Fig. 8-4). Other media are askarels (where fire hazards may be high), air (Fig. 8-3), vacuum (Fig. 8-5), and sulphur hexafluoride (Fig. 8-6). Arcs resulting from opening breakers while current is flowing may be broken into a series of small arcs, called deionization, that may be more easily quenched.
5. The material, shape, method, and speed of operation of the contacts are very important as all these factors determine the rating of breakers, including their interrupting duty. Contacts may be of the butt, (Fig. 8-9), wedge (Fig. 8-10), brush (Fig. 8-11), or bayonet types (Fig. 8-12).
6. Tanks are usually of steel and conductors are carried through by means of bushings, insulation surrounding the conductors. The insulation of the internal conductors and of the bushings is coordinated so that failure may be more apt to occur outside the tank where repairs or replacement can be more easily accomplished. All insulation must withstand line or lightning surges of voltage. For greater safety, tanks and supporting structures should be connected to one or more grounds.

STUDY QUESTIONS

1. What is the function of a circuit breaker?
2. What is meant by a short circuit?
3. What happens when short-circuit current flows through a circuit breaker?
4. What is meant by the interrupting duty of a circuit breaker?
5. How are circuit breakers rated? Why?
6. What is the function of the oil in an oil circuit breaker?
7. What other types of breakers are there besides oil circuit breakers? What are some of their advantages? Disadvantages?
8. What are the principal types of circuit breaker contacts?
9. What are explosion chambers? De-ion arc quenchers? How does each function?
10. What is an oil circuit recloser? How does it operate?

9

CIRCUIT BREAKERS: OPERATION AND MAINTENANCE

CONTROLS

Although circuit breakers may be opened and closed directly by manual means, more often they are opened and closed by remotely controlled electrical means.

Trip-Free Feature

The operating mechanisms of most circuit breakers are designed to prevent the breakers from staying closed on a fault or overload condition. This characteristic is referred to as "trip-free." If a breaker trips open and the abnormal condition that caused it to do so persists, any attempt to close the breaker will cause it to trip again immediately. The tripping device, as long as it is activated, will prevent the mechanical linkages of the mechanism from engaging to close the breaker. Even if an attempt is made to close the breaker manually, it cannot be kept closed so long as the abnormal condition exists. (Fig. 9-1).

Antipump Feature

Electrically operated breakers have "antipump" provisions included in their control circuits. If an attempt is made to reclose a breaker by holding the control handle in the "close" position while the fault or overload persists, the breaker will attempt to trip and close, trip and close, indefinitely (since the

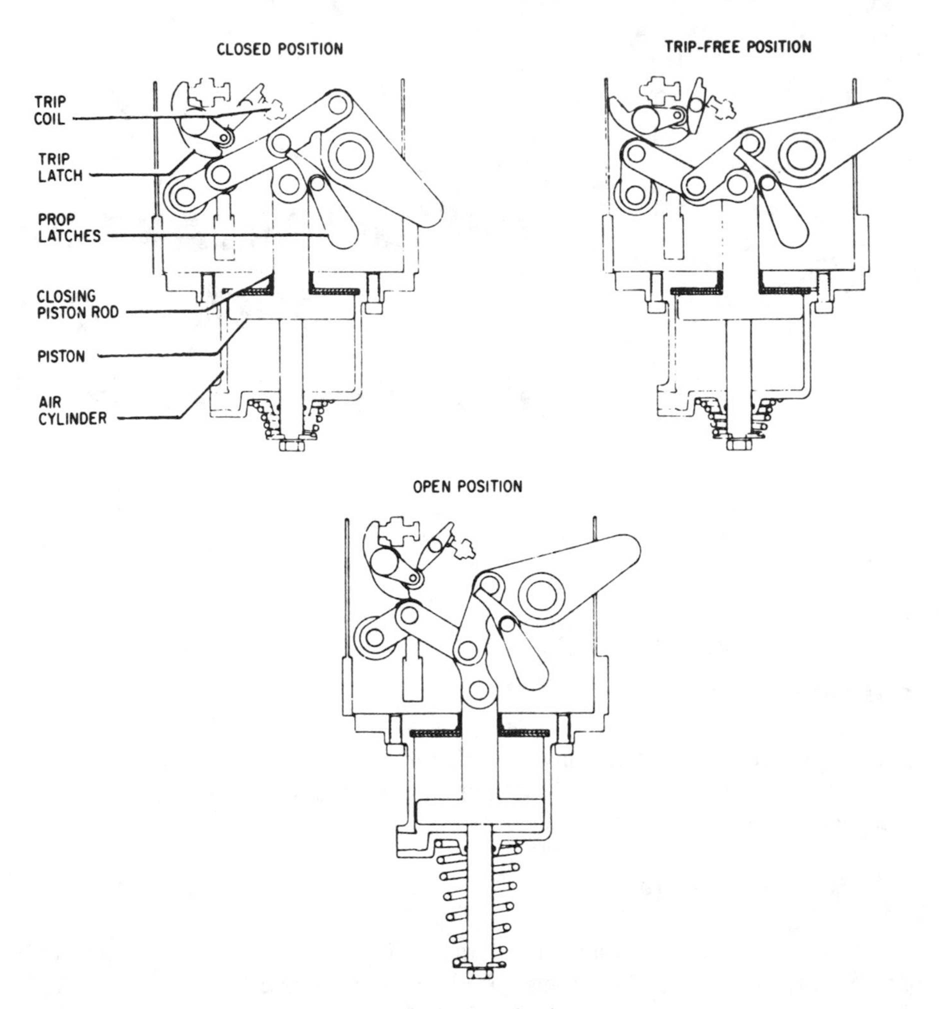

Figure 9-1 Trip-free mechanism system.

"close" contacts will still be made up). The "antipump" feature prevents this situation by requiring an auxiliary relay to be reset (by moving the control handle to the neutral position) before another "close" operation can be achieved.

Automatic Reclosing Feature

Some circuit breakers can be automatically reclosed in accordance with a predetermined pattern, usually by electrically operated mechanisms. A definite but different time delay is inserted in the closing circuit, which is automatically energized a predetermined number of times before becoming locked in a de-energized or open position. For example, the breaker may be set to reclose the first time immediately after opening, then wait for one or two seconds before reclosing again, then perhaps five seconds before a third, and ten seconds before a final attempt to reclose. This pattern takes care of the so-called "transient" or temporary faults or overloads that may cause the breakers to trip. If the abnormal condition is removed or disappears after tripping, the circuit may be re-energized without a long interruption of service. This operation is similar to that of the circuit recloser described in Chap. 8, except that the controls are not self-contained as they are in the recloser.

Storage Battery Supply

To insure the continuity and reliability of the control circuits and other devices associated with the operation of circuit breakers, the electric supply is obtained from storage batteries "floating" on a commercial or distribution supply line; that is, the batteries are permanently connected through suitable transformers, rectifiers, and control devices to the normal electric supply line and are thus kept fully charged at all times. Storage batteries will be discussed at greater length in Chap. 15.

METHODS OF OPERATION

Almost all circuit breakers are opened by means of a spring. The spring is compressed during the closing operation and maintained in that position by some form of latching device. When the breaker is called upon to open, the actuating controls trip the latch, thereby permitting the spring to decompress, opening the breaker contacts very rapidly and positively.

Circuit breakers may be installed and operated in several different ways: (1) Mounted on a switchboard and direct-operated manually; (2) remote mechanism manually operated; (3) electrically operated; or, (4) pneumatically operated.

Switchboard Mounting

Circuit breakers of small current-carrying and interrupting capacity for voltages to about 5,000 V may be mounted on a panel and operated directly by a lever; if mounted on the back side of the panel (for greater protection of the operator), the lever can extend through the panel (Fig. 9-2).

Remote Mechanisms

Circuit breakers arranged for manual operation may also be mounted at some distance from the operating point and actuated by means of linkages or bell-crank mechanisms (Fig. 9-3). This method of operation is subject to

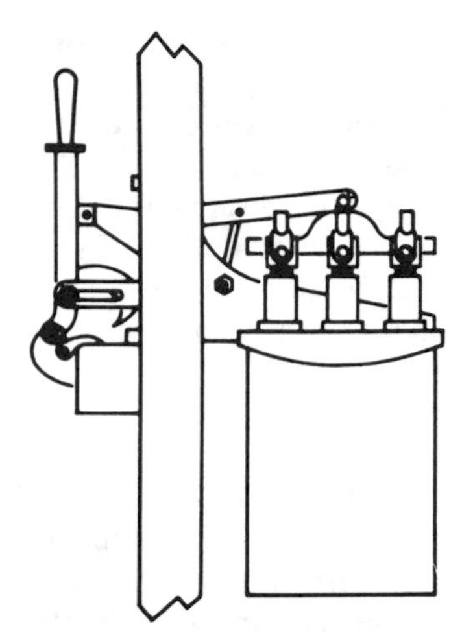

Figure 9-2 Direct mounting of manually operated oil circuit breaker.

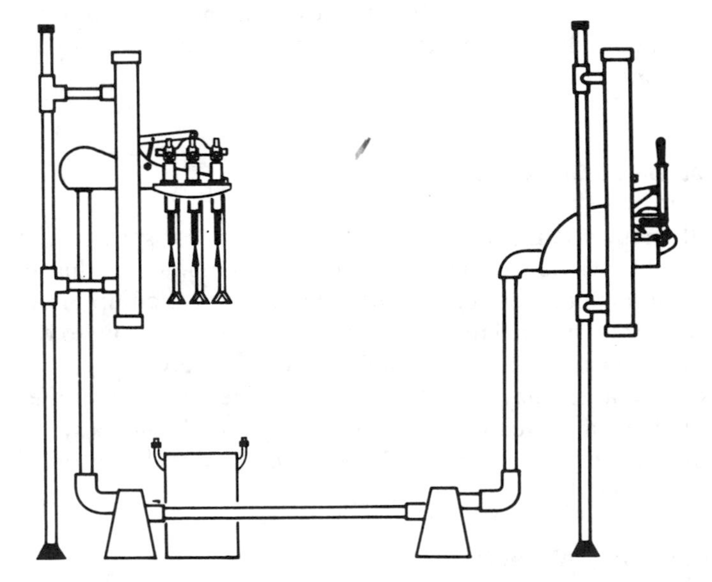

Figure 9-3 Manually operated oil circuit breaker for remote operation.

mechanical limitations because the weight and friction of the mechanisms not only tend to retard the speed of operation of the breaker but also may be beyond the physical capacity of some operators.

Electrically Operated Mechanisms

With distantly located or large circuit breakers, manual operation becomes impractical. Such breakers, therefore, are electrically operated and actuated by either solenoid (coil) or motor mechanisms.

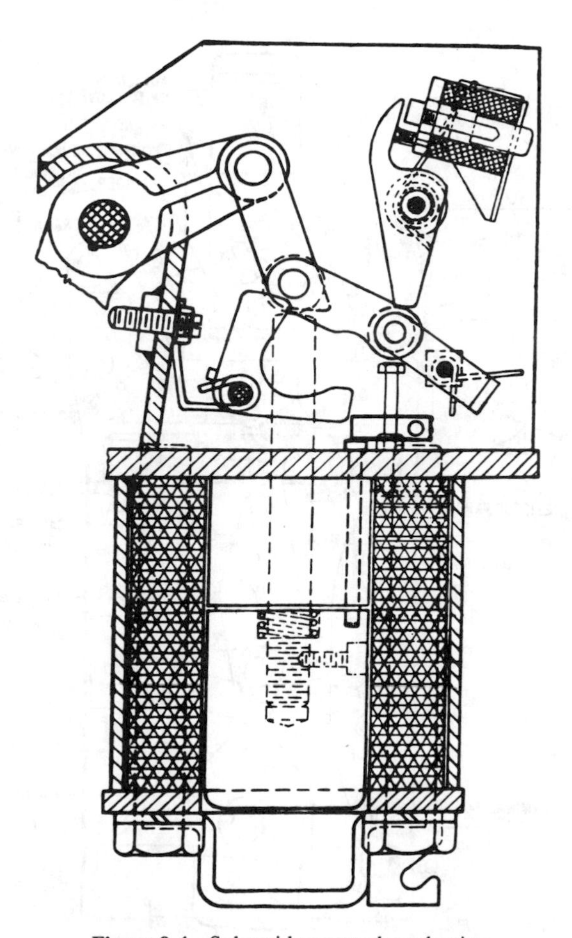

Figure 9-4 Solenoid-operated mechanism.

Solenoid type. This type of mechanism consists of a powerful closing coil, a relatively small trip coil, a latching device, an auxiliary switch and system of levers, and links and toggles for multiplying the short stroke of the closing coil armature (Fig. 9-4). The breaker is closed by energizing the solenoid so that it pulls the breaker rod upward and closes the contacts; at the same time, it may also compress a spring. The breaker (and spring, if so equipped) is latched in this position and the solenoid de-energized by means of the auxiliary switch. Opening is accomplished by energizing the small trip coil so that it releases the breaker contacts; these open through the action of gravity, or a spring, or both. Solenoids may be operated either on ac or dc.

Motor type. Motor operated mechanisms are usually of two types. One type utilizes the principle of centrifugal force (Fig. 9-5). The motor revolves two fly-weights at high speed, and the centrifugal force thus devel-

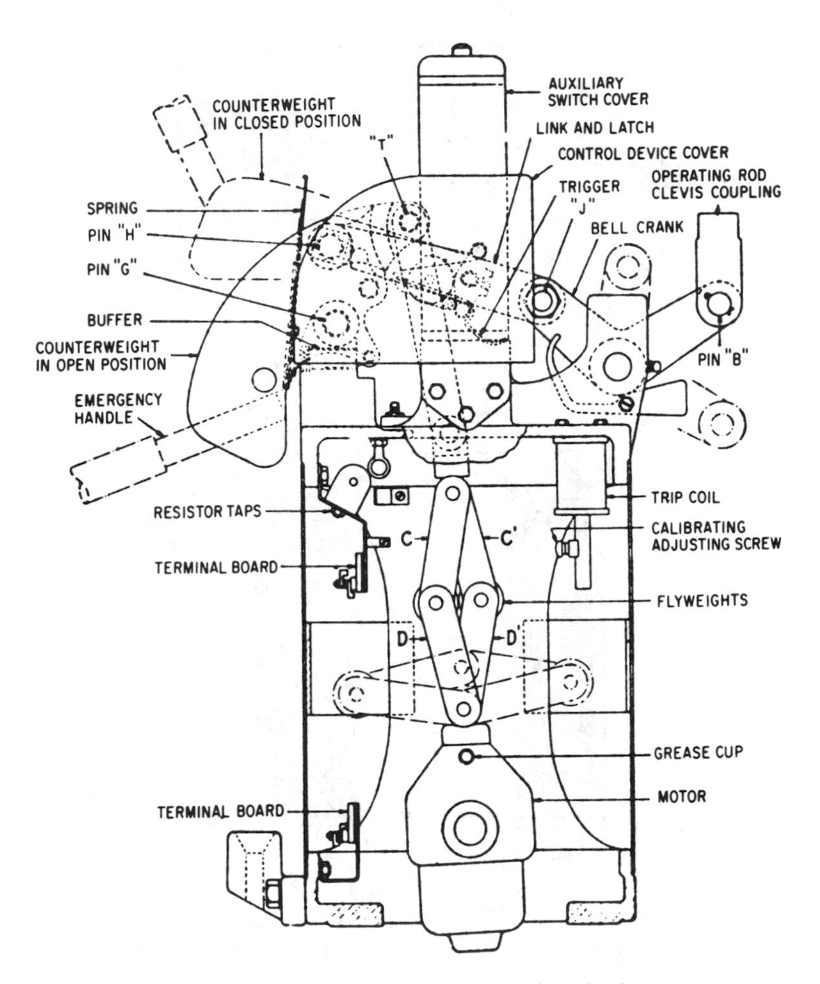

Figure 9-5 Centrifugal-motor-operated mechanism.

oped closes the breaker through a system of linkages (and compresses a spring, if the mechanism is so equipped). The breaker and spring are latched in this position, and the motor is de-energized by means of the auxiliary switch. Opening is accomplished in the same manner described for solenoid operation. The motor may be operated either on ac or dc. This type of mechanism is often used in automatic reclosing equipment.

Pneumatically Operated Mechanisms

The pneumatic or compressed air method of operation is used for heavy duty or very large breakers (Fig. 9-6). It has an unusually fast action. Since compressed air is used to close the breaker contacts, an air compressor and

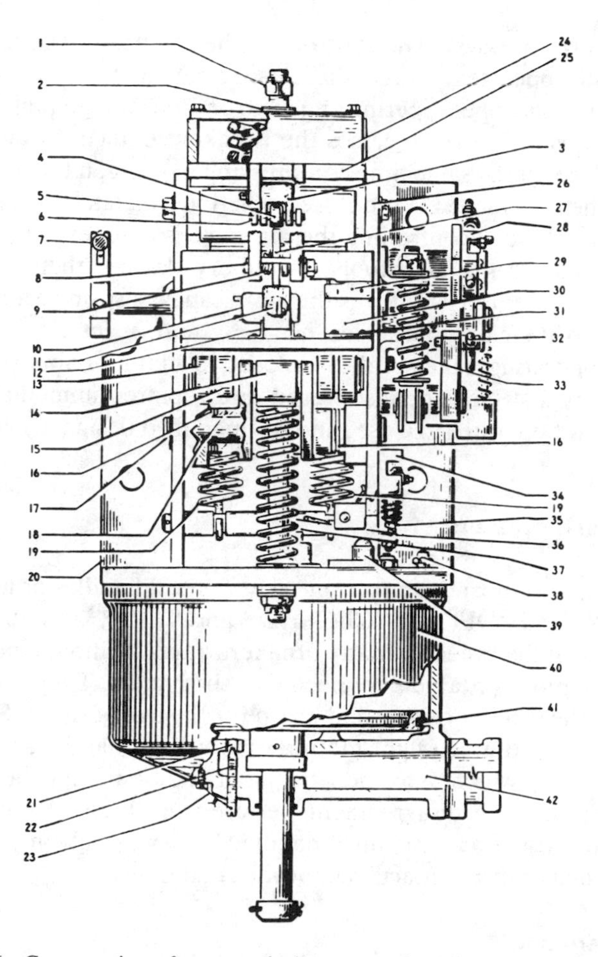

Figure 9-6 Cross section of pneumatically operated mechanism: (1) Buffer nut, (2) buffer adjusting washers, (3) front crank, (4) front trip roller, (5) trip assembly link, (6) trip latch roller pin, (7) auxiliary switch operating linkage, (8) hardened front trip latch pin, (9) front trip latch return spring, (10) trip coil, (11) trip coil plunger, (12) trip coil frame, (13) intermediate prop roller, (14) linkage reset spring guide, (15) lever arm, (16) antifriction shims, (17) crank pin, (18) linkage return crank, (19) linkage return spring, (20) mechanism frame, back plate, (21) closing speed orifice, (22) orifice adjusting screw, (23) orifice adjusting screw locking wire, (24) buffer housing, (25) mechanism frame lever support, (26) mechanism frame, side plate, (27) front trip latch, (28) "ac" switch, (29) latch checking switch ("lc"), (30) latch checking switch support bracket, (31) "lc" switch wipe adjustment shims, (32) manual trip linkage, (33) manual trip lever return spring, (34) piston cutoff switch ("aa"), (35) "aa" switch adjustment screw, (36) operating lever for the "aa" switch, (37) "aa" switch adjustment screw locking nut, (38) oil fitting, (39) "aa" switch operating lever buffer, (40) main operating cylinder, (41) main operating piston ring, and (42) air supply to cylinder.

storage tank are necessary. The controls on the compressed air lines, however, are electrically operated. When the closing coil in the control air valve is energized, the valve opens, permitting air from the storage tank to flow into a cylinder, the piston of which closes the breaker contacts by means of a mechanical linkage (at the same time compressing a spring, if the mechanism is so equipped); they are locked in the closed position by a latch. When the breaker contacts close, the coil controlling the air valve is de-energized and the valve closes and cuts off the air supply. Auxiliary devices then permit any air trapped in the system to escape; as the air escapes, a spring returns the piston and linkage to their original positions. A system of interlocks prevents the piston from operating if the breaker is closed in a fault or abnormal condition. It is important that leaks be discovered and repaired immediately since the breaker will not operate (to close, and to open if so designed) without proper air pressure.

Control Indicators

Located at or near the circuit breaker control handles or levers are signs labeled "ON" and "OFF" or "CLOSED" and "TRIP" to indicate the operating position of the breaker. Many breakers have additional indicators in the form of semaphores or targets connected to the handle. They usually show the color red for the closed position and green for the trip position. Some breakers also have a red and green light mounted above the handle and governed by auxiliary breaker switches to indicate position. Sometimes a white light is included to indicate a disagreement between the breaker position and the semaphore or targets at the control handle. It is extremely important that the position of the breaker contacts be known at all times.

Maintenance

Before any work is performed on circuit breakers, care must be taken to isolate them completely from the circuit. They may be provided with disconnecting switches or devices on both sides, and these should be locked open and grounded if necessary. When such devices are not available, it may be necessary to disconnect the breakers physically from the circuit.

For most breakers, the tank must be removed before contacts and other parts can be inspected and maintained. In oil circuit breakers, the oil must be removed before the tanks are lowered. Such breakers, especially large outdoor types, are equipped with tank lifters, either as part of the mounting frame or as an accessory device, because of the weight of the tanks. These lifters may be operated manually by handcranks or driven by electric motors. Tanks are usually equipped with a valve at the bottom through which the oil may be drained.

Three main elements of circuit breakers require special attention:

1. Contacts and associated arc-quenching devices
2. Insulation, including oil, insulating partitions or spacers, and bushings
3. The mechanisms, including closing and opening devices and control circuitry.

Contacts and Arc Quenchers

Contacts should be properly aligned and adjusted, if necessary, so that contact surfaces bear with firm, uniform pressure. One of the best ways to check contact impression is to close the contacts on a piece of thin tissue paper and a piece of thin carbon paper, with the carbon next to the tissue. When the breaker is closed and opened, it is a simple matter to determine the amount of impression made on the paper (Fig. 9-7).

Silver-to-silver contacts normally require little maintenance but should be resilvered and dressed to maintain a line contact. Badly pitted or burned contacts should be replaced before causing damage to other parts. If the contact surfaces are only roughened, they may be smoothed down with clean fine files or sandpaper. Arcing tips are normally replaced while the main contacts are being maintained. Arc-chutes, explosion chambers, and de-ion grids should be cleaned. Badly projecting tips and scale (the result of pitting) should be filed or scraped off.

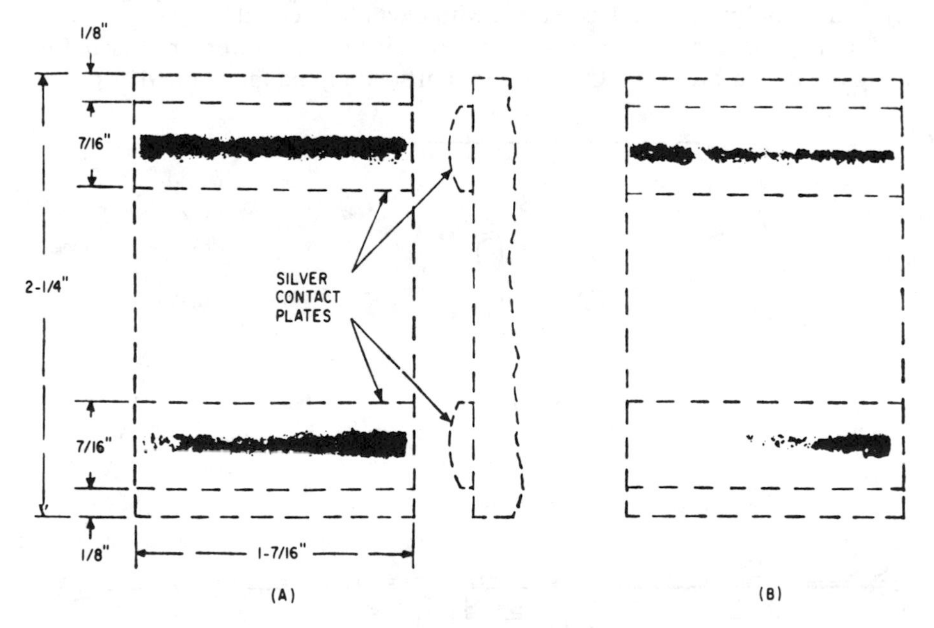

Figure 9-7 Checking contact impression: (A) This contact needs no adjustment since the impression of each bar is satisfactory (75 percent or more of contact length); (B) adjustment needed since impression shows that contacts are not closing properly.

In the case of large circuit breakers, contacts and arcing devices may be reached through a removable cover plate after sufficient oil has been drained to uncover them.

Oil

The oil in a circuit breaker should be frequently checked to ascertain its condition and that of the breaker. The condition of the oil may be determined on the site by simple tests. Oil samples may be obtained through the drain valve at the bottom of the tank. If the oil shows signs of moisture, carbonization, or dirt, it should be filtered through a paper press or centrifuge before being put back in the tank.

The task of draining oil from circuit breaker tanks may be a long and cumbersome one since some of the larger circuit breakers may require thousands of gallons of oil. For smaller breakers, oil may be drained into drums or barrels; for larger breakers, oil may be drained into plastic storage "bags" of rather large dimensions or pumped into truck-mounted tanks such as those used for oil or gasoline deliveries (Fig. 9-8).

If there is evidence of carbon, the tank and other parts of the breaker, insulating barriers, partitions and spacers, lift-rods, and the like, should be washed down with clean hot oil under pressure and checked for cracks or deformities before the tank is refilled with clean filtered oil.

It is good practice to pass even new oil through filter presses before filling the circuit breakers. Oil levels in tanks should be maintained at proper

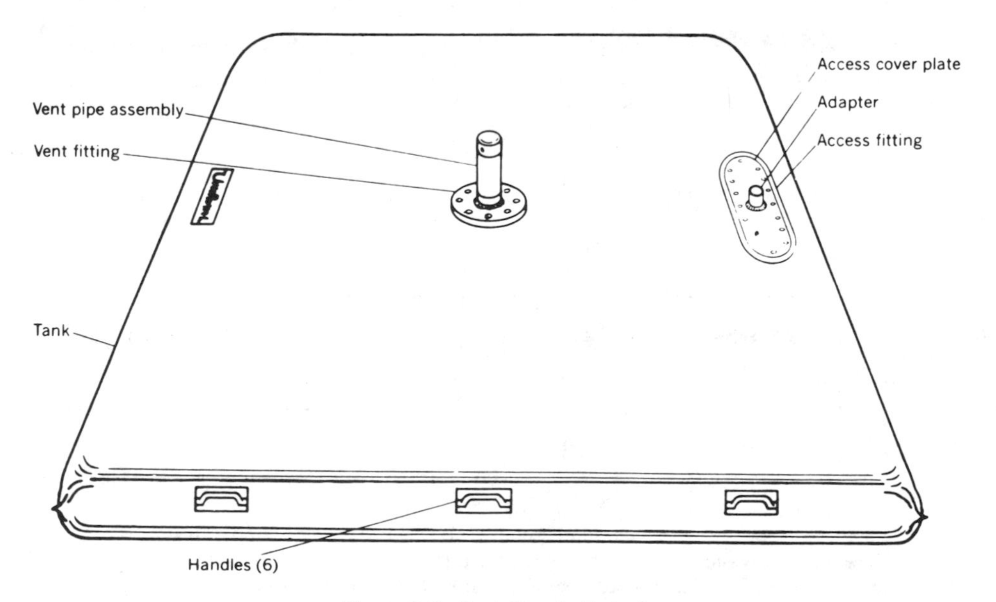

Figure 9-8 Portable plastic tank.

heights to allow for expansion under normal and abnormal conditions. Any gauges should be checked to make sure that they indicate actual oil levels properly.

Oil valves should be checked for leaks as well as for their ability to open and close readily without sticking.

Bushings

The external surfaces of bushings should be cleaned to remove any accumulation of dirt or other deposits (cement, dust, metallic particles, salt, and so forth), and the bushings should be checked for cracks and breaks. The dielectric strength of the bushings may be determined by a power-factor test or by a megohmmeter (Megger).

Bushing supports should be inspected to ascertain that vibrations from the breaker operations have not caused the bushings to move, resulting in misaligned contacts.

Mechanisms

The mechanisms should be operated several times to make sure that they (and the breakers) are working smoothly and freely without binding. If the breakers are normally operated electrically or pneumatically, they should first be operated manually to determine the absence or presence of binding. Toggles, buffers, stop clearances, dashpots, and other auxiliary devices should be cleaned of dirt and grime. The length of the breaker stroke (lift rod) and its opening and closing speed should be measured and adjusted according to the manufacturer's specifications. Opening and closing times should be rechecked after the breaker has been properly filled with oil and is ready to be restored to service.

Electrical Operation

Operating voltage at the mechanism terminals with full operating current should be checked to determine its adequacy. Closing relays, indicating lamps, trip coils, tripping control switches, and the like, should be checked to make sure that they are operating properly, especially that the (usually red) indicating lamp is actually in the trip coil circuit.

Pneumatic Operation

Air pressure should be checked to ascertain its adequacy for correct operation and to make sure that it is restored after each breaker operation. Air connections should be checked for leaks. Air valves should be operated to make sure that they open and close the breaker properly. Opening and closing time and tripping arrangements should be checked as they were for electrical operation.

Accessories

Undervoltage devices should be checked to make sure that tripping will be positive if the voltage drops to, or below, a predetermined value. If a time delay is included, the time setting should be checked. Auxiliary switch adjustments and contacts should be checked. Grounding connections on breaker frames should be checked to be sure that they are intact and directly connected to a dependable ground for safety. Mechanical and electrical interlocks should be checked to make sure that they are functioning properly and that all adjustments are correct. Cyclometers, or operation counters, should be checked for proper operation. Heaters, should they exist, should be checked to see whether they are in good operating condition. If thermostat controls are associated with them, the settings and operation of the thermostat should be checked.

General

Bolts, nuts, pins, cotter pins, pipe plugs, pipe fittings, as well as rods, levers, links, and other items should be in place and tight. Broken, worn, or damaged parts should be replaced. Gaskets and seals should be checked and replaced if worn or leaking. Connections, splices, and terminals should be checked to make sure that they are clean and tight. Painted surfaces of oil tanks, housings, and the like, should be checked, and damaged, scarred, or scratched places touched up to avoid corrosion.

TESTING

The frequency of maintenance depends on many factors—local conditions, importance of the equipment being protected, and, to a great extent, on the number of operations engaged in by the breaker and the duty imposed on it. In certain cases of severe duty, such as the interruption of a heavy short-circuit current, a special inspection and maintenance may be advisable. On the other hand, if operating conditions leave a breaker inoperative for long periods of time, it may be well to open and close it periodically to make sure that all parts are in proper operating condition.

Since circuit breakers are not often called upon to operate under severe conditions for long periods of time, lengthy, expensive maintenance procedures can be reduced substantially by substituting simple, relatively inexpensive tests that can be performed in short order. Such tests can be made periodically, or following an operation under severe fault conditions, and their results used to determine the extent of maintenance required. Full maintenance schedules based on regular time periods can thus be extended considerably.

Contacts are tested by taking resistance measurements across the terminals of each pole of a breaker. Oil tests determine the condition of the insulation generally. The working condition of the operating mechanisms can be determined by recording the time and smoothness of the opening and closing of the lift rod. All these tests can be accomplished in a relatively short time, perhaps even less than an hour, by a small crew, without undertaking the long and tedious process of draining oil from the tanks so that they can be dropped and then the equally long process of raising the tanks, filtering the oil, and refilling the tanks. Simple tests like these will be described below. The breaker under test, however, must still be isolated from the circuit by air-break switches or by physically removing its connections.

Resistance Test

This test employs a dc source (such as a battery, a voltmeter, and an ammeter (Fig. 9-9). The resistance of the contacts is derived from Ohm's Law, using the equation,

$$R = \frac{E}{I}$$

where

R = resistance, in ohms
E = voltage, in volts
I = current, in amperes

A record of these values of resistance is kept, and any significant change in later tests determines the time for maintenance.

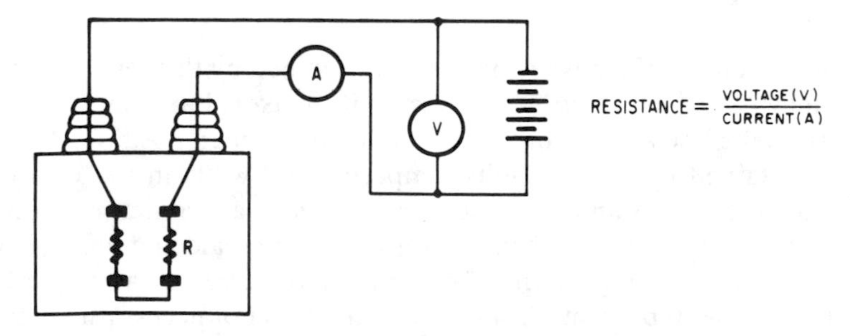

Figure 9-9 Voltmeter-ammeter method of resistance testing.

Oil Test

The condition of the oil may be determined in the field by two simple tests. These are similar to those for transformers and are described in Chap. 2.

Megohmmeter Test

This test is also known as a Megger test. A Megger (trade name) is a small dc generator, turned by hand or motor, that is connected to the terminals of the circuit through a meter calibrated to read directly the megohms (millions of ohms) in the circuit. See Chap. 2.

In testing the bushings of a breaker, the breaker contacts are in the open position, and the Megger is connected between the terminal of one bushing at a time and the tank. A comparison of readings of the several bushings over a period of time will give an indication of the condition of the bushings.

The Megger can also be used for an overall check of the breaker. Again, with the breaker in the open position, the Megger is connected across the breaker pole, that is, one lead of the Megger is connected to the incoming bushing terminal and the other lead to the outgoing bushing terminal. The reading of the Megger will then indicate the insulating condition of the two bushings and the condition of the oil in the breaker (Fig. 9-10).

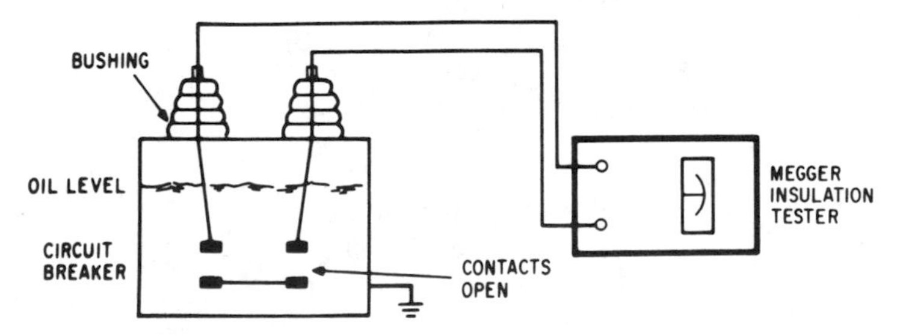

Figure 9-10 Megger testing of circuit breaker insulation.

Power-Factor Test

The condition of a bushing may be determined by this test. The bushing may or may not be removed from the circuit breaker, but is must be disconnected from the line and the breaker must be in the open position. The power factor is obtained by reading volts, amperes, and watts in the circuit (Fig. 9-11). The instruments and instrument transformers are sometimes contained in one package, with only a power factor indicating scale and voltage-control device mounted on the package. Power-factor values cover a range of a few percent, perhaps 10 percent. A single reading does not necessarily determine the condition of the bushing insulation. As it deteriorates, more current will

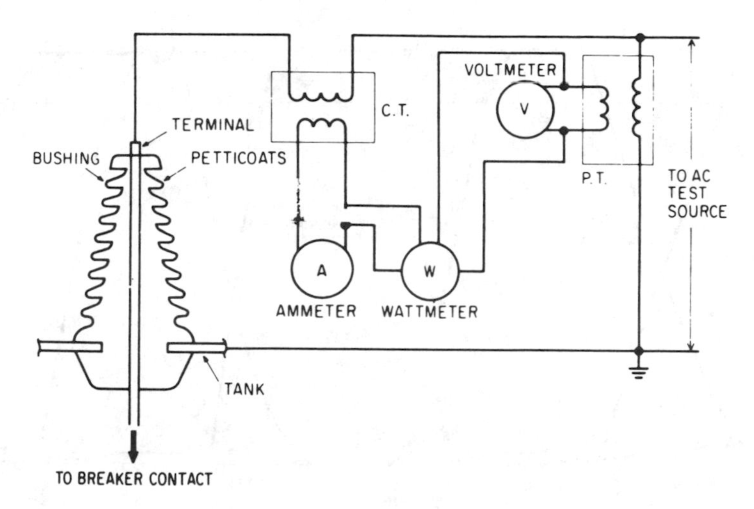

Figure 9-11 Connection for power-factor test.

leak through it, and the power factor will become increasingly greater, approaching 100 percent.

Power-factor readings are taken periodically as part of maintenance programs. These readings are compared, not with readings of other bushings, but with previous readings on the same bushing. To what extent power factor increases may be acceptable is a matter of judgment. "High" values may depend on the importance of the installation and other factors. Values will vary with temperatures, and excessive readings should be compensated for by any differences in temperatures at the times the tests were made.

Breaker Travel Indicator

This instrument records in graphic form (a chart wrapped around a turning cylinder) the time and "smoothness" of the lift rod—that is, its acceleration and any hindrance to its motion when opening and closing the contacts in the breaker. The travel indicator is a mechanical device attached to an extension of the lift rod through a fitting provided for checking the rod. The shape and timing shown by the curve produced on the chart will determine the need for maintenance (Fig. 9-12).

VACUUM AND SULPHUR HEXAFLUORIDE BREAKERS

Although the practices previously described apply specifically to air and oil circuit breakers, most of the details also apply to vacuum and sulphur hexafluoride breakers. In the latter, the contacts and operating rods operate in a

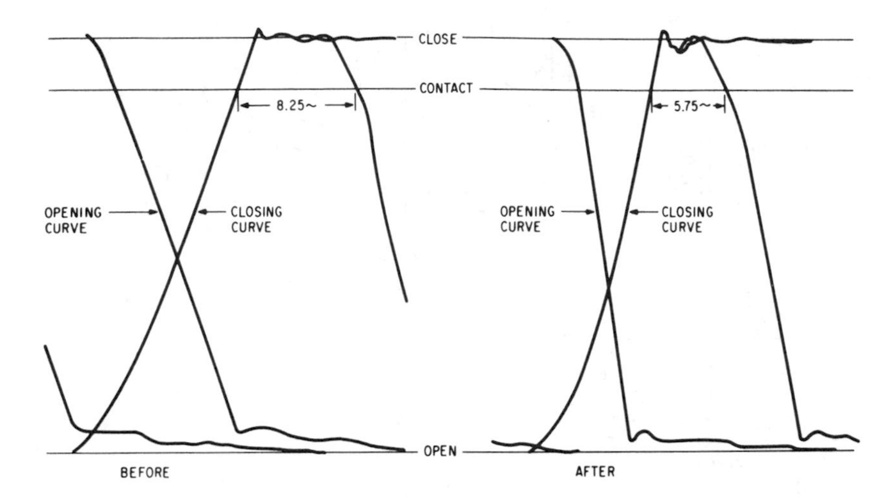

Figure 9-12 Lift-rod travel indicator.

separate module (sometimes called a *bottle*) containing the vacuum or gas, and maintenance is accomplished by the removal of the module and its replacement with a new one. Repair or replacement of internal parts and restoration of the vacuum or gas is performed in the shop under more controlled conditions, using the more specialized equipment available. The gas system of the sulphur hexafluoride breaker is inspected and maintained in a manner similar to that used for the compressed air system in a pneumatic operated breaker. A schematic diagram of this gas system is shown in Fig. 9-13.

TROUBLESHOOTING

The principal troubles experienced in circuit breakers are overheating; failure of the breaker to trip, to close, or to latch close; and oil problems. The causes and remedies are indicated in Table 9-1.

REVIEW

1. Circuit breakers may be operated directly by hand (Fig. 9-2), by remote mechanism (Fig. 9-3) and electrically by motors (Fig. 9-5) or solenoids (Fig 9-4). To insure continuity and stability of electric supply, storage batteries may be employed that may be continuously charged by "floating" on a distribution supply line.
2. Some breakers may be automatically reclosed a predetermined number of times for predetermined periods of time (usually several cycles or seconds) before becoming "locked out" in an open position. This is done

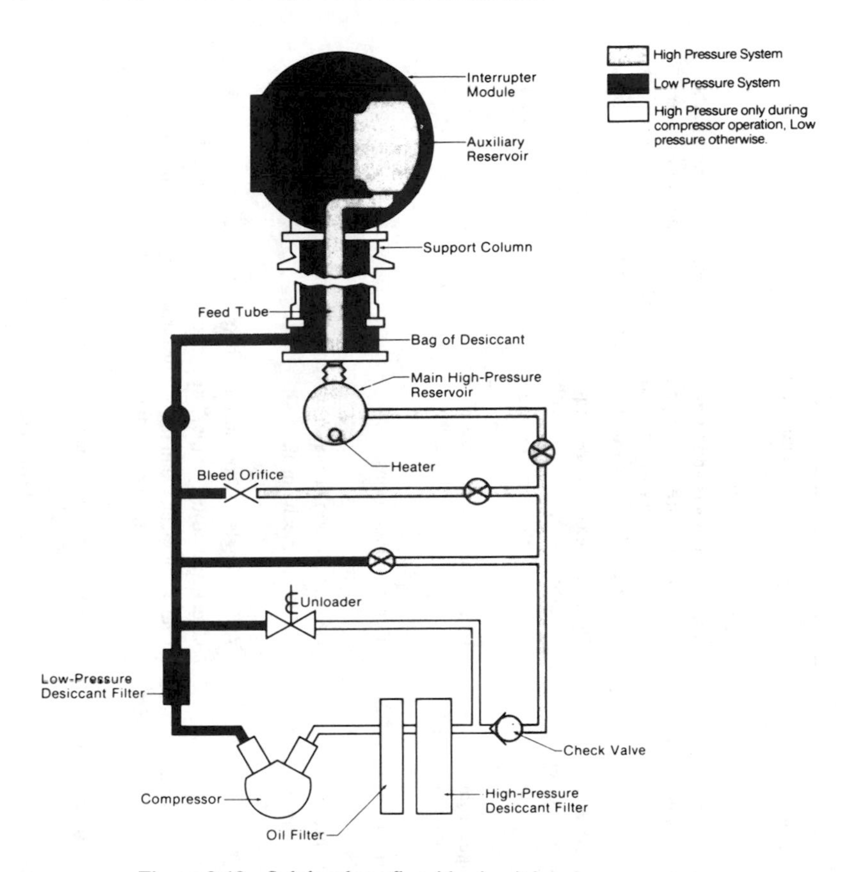

Figure 9-13 Sulphur hexafluoride circuit breaker gas system.

so that temporary fault conditions need not interrupt service unnecessarily.

3. Contacts need to be aligned, adjusted, and dressed (Fig. 9-7) to insure proper connection between incoming and outgoing electrical conductors. Arcing tips, explosion chambers, and de-ion grids should be thoroughly cleaned.
4. Oil, or other fluid insulation, should be checked for carbonization, sludge, or other impurities. This may be done by visual observation of color and clarity, and by electrical tests on samples. External surfaces of bushings should be cleaned and checked for cracks and breaks. Electrical tests may include resistance and power-factor checks, or by the use of a Megger.
5. Mechanisms and devices associated with the breaker should be checked for tightness, rust, and other signs of deterioration; tanks should be sanded and painted when necessary, gaskets replaced, auxiliary switches and other items checked for proper operation.

TABLE 9-1 Troubleshooting Circuit Breakers

Condition	Possible Cause	Suggested Remedy
Overheating	Contacts out of proper alignment and adjustment	Line up contacts and adjust properly.
	Contacts dirty or greasy	Remove accumulated dirt.
	Contacts burned and pitted from too frequent operation or many heavy interruptions	Dress up contacts and fit properly, or replace with new parts. (High-pressure, butt-type contacts usually do not require dressing; silver-to-silver contacts should be dressed very carefully but only when actually required.)
	Breaker kept open or closed for too long a period; copper contacts oxidized	Operate breaker more frequently to wipe contacts clean; perhaps replace them with silver-to-silver contacts.
	Overloading (continuous or prolonged current in excess of rating)	Replace with adequate rating breaker or remove excess load.
	Heat transmitted to breaker from overloaded or inadequate connecting cables or bars	Increase number or size of bars, or reduce loads.
	Loose connections on terminal connectors	Tighten connections.
	Location too hot an ambient	Relocate to cooler place, or ventilate and cool.
Failure to trip	Mechanism binding or sticking: Lack of lubrication	Lubricate mechanism.
	Out of adjustment	Adjust toggles, stops, springs, links, etc.
	Failure of latching device	Check and adjust "wipe," or replace if worn or corroded.
	Auxiliary contacts and interlocks not making proper contact; damaged trip coil	Check contacts; clean if dirty or tarnished; adjust if out of adjustment. Replace damaged parts or trip coil.
	Damaged or dirty contacts on tripping device	Dress or replace damaged contacts, or clean dirty contacts.
	Faulty connections in trip circuit	Repair loose or broken wires. Tighten binding screws.
	Blown fuse in control circuit	Replace blown fuse. Check for grounds or short circuit.

Unnecessary tripping (when tripping should not occur)	Calibration set too low	Set device or relay for proper value according to ampere load of circuit.
	Worn latches	Replace with new parts.
Failure to close or to latch close	Mechanism binding or sticking: Lack of lubrication	Lubricate mechanism.
	Improper adjustment of breaker mechanism	Adjust toggles, links, stops, buffers, springs, etc.
	Burnout of closing coil: Operator holding control switch closed too long	Replace coil and educate operator, or rewire to include an auxiliary switch that automatically cuts off closing coil as soon as breaker closes.
	Closing relay sticking	Check or adjust closing relay.
	Cut-off switch operating too soon	Adjust switch to delay cut-off and allow breaker to close fully.
	Cut-off switch operating too late, causing the breaker to "bounce" open	Readjust switch to reduce power at end of stroke and eliminate "bounce."
	Insufficient control voltage: Too much drop in leads	Install larger wires. Improve contact at connections.
	On ac control: poor regulation	Install larger control transformer. Check rectifier to be sure it is delivering adequate dc voltage from adequate ac supply.
	On dc control: battery not fully charged or in poor condition	Give battery a sustaining charge or replace cells.
	Blown fuse in control circuit; faulty connection or broken wire in control circuit; damaged or dirty contacts in control switch	Replace blown fuse. Repair faulty connections or broken wire. Dress or replace damaged contacts or clean dirty contacts in control switch.
Insufficient oil (in tanks)	Leakage of oil	Find leak and repair. Tighten up joints in oil lines.
	Oil thrown during operation	Fill tanks to proper level.
Dirty oil	Carbonization from many operations	Drain dirty oil, and filter or replace with new oil. Clean inside of tank and internal parts of breaker.
Moisture in oil	Condensation of moist atmosphere	Drain and filter oil or put in new oil. Check mufflers.
	Entrance of water from rain or other source	Locate and repair source of water entrance.

TABLE 9-1 (cont.) Troubleshooting Circuit Breakers

Condition	Possible Cause	Suggested Remedy
Sludging of oil	Overheating	Filter or put in new oil. Locate and repair source of overheating.
Gaskets leaking	Oil saturation or improper installation	Replace with new gaskets, properly installed.
Insulation failure	Absorption of moisture or accumulation of dirt, grime, carbon, etc., on bushing and insulation parts	Thoroughly clean all insulated parts. Bake or dry out water-soaked parts, or replace them.
Tanks or supporting structure energized (leakage or induced voltage)	Ground connection broken	Repair connection.
	Grounding inadequate	Add additional grounds. Bond as many structures together as practicable.
	Dirty or cracked bushings	Clean bushing, or replace it if cracked. Check bushing for flashover and lightning protection.
Noises due to vibration	Loose bolts or nuts, permitting excessive vibration	Tighten bolts or nuts.
Rust and paint deterioration	Weather; pollution or salt contamination; overheating	Remove rust and deteriorated paint. Clean surfaces. Repaint with proper paints and coatings.
Broken bushing or parts	Strain on terminal connections	Install flexible connection between terminals and cables or busses to remove strain.

6. Before draining oil (or other fluid) to maintain elements of a breaker, its condition may be determined by checking the resistance across each set of contacts, by visual and dielectric test of the oil, and by the speed and smoothness of the closing of the contacts using a graphic travel indicator. Resistance (Fig. 9-9), Megger (Fig. 9-10), and power-factor tests (Fig. 9-11) may be used to check the condition of the bushings.

STUDY QUESTIONS

1. What is meant by the trip-free feature of a circuit breaker?
2. What is meant by the antipump feature of a circuit breaker?
3. What type of mechanisms are employed in the operation of circuit breakers?
4. How are circuit breakers opened? How are they closed?
5. What three elements of a circuit breaker require special care in their maintenance?
6. What is meant by the "throwing" of oil?
7. How should the oil in a circuit breaker be checked?
8. What three tests should be made to determine the condition of a circuit breaker?
9. What precautions must be taken to insure a continuous electric supply for the reliable operation of circuit breakers?
10. What is a "Megger" and where is it used?

10

DISCONNECTING DEVICES

SWITCHES

Less expensive equipment than circuit breakers can be used to accomplish many switching operations. Such equipment, commonly referred to as *switches,* also has the dual function of providing for physical separation in an electric circuit and restoring its continuity.

Switches serve many purposes. They may be used to

1. De-energize equipment or portions of circuits upon which work is to be performed
2. Isolate faulted equipment or faulted portions or circuits
3. Provide electric supply from alternate circuits
4. Distribute loads throughout an electric circuit to avoid or eliminate overloading of portions of the circuit
5. Ground circuits
6. Serve temporarily in place of circuit breakers
7. Serve for a variety of purposes to meet particular situations.

Switches are of many types, shapes, and designs, ranging from a small snap switch used to control one or more lamps or signals to large electrically operated devices with some arc-handling capability. A variety of small specialty switches widely used in control and communication circuits are not normally considered as power equipment and are therefore not included here.

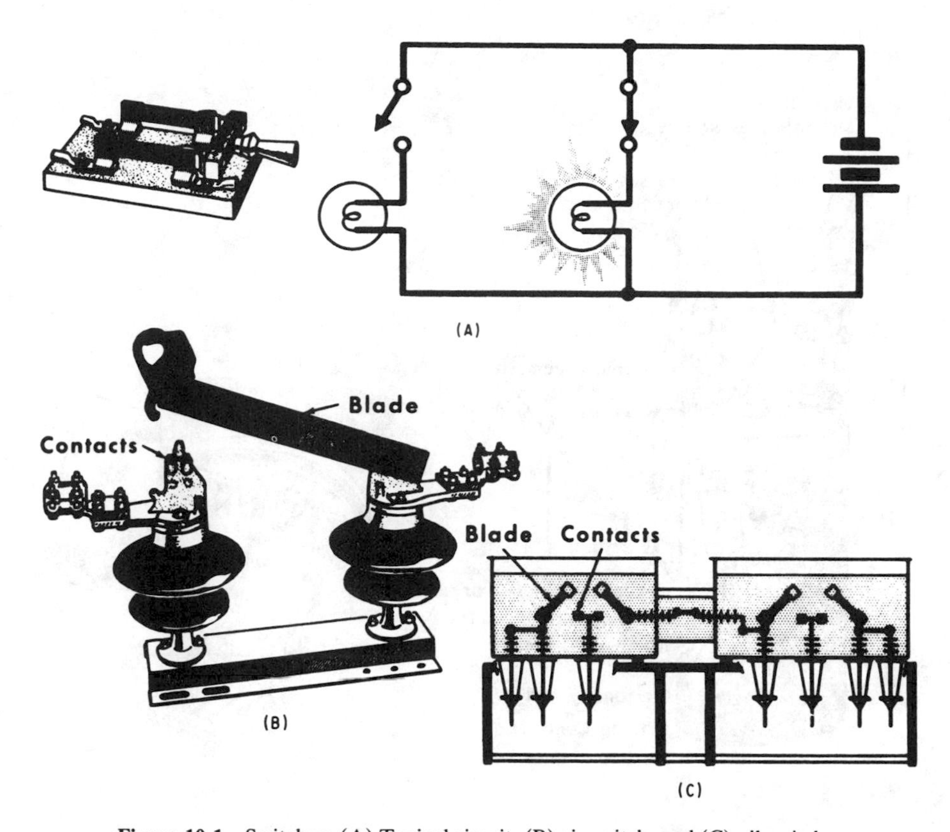

Figure 10-1 Switches: (A) Typical circuit, (B) air switch, and (C) oil switches.

Switches are almost always used in series with a circuit breaker or fuses (to be discussed later), both of which can safely interrupt short-circuit currents.

The two basic types of switches are oil switches and air switches. As their names imply, oil switches are those whose contacts are opened in oil, and air switches are those whose contacts are opened in air (Fig. 10-1).

OIL SWITCHES

Oil switches suitable for switching load currents but not for interrupting short-circuit currents are widely used in high-voltage circuits (such as utility primary circuits). They are also used where low-voltage currents may be rather sizable, that is, where the arc that may form, though not as great or persistent as that from a short circuit, may not be successfully handled by other devices (Fig. 10-2).

Oil switches are one of the more dependable and convenient discon-

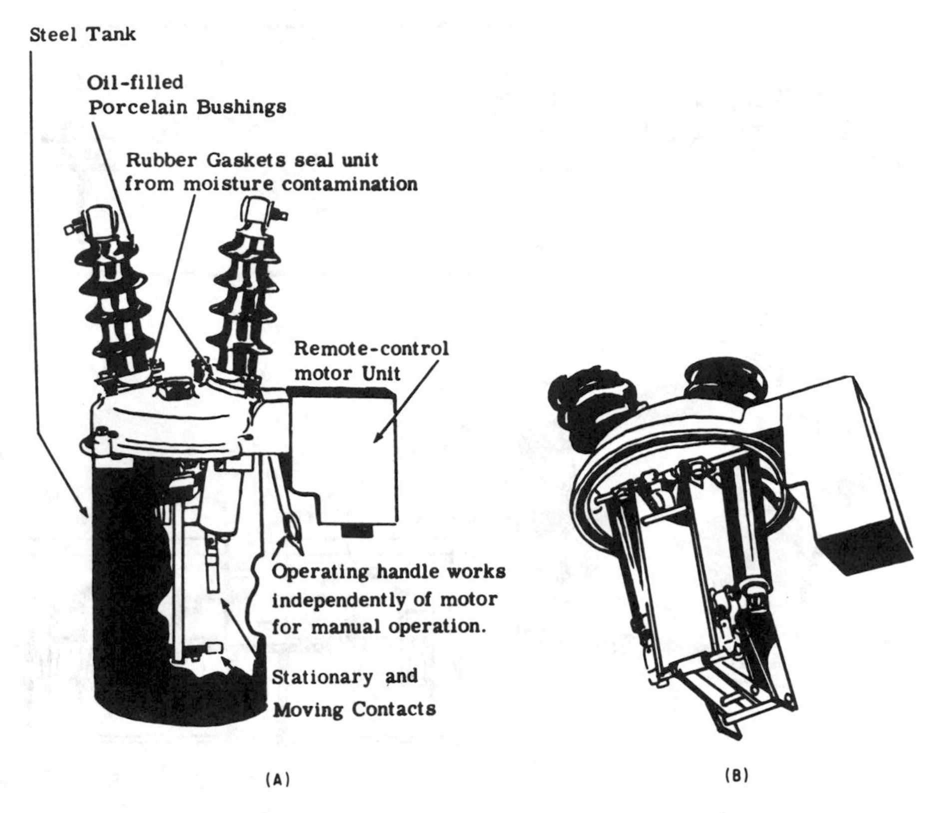

Figure 10-2 Oil switch: (A) Open position, and (B) closed position with tank removed.

necting devices available, but they are also one of the more costly. Size for size, they are considerably more expensive in first cost than air switches and also in operating and maintenance expense.

Oil switches are especially suitable for underground application and any situation where water, moisture, salt atmosphere, or chemical or other pollution makes air switches impractical.

The tanks and operating handles of oil switches should be effectively grounded. Provisions should also be made for locking the operating handles in the open or closed positions.

AIR SWITCHES

Switches of this type can be classified into two categories: disconnects and air-break switches.

Disconnects

Switches used merely to insert a physical separation in a circuit are associated with almost every type of major circuit or piece of electrical equipment as a safety measure to ensure that the line or equipment upon which work is to be done cannot be energized accidentally. To differentiate them from air-break switches, they are often called *disconnects* (Fig. 10-3A). They are not designed to be opened while the circuit in which they are connected is energized. While they are also not designed to close circuits, they may, under some circumstances and with special care, be closed while a circuit is energized.

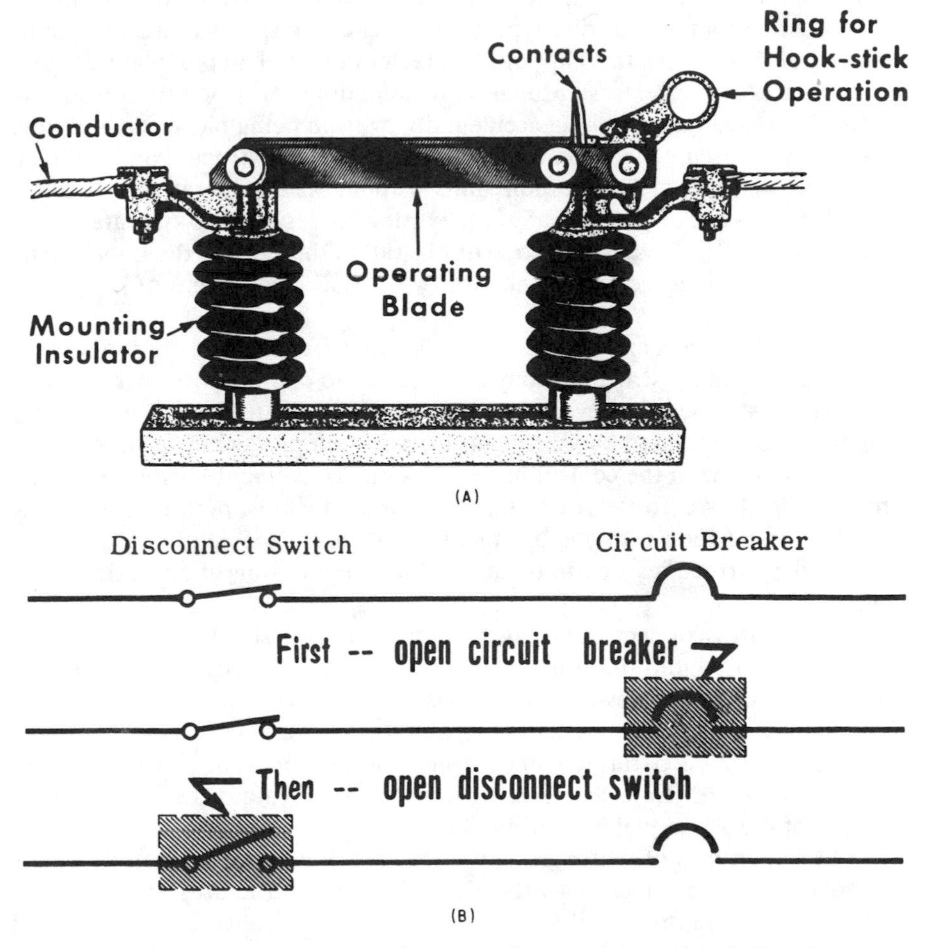

Figure 10-3 Disconnect switch: (A) Switch in closed position, and (B) switch never opened under load.

Disconnects are connected in series on both sides of a piece of electrical equipment (including circuit breakers) or at some point or points in an electric supply line. They should never be opened until the oil circuit breaker or air-break switch in the circuit has been opened (Fig. 10-3B). They thus provide a second (and usually visible) physical break in the circuit in the event the breaker or air-break switch is accidentally closed. The disconnect should always be closed before the breaker or air-break switch has been closed to energize the equipment or circuit from which workmen have been cleared.

Operating in the same manner, disconnects may also be used to sectionalize circuits when, for example, it is necessary to locate a faulted section of a circuit. They may also be used to transfer loads from one circuit to another when, for example, it is necessary to eliminate or avoid overloading a circuit.

Most disconnects or disconnecting switches consist of hinged single or multiple blades and contact clips or receptacles mounted on porcelain or other insulation. Locking devices or latches are sometimes provided to prevent the disconnect blade from opening accidentally or from being blown open during a heavy short-circuit condition. Disconnects may be constructed and installed as single-pole units or as multipole units with all blades operated by a single actuating device. For greater safety, they may be designed to operate with a switch stick made of insulating wood or plastic. In this design, the hook on the end of the stick is placed in the hole or eye at the opening end of the switch blade.

Both disconnect and air-break switches that are mounted outdoors at the top of poles or other structures may be designed to operate with such a switch stick. In most cases, however, such switches are located at a considerable height and are operated by a mechanism, a system of linkages actuated by a crank or motor with the control handle at some convenient location near the ground (Fig. 10-4). To avoid unauthorized or erroneous operation, they may be locked in position with one or more special locks capable of being opened only by the correct key or combination in sole possession of authorized personnel.

When a disconnect is closed, the switch blades should be rapidly and completely closed into the contact clips or receptacles. Because arcs are usually not formed when switching contacts are closed and held closed, the closing of a de-energized disconnect blade onto an energized clip of the disconnect may be permitted under very special conditions, usually of an emergency nature. In such cases, extraordinary precautions must be taken because of the additional hazards involved.

In some types of switch gear, such as metal-clad distribution modules, disconnects are so attached to the sliding breaker structures that they are disconnected from the circuit when the breaker unit is rolled out. Mechanical interlocks prevent the moving of the breaker unit and the disconnects into or out of the operating position unless the breaker is open (Fig. 10-5).

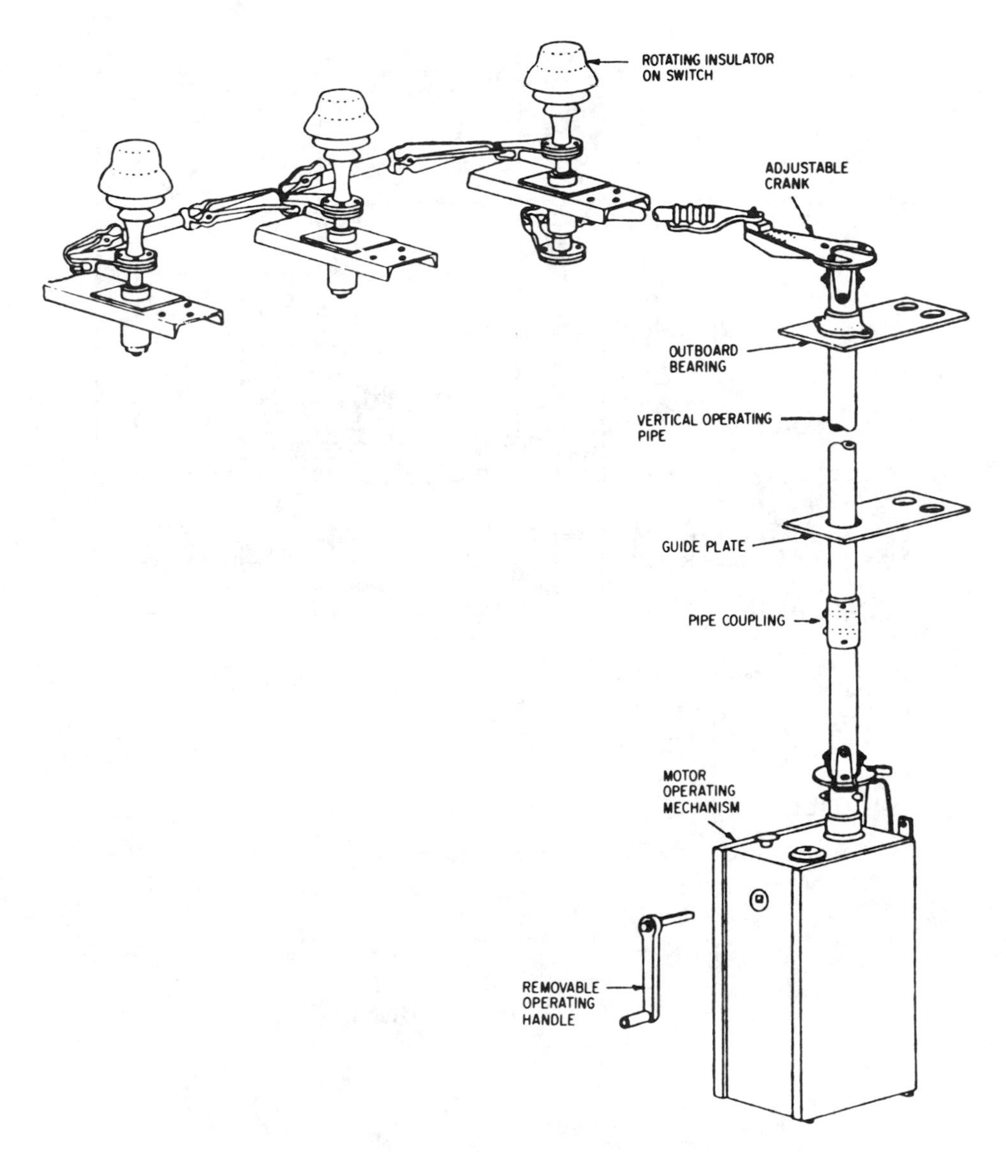

Figure 10-4 Air disconnect switch with motor-operating mechanism.

Air-Break Switches

Air-break switches are essentially disconnect switches to which arcing "horns" or other arc-suppressing devices are added (Fig. 10-6). They are designed to interrupt limited amounts of current but not short-circuit currents. The interrupted currents may represent load currents or magnetizing (or

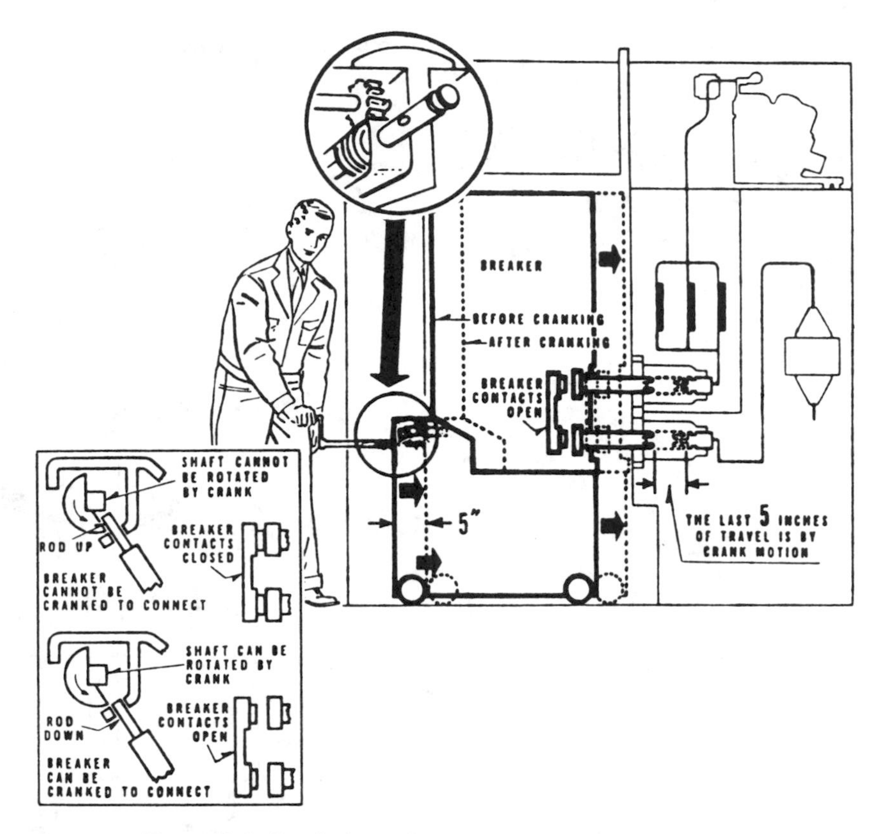

Figure 10-5 Interlocks on disconnects in metal-clad switch gear.

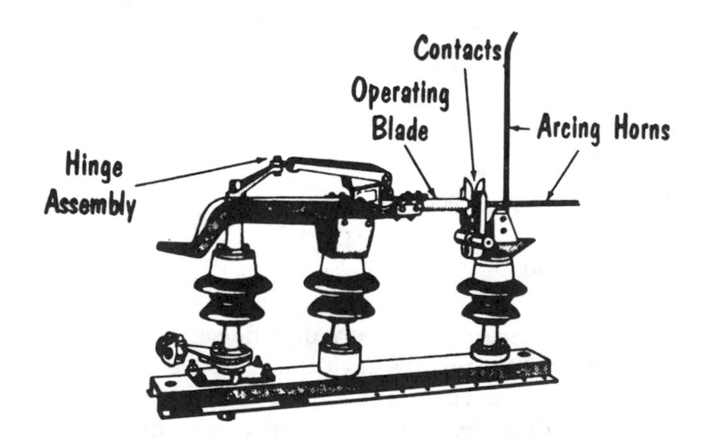

Figure 10-6 Air-break switch in closed position, with arcing horns.

Figure 10-7 Pole-mounted air-break switch in open position.

exciting) currents from transformers connected to the circuit, or they may represent charging currents from capacitors connected to the circuit. If these currents are large, oil switches may be advisable.

As with disconnects, air-break switches may be operated as single-pole units or "gang" operated in multipole units actuated from a single device (Fig. 10-7). They may be operated by insulated switch sticks or by manual or motor-operated mechanisms.

Arcing horns are metal rods attached to the clip or stationary end of a switch but bent at an angle from the path of the blade. As the blade of the switch is opened, the arcing horns remain in contact until after the main contacts have separated. As the switch blade continues to open, an arc is permitted to form between a part of the blade not designed to make contact with the stationary clip and the arcing horn. As the spread between these horns and the blade becomes larger, the arc continues to lengthen until it is finally extinguished through inability to span the gap. Any pitting or burning that may occur takes place on the horns and parts of the switch blade where it is not critical. Although the blade of most air-break switches operates in a vertical plane (though they may be mounted horizontally or vertically), the blade of others operates in a horizontal plane. The advantages claimed for the latter are that they require fewer insulators and have less trouble operating under conditions of freezing rain, snow, and ice, the reason being that stationary contacts, because of their position, are partially protected from the elements.

Auxiliary Interrupter Devices

To permit air-break switches to interrupt larger currents, several devices can be added to them to reduce the effects and duration of the arcs formed.

One simple device replaces the rigid arcing horns with a whip attached to the blade. As the switch opens, the whip remains in contact until the blade is part way open. The whip is then released, and its rapid swing extinguishes the arc very rapidly (Fig. 10-8).

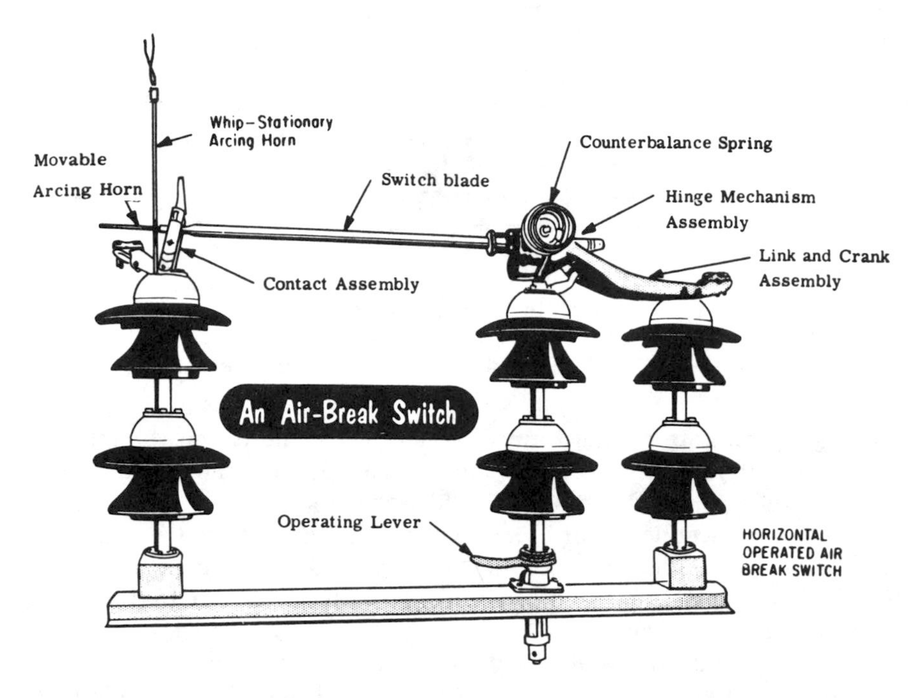

Figure 10-8 Air-break switch with whip arcing horn.

Other devices employ interruption chambers or arc tubes and vacuum modules (Fig. 10-9). After the main contact is open, the arc is extinguished (1) within the interruption chamber or arc tube or (2) within the vacuum module, both of which are mounted on the end of the switch blade. The theory behind the former of these devices is similar to that behind the explosion chamber in a circuit breaker. The action of the latter device is similar to the action of the vacuum module in a circuit breaker. In both these types of units, the arc is contained and extinguished within the interrupting device at the time of circuit interruption. Supposedly, these units are faster, more positive, have reduced fire hazard, and are less affected by weather.

CONTACTS

Contacts for air switches fall generally into two types, one based on line contact and the other on area contact.

Line contact (Fig. 10-10) is usually achieved by the switch blade's being rotated within the stationary contacts and essentially resting under pressure at right angles to the inner side of the clips. The wiping action as the blade turns not only cleans the contacts but clears them of ice formations. The line contacts established are further enhanced by silvering the contact areas of

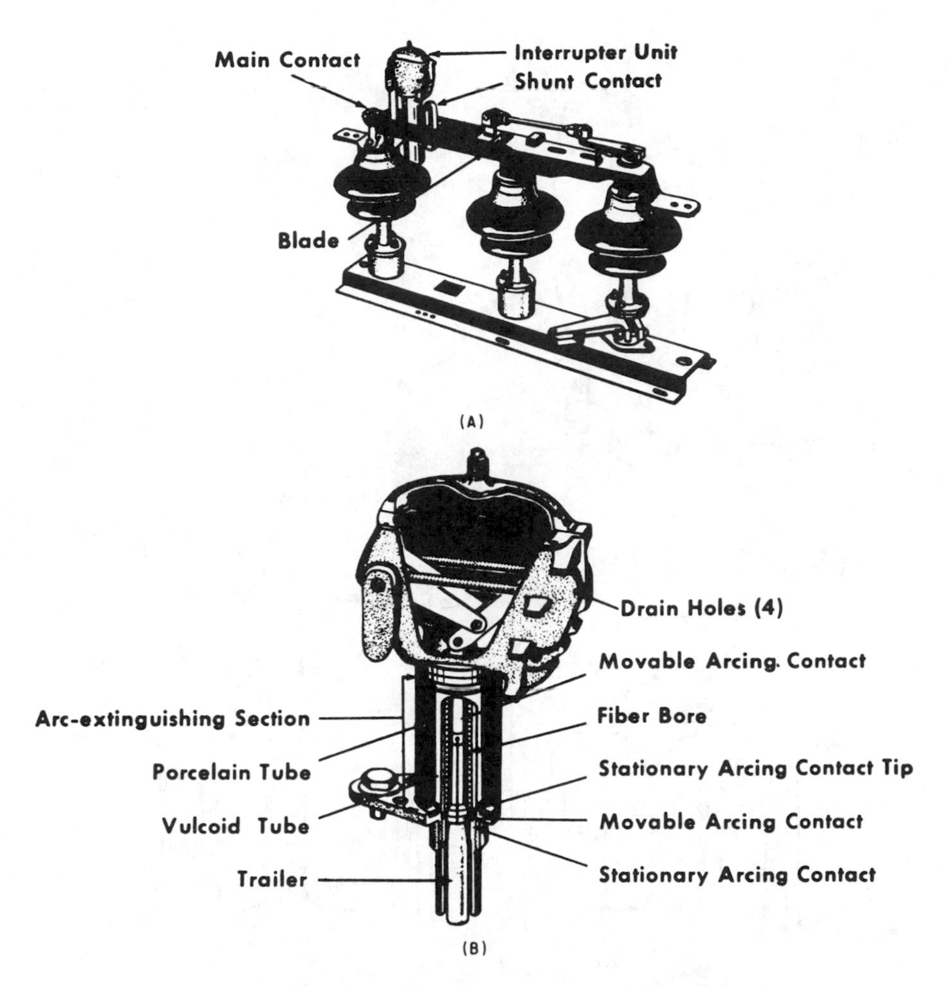

Figure 10-9 Air-break switch with interrupter unit: (A) Closed switch, and (B) cutaway of interrupter unit.

both the blade and the inner surfaces of the stationary contacts. This type of contact is usually associated with air-break switches.

Area contact is usually achieved by the switch blade's slipping between stationary contacts on both sides, producing a wiping motion that maintains a clean surface on both the blade and the stationary contacts. The stationary contacts may consist of the inside surfaces of the clip or jaws and exert pressure through the elasticity of the shaped metal. The contacts may also be separate, in which case the pressure is supplied by springs (Fig. 10-11). Area contact types of stationary contacts are usually employed in disconnect switches.

Whatever type is used, it is necessary to hold the electrical resistance of all contacts to low levels to keep them from overheating.

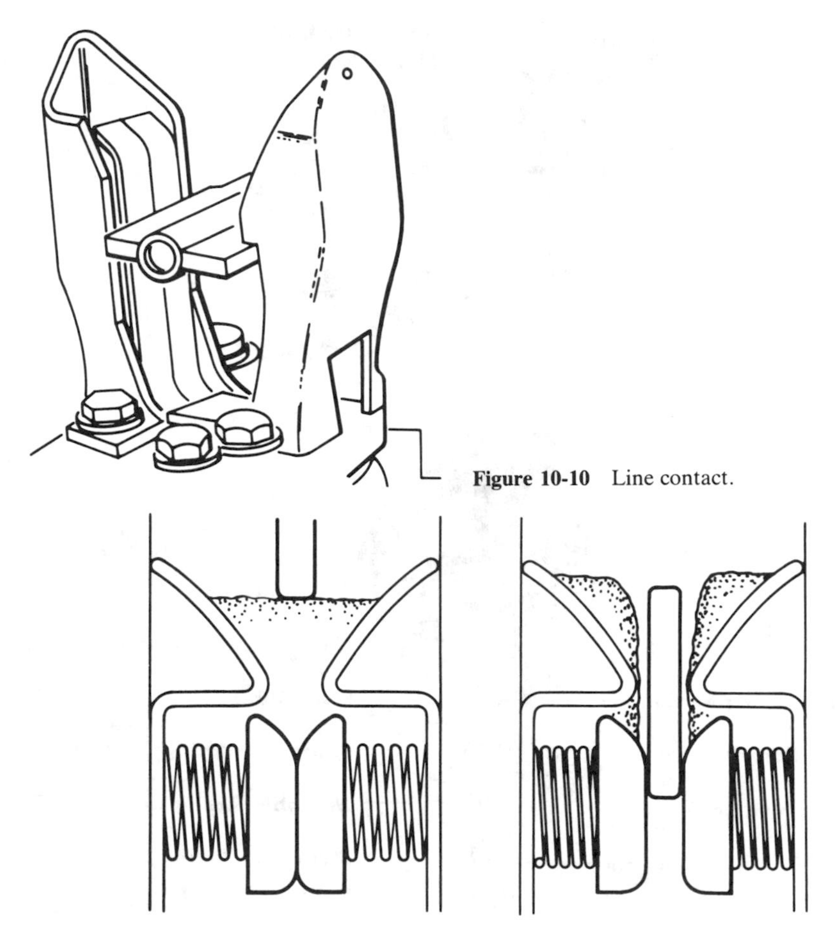

Figure 10-10 Line contact.

Figure 10-11 Area contact.

OTHER DISCONNECTING AND SWITCHING DEVICES

Disconnect blades often consist of, or incorporate, a fuse element. For voltages of about 5,000 V or less, these fuses may be enclosed in a porcelain enclosure known as a *fuse cut-out.* A solid bar or link sometimes takes the place of the fuse, resulting in a disconnecting cut-out (Fig. 10-12). Similarly, oil fuse cut-outs may perform the same function, with the fuse element or solid link operating under oil. When used as disconnecting devices, these units are generally operated in the same manner as other types of disconnects. The function and application of fuses will be discussed in the next chapter.

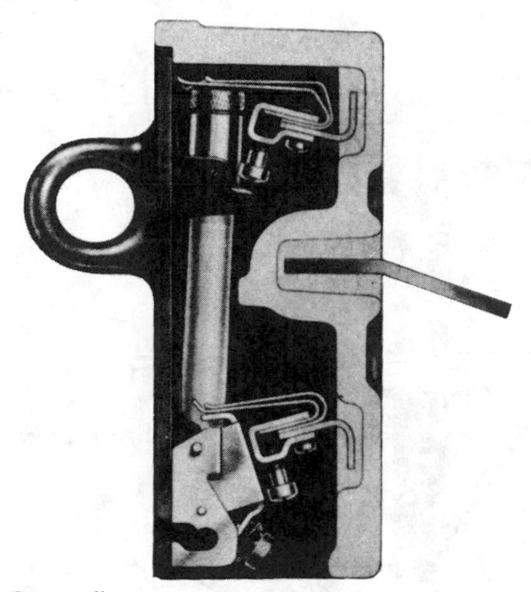

Figure 10-12 Cut-out disconnect.

Potheads

Disconnecting potheads are customarily used whenever the conductors of an underground cable are connected to overhead conductors (Fig. 10-13). They are essentially single-pole disconnects whose contacts are usually totally enclosed in porcelain when the circuit is closed. They are used whenever the currents to be broken are relatively small. Although obsolete, many will remain in service for a long time.

Clamps

Several types of clamps, especially so called *hot-line clamps,* are sometimes used as disconnects, their purpose and operation being essentially the same as those of disconnects (Fig. 10-14).

Jumpers

Jumper cables, with clamps at either end, are designed for other purposes but can be used incidentally for performing a disconnecting function. Usually handled at the end of insulating wooden or plastic "hot-sticks" (similar to switch sticks), they must be manipulated with extreme care (Fig. 10-15).

The current that may be interrupted by these devices is very much smaller than that which can be managed by other disconnecting devices designed specially for such a purpose.

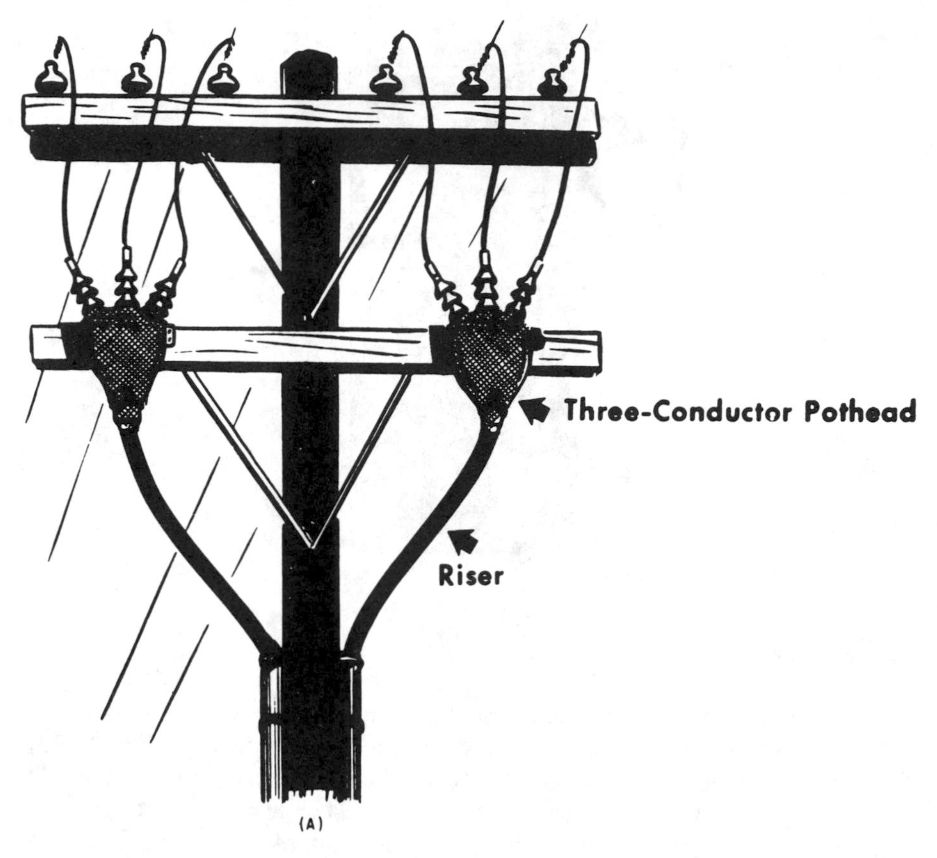

Figure 10-13 Disconnecting potheads: (A) two potheads connecting two underground cable lines with two overhead lines.

Load-Interrupt Device

Another device enables disconnect switches and fuse cut-outs to interrupt circuits safely, essentially by converting them to air-break switches (Fig. 10-16). This accessory device provides the capability for the disconnect to be used for interrupting load and fault currents up to the rating of the device.

Handled at the end of an insulated hot-stick, this device is connected between the movable blade and the stationary contact while the disconnect is closed. The device is pulled away with the same stick, thereby opening the disconnect and charging an internal spring at the same time. At a predetermined point in the opening stroke, its internal trigger trips, releasing the spring and separating the internal contacts, thereby interrupting the circuit. An external latch keeps the device in the open position until it is removed from the disconnect or cut-out. Release of the latch prepares the device for the next operation.

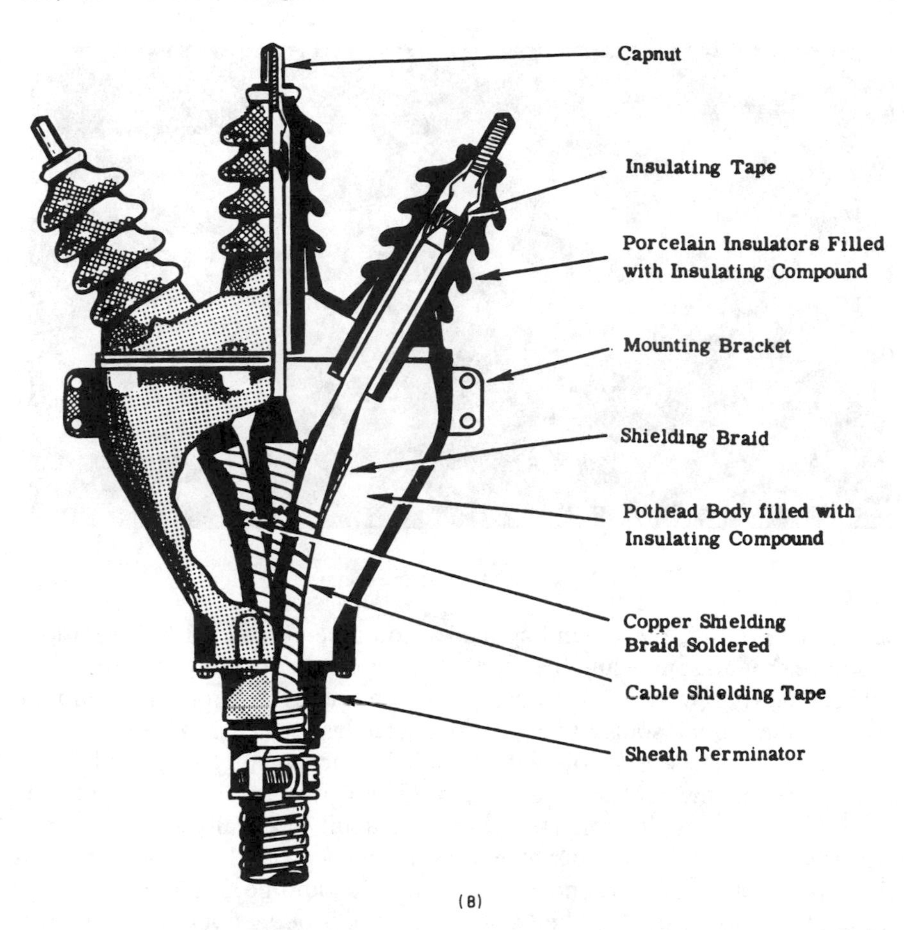

Figure 10-13 (cont.) Disconnecting potheads: (B) cross section.

Switch Operation

The operation of both air and oil switches has been described previously. Emphasis should be placed on the operation of disconnects only when the circuit of which they are a part is de-energized and only when limited amounts of current (very much less than short-circuit currents) are to be interrupted. These devices should also be operated by insulated switch sticks, preferably handled with rubber gloves.

The handles of air-break switch levers and of oil switches should be connected to one or more grounds to protect the operator of the switch.

When work is done on an air switch to repair or maintain it, it is essential that it be completely and physically separated from the energized circuit. If physical disconnection is impractical, the circuit should be de-energized and the terminals on both sides of the switch must be effectively grounded. In

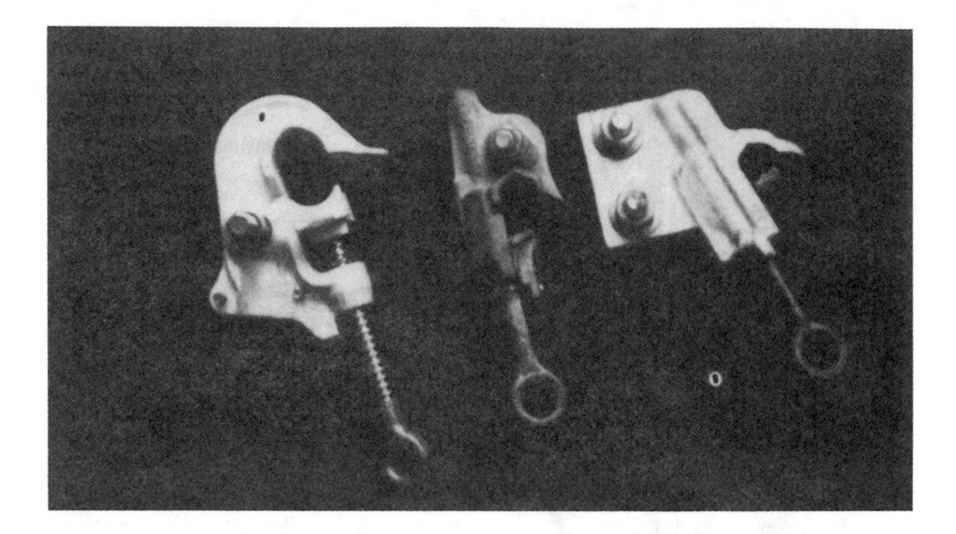

Figure 10-14 Hot-line clamp.

placing the ground, the grounding leads should be connected first to a low-resistance ground source and then to the circuit conductors or terminals of the switch requiring work. In removing the ground, the process should be reversed, the ground source being disconnected last.

The tank or case of an oil switch should be permanently grounded. If this is not practical, the tank should be treated as if it were energized until the switch is completely disconnected from the circuit physically and the circuit terminals and all parts of the switch are effectively grounded. If the switch is taken from service and grounded, the grounds should be so placed that they will not be removed or disturbed as long as work is being done on the switch.

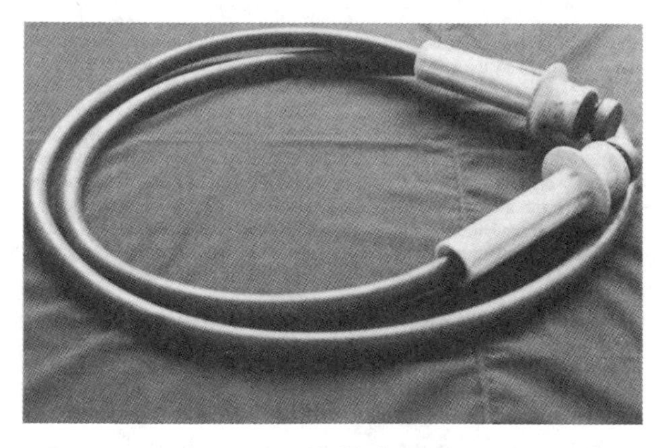

Figure 10-15 Jumper.

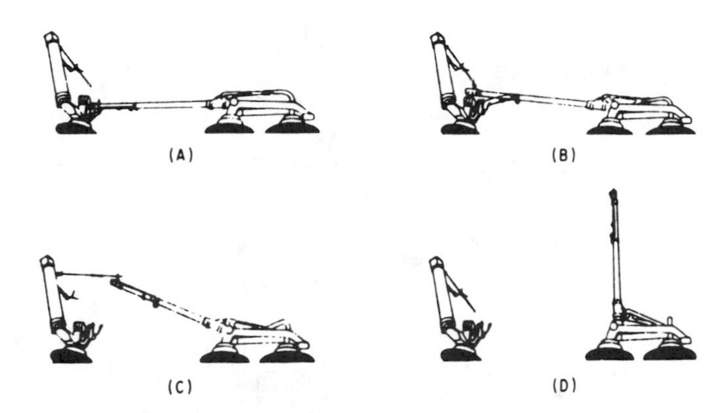

Figure 10-16 Load-interrupting device for air-break switch: (A) Switch in closed position; (B) beginning of opening cycle with by-pass switch still engaged to prevent arcing; (C) by-pass switch disengaged with current path now directed through activating arm; (D) completed load-break cycle with switch blade open.

The ground leads should be placed and removed with the use of rubber gloves, insulated switch sticks, or both.

As a further safety measure, it might be wise to check all the de-energized circuit conductors, switch terminals, and other parts involved in the work with a voltage detecting device prior to applying grounds (Fig. 10-17).

Both air and oil switches should be capable of being locked in either the open or closed position, and provision is often made for more than one lock to be employed at the same time (Fig. 10-18).

Equally important as all of the physical and electrical preparations involved in operating these switches is the supervision of the activities of the personnel associated with them. Work on lines or equipment, especially the operation of air and oil switches as well as the placing and removing of protective grounds, should always be coordinated by one responsible person. Their job is to see that all necessary steps are executed properly and in correct sequence. It is they who give the orders for circuit breakers and switches to be opened or closed and to be locked in the appropriate position. They must make sure that proper notices are displayed at the breakers, switches, lines, and equipment involved (Fig. 10-19). The workers responsible for the work to be done on the lines or equipment not only receive their orders from the coordinator but must also inform him of the completion of each instruction. As a further safety precaution, orders and instructions must be put in written form. Those given by telephone must be written down by both coordinator and worker and repeated back to the other to avoid error. These written orders and instructions, as well as reports of the completion of each order, become part of a permanent record that can be analyzed for possible improvements in safe operating practices and for investigating events should something go wrong despite the precautions taken.

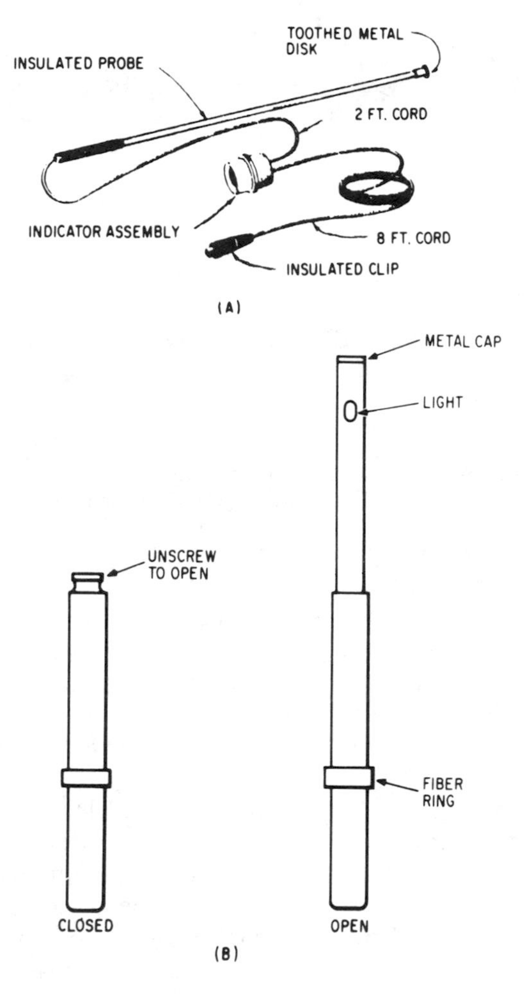

Figure 10-17 Voltage-detecting device: (A) Voltage tester (60 to 7,500 V), and (B) statiscope overhead type.

MAINTENANCE

For air switches, maintenance work centers mainly on the contacts. It must make sure that the contacts between blades and stationary contacts are made smoothly and evenly, that the pressures exerted meet the specifications, and that the movable parts move freely. Mechanisms should be adjusted to maintain full contact on all poles of a multipole switch as simultaneously as possible. Hinges should be lubricated but kept sufficiently stiff so that when the blades are open, they will not fall back on the live stationary contacts. Other parts that rotate, linkages and so forth, should also be lubricated. Contacts should be cleaned of dirt and dressed; oxides and pittings and line contacts

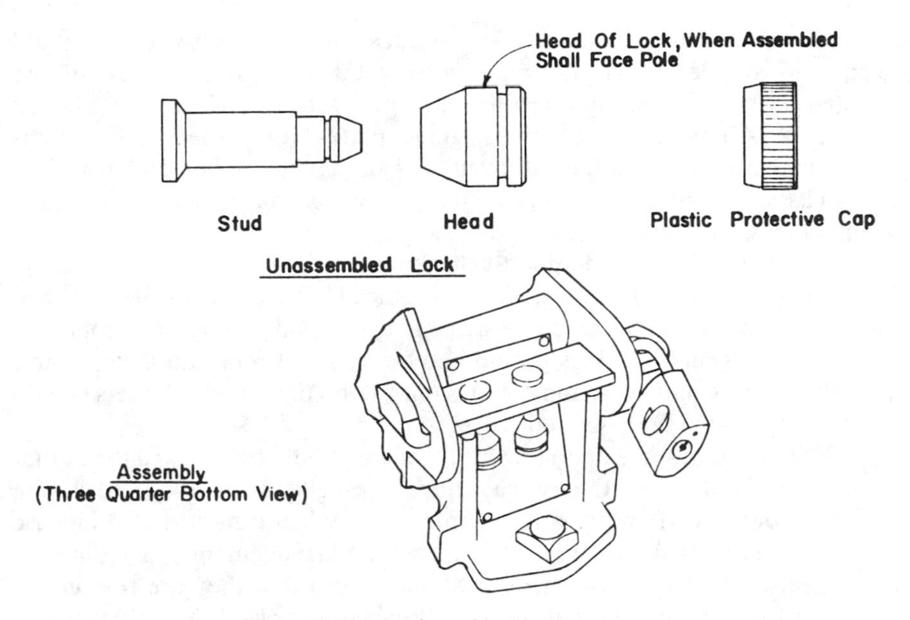

Figure 10-18 Locks and locking arrangements.

_____ TAGS NO. _____

LIGHTING COMPANY
SYSTEM OPERATORS

HOLD OFF

SWITCH No. _____

AT _____
Location

HAS BEEN OPENED AND TAGGED BY

_____
Signature

AT _____ ON _____
Time Date

TO CLEAR _____

_____

FOR _____
Authorized Employee

PER ORDER OF: _____
System Operator

LIGHTING COMPANY
SYSTEM OPERATORS

HOLD OFF

TAG REMOVED BY _____
Signature

AT _____ ON _____
Time Date

PER ORDER OF _____
System Operator

Figure 10-19 Typical hold-off card.

should be resilvered, if indicated. Arcing horns of air-break switches, should be cleaned and cleared of pittings, or replaced if necessary. They should be adjusted so that they barely touch the movable switch member when opened and closed. All bolts, nuts, clamps, guide plates, and other similar items should be tightened and adjusted. Insulators should be cleaned and inspected for cracks and signs of flashover or fatigue (when the insulator and blade assembly move as a unit).

For oil switches, in addition to checking and adjusting contacts, the oil should be changed, filtered, tested, and replaced. The inside of the tank and the mechanism parts should be cleaned, and the tank refilled to proper oil levels. Bearings and bushings should be checked and replaced if necessary. Loose bolts, nuts, linkages, and the like, should be tightened. Gaskets should be replaced and the tank painted if necessary.

For all switches, the operating mechanisms should be checked to see that they operate freely and with positive action, especially the opening and closing of switch contacts. If switches are motor operated, the motors and electric supply and controls should be checked. Locking arrangements, including the locks themselves, should be checked to make sure that they are reasonably safe from being tampered with by unauthorized people. Electrical grounds should be checked for continuity and tightness, and ground sources should be checked to be sure that their resistance is sufficiently low.

Since some of these switches may not be used over long stretches of time, it is good practice to open and close them periodically (while de-energized, of course) to make sure that they are in good mechanical operating condition. The ideal frequency of such testing depends in great part on the geographical location of the switches and their exposure to cold, rain, ice, salt, polluted atmosphere, and other local adverse conditions.

Table 10-1 provides a detailed breakdown of troubleshooting procedures for disconnecting devices.

REVIEW

1. Less expensive equipment, of lesser circuit breaking capability, can be used to accomplish many of the circuit breaking operations. These are generally referred to as switches, and provide for physical separation in the electric circuit and the restoration of its continuity.
2. Switches may be used to: de-energize equipment or portions of circuits upon which work is to be performed; isolate faulted equipment or faulted portions of circuits; provide electric supply from alternate circuits; distribute loads throughout an electric circuit to avoid or eliminate overloading or portions of the circuit; ground circuits; serve temporarily in place of circuit breakers; and to serve and meet particular situations.
3. Switches range from small snap switches to large air or oil switches

TABLE 10-1 Troubleshooting Disconnecting Devices

Condition	Possible Cause	Suggested Remedy
Discoloration, signs of over-heating	Overload	Replace with proper rated unit.
	Contacts poor	Check and dress contacts; resilver line contacts if necessary. Check alignment of contacts and operation of closing mechanism.
	Loose connection	Check and tighten all connections, clamps, etc.
	In an oil switch, dirty oil or low oil level	Check oil, filter it, and refill tank to proper level. Clean all parts inside tank with clean oil.
Noisy switch	Loose bolts, pitting, linkages, etc.	Tighten all parts, bolts, links, etc., and replace if necessary.
Sluggish and uneven closing of gang switches	Loose linkages in control devices	Check alignment and tightness of linkages and the condition of bolts, cotter pins, guide plates, etc. Tighten or replace if necessary.
Cracked or chipped insulators	Flashover from dirt or contamination; lightning	Clean insulators, replace with larger insulators if necessary. Check grounds and lightning arrester installation.
Operating handles energized	Leakage from dirty insulators; poor grounds or ground connections	Clean insulators or replace as above. Check ground resistance and install additional grounds if necessary. Tighten all ground connections. Bond to other structures if practical.
Oil leakage (oil switches)	Leaks; improper oil level; oil "thrown" during transient short circuit	Check and repair leaks. Fill tank to proper level. Check internal parts for damage Replace with larger unit if necessary.
Rust and paint deterioration	Weather; polluted or salt atmosphere; overloads	Remove rust and deteriorated paint. Clean surfaces. Repaint with proper paint and sufficient coatings.

which, under some conditions, can safely interrupt relatively small currents. Switches, however, are almost always operated in series with circuit breakers or fuses.

4. Switches are generally used to insert a physical separation in a circuit associated with circuits or equipment as a safety measure to ensure the

line or equipment being worked upon cannot be energized accidentally. To differentiate them from air-break switches, they are called "disconnects" (Fig. 10-3).

5. Air switches, equipped with arcing horns or arc suppressors, are known as air-break switches, and are capable of interrupting relatively small load currents (Fig. 10-6).
6. Contacts, links, levers, and supporting structures must be maintained in a fashion similar to those of other circuit breaking devices.
7. Potheads are disconnecting devices designed to connect or disconnect overhead conductors with conductors of underground cables (Fig. 10-13). They are usually de-energized before operating them, but under special conditions, may break only relatively small load currents.

STUDY QUESTIONS

1. What is a switch? How many types are there? What are their functions?
2. What is the difference between an air-break switch and a disconnect?
3. When and how are disconnects connected in a circuit? What precautions should be used in operating a disconnect?
4. What are arcing horns? What is their function?
5. How many air-break switches have their interrupting capacity increased?
6. What is the advantage of a load-break switch?
7. What other types of disconnecting devices are there besides the air-break switch and disconnect?
8. How are some of the larger air-break switches operated? What precautions should be taken before working on such a switch?
9. Where and why are barriers sometimes used?
10- What advantages do air-break switches have over circuit breakers? What are their disadvantages?

11

FUSES

Fuses constitute a very important element in the protection of electric circuits. They are devices designed to melt at a predetermined current value and thereby serve as protection against abnormal current by opening the circuit in which they are installed. Essentially "one-shot" circuit breakers, they are perhaps the most common protective device now in use.

PRINCIPLE OF OPERATION

All fuses work on the same principle: a metallic fusible link that melts when more than a rated current flows through it. The factors affecting the rating of a fuse include not only the composition and dimensions of the fusible link but also the characteristics of its mounting and enclosure. The former determines the magnitude of the current and the minimum time necessary to melt the link. The latter also determines the melting time, but equally important, it determines the arcing or arc-clearing time. The sum of these two times is known as the total *clearing time* of the fuse (Fig. 11-1).

Fuses come in a variety of sizes, shapes and types, from the common, small household type that screws into a socket, to the larger cartridge types found in industrial plants, to the power fuses that are considerably more complex in construction and operation and greater in cost. The basic reasons for these differences are the amperage of the short-circuit or fault currents that need to be interrupted and the voltage at which they operate. Typical cartridge fuses operate at voltages under 600 V and can interrupt currents of

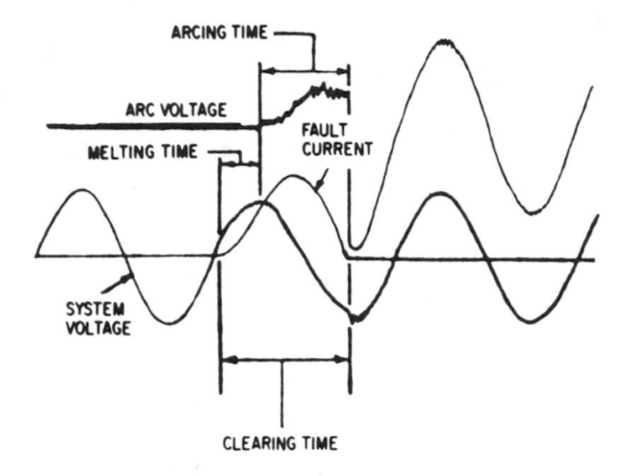

Figure 11-1 Oscillogram showing clearing time.

several hundred amperes. Power fuses are designed for voltage values in the thousands of volts and short-circuit currents in the thousands of amperes. Only high-voltage or power fuses will be discussed here.

POWER FUSES

Time-Clearing Characteristics

Power fuses are rated in terms of voltage, normal current-carrying ability, and interruption characteristics expressed as time-current relationships. It is obvious that a fusible link will take less time to melt with a larger current than with a smaller one, and the relationship between the minimum melting time and the current applied is usually portrayed in a curve. To this curve must be added arcing times for the various current values. The melting time and arcing time values are obtained from tests carried out by the manufacturer and represent the average values of a number of tests. Moreover, the minimum melting-time curve is adjusted by increasing the minimum time values by a predetermined percentage to allow for manufacturing tolerances and to assure a safe and positive functioning of the fuse by providing adequate clearing time. The second curve thus developed will indicate the maximum clearing time for that particular fuse size and type (Fig. 11-2).

A similar set of curves is obtained for each of the different ratings and types of fuses (Fig. 11-3).

FUSE COORDINATION

In an extended electric circuit, such as those that may be found in large commercial and industrial installations or in electric utility systems, it is desirable, if not essential, that the outages caused by a fault, short circuit, or

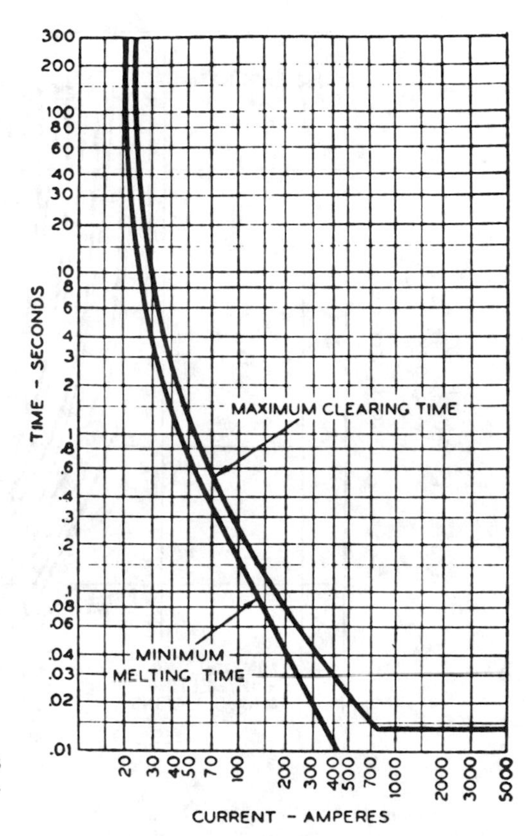

Figure 11-2 Typical time-current curves for a 10-k link. (*Courtesy*, McGraw-Edison Co.)

malfunction be restricted to the smallest practical section of the system for the shortest time. Systems, therefore, are divided into smaller sections, the number and size of each of the sections being determined by its importance and general economic considerations.

Each of the sections may be protected by an interrupting device that operates to de-energize that section before the short circuit or fault occurring within it can affect the rest of the system. The number and type of interrupting devices depend on several factors: the system voltage, usual load current, maximum possible fault current, number of branches, lengths of the sections, the equipment connected to the lines in each of them, and other local or particular conditions. The locations of these protective interrupting devices, known as *coordinating points,* are usually key points and frequently at the branch intersections. If two or more such devices are installed in a circuit, it is essential for them to be coordinated so that only the section in trouble is de-energized. Needless to say, these devices must be of the correct rating for carrying the normal load current and for responding properly to likely faults.

Although these protective devices could be circuit breakers or reclosers, economics generally dictates that many of them be the less expensive fuses. In

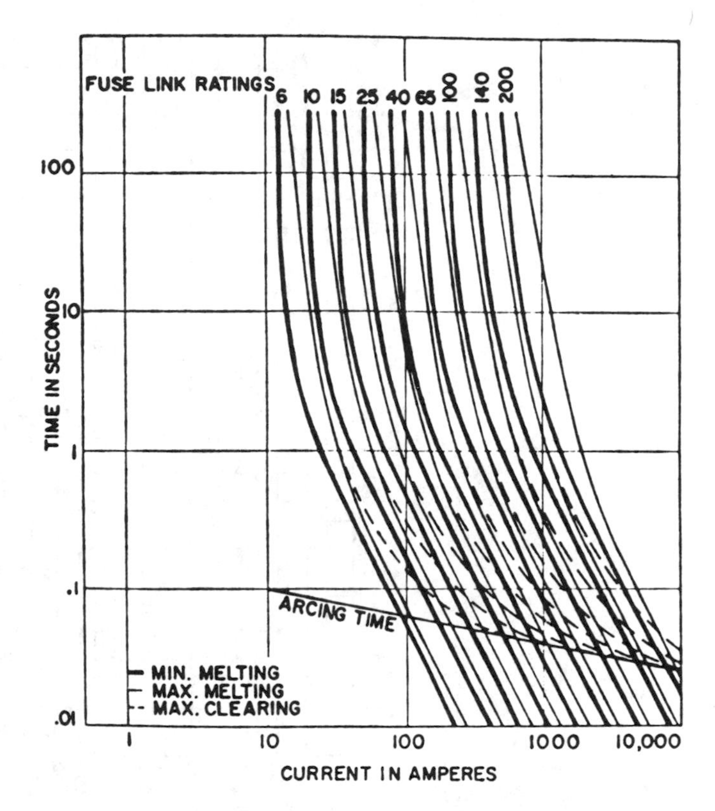

Figure 11-3 Minimum and maximum time-current curves for type "K" fuse links. (*Courtesy*, McGraw Edison Co.)

Fig. 11-4, fuse J must clear before fuse G, or, expressed another way, fuse J protects fuse G. Following this train of thought, G protects E; E protects D; D protects C; and C and B protect A. At the load location, L protects K, while K protects E. All these devices, from A to L, must be coordinated, that is,

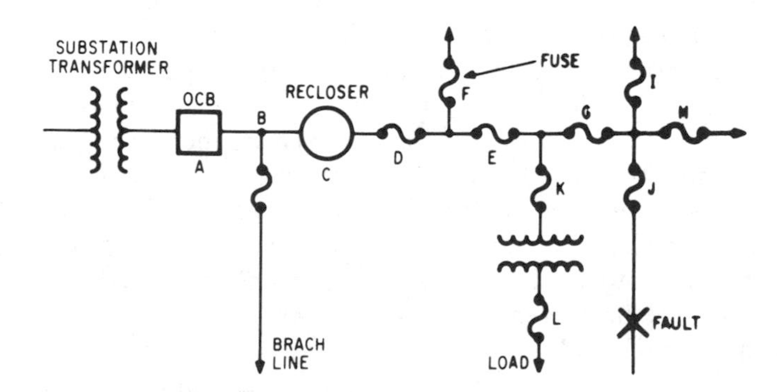

Figure 11-4 Single line showing coordination points and devices.

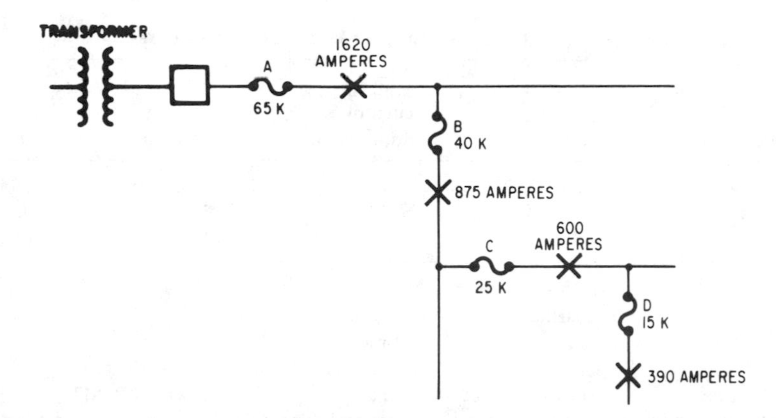

Figure 11-5 Typical fuse-coordinated system.

they must have the correct ratings for carrying normal load current and for responding properly to a fault.

If a fault occurs in a branch line, as shown in Fig. 11-4, the fault current flows through the main part of the circuit back to the source. The magnitude of current caused by a fault at the different coordinating points depends on the resistance (or impedance for ac circuits) between the source and the point of fault. When a fault is distant from the source, the impedance of this part of the circuit is high and the magnitude of the fault current is low. Conversely, when a fault is close to the source, impedance is low and the magnitude of the fault current is high.

Fuses at the coordinating points furthest from the source will thus have the lowest rating consistent with the normal load at this point, whereas fuses at other coordinating points along the main line will increase in rating as they approach the source (Fig. 11-5). The characteristics of these fuses must also coordinate with those of the circuit breaker at the source and with other protective devices in their path.

The time-current values of the several fuses are indicated in Table 11-1. It will be noted that for the faults and fault currents (X) indicated in Fig. 11-5, the fuse at point D will clear before that at point C, that at point C will clear before that at point B, that at point B before that at point A, and that a fault close to the source at point A would clear before the circuit breaker opens. Further reference to the time-current characteristics of fuses will be made in Chap. 13.

FUSE CONSTRUCTION

Probably no part of an electrical circuit is subjected to more severe conditions than the fuse. Power fuses are specially constructed so that they will not only operate safely at voltages greater than 600 volts but must be able to operate

TABLE 11-1 Time-Current Values of Various Fuses

Position	Fault current amperes	Load current amperes	Protecting link	Protected link
D	390	14	15 k	25 k
C	600	23	25 k	40 k
B	875	35	40 k	65 k

Position	Maximum clearing time (protecting link)	Minimum melting time (protected link)	Percent CT/MT
D	0.061 sec.	0.084 sec.	72.6 (.061/.084)
C	0.666 sec.	0.093 sec.	71.0 (.066/.093)
B	0.078 sec.	0.120 sec.	65.0 (0.78/.120)

under very heavy loads and must interrupt without damage high energy flow of power amounting to thousands of kilowatts, although only for a very short period of time.

The interruption of such large currents creates a great amount of heat that must be dissipated, and the blowing of a fuse can be compared to a violent explosion. The heat causes a rapid expansion of the surrounding air, and the gaseous metallic vapors created by the arc, if present for more than a few cycles, may destroy the fuse holders and nearby objects. That is why particular attention is given to coordinating points in the design of fuses and fuse holders.

When fuses are installed, they are usually regarded as part of a circuit and, with few exceptions, are given no further attention. The fuse, therefore, must not only be able to withstand extreme weather conditions and all kinds of varying loads but, by its very nature, must open the circuit satisfactorily whenever called upon.

There are three principal kinds of power fuses in general use:

1. Open or link fuses
2. Enclosed fuses
3. Expulsion fuses.

Although there are many types of each kind, the ones described here are the most typical and illustrate the factors involved in the construction of the several kinds.

The fuse element is usually made of tin, lead, zinc, aluminum, or of silver, often alloyed with tin or lead. Its length and cross section generally determine the characteristics of the fuse.

Open or Link Fuses

The open or link fuse is merely a piece of metal ribbon or fuse wire connected between two terminals. An open-mounted fuse is subject to air currents and will itself cause some movement of air surrounding it as it heats up. The fact that the fuse rating will change as the ambient temperature changes has the effect of decreasing the accuracy of fuse operation. Since open fuses must vaporize metal in their operation, they may smudge or burn surrounding surfaces. Copper-terminal clips are sometimes attached to the fuse strip, which is often also enclosed in insulating boxes of porcelain, asbestos tubes, or other casings of fire-resisting material that confine the explosion and increase the accuracy of operation. These can be readily disassembled and the fuse element replaced.

Enclosed Fuses

In enclosed fuses, the fuse element is enclosed within a tube of insulating material, such as glass, fiber, or Micarta (Fig. 11-6). For small-capacity fuses at voltages to about 15,000 V, the fuse wire or element is fastened to caps at each end of the casing that are usually made of brass. The tube is filled with a material, usually in powdered form, that acts to absorb the heat and the gases liberated when a fuse is blown. It also acts to condense the vapor of the molten metal and to chill the arc, thereby breaking the continuity of the circuit by filling the gap that is formed when the fuse melts and preventing the arc from restriking. These fuses are usually used with potential transformers or other accessory equipment.

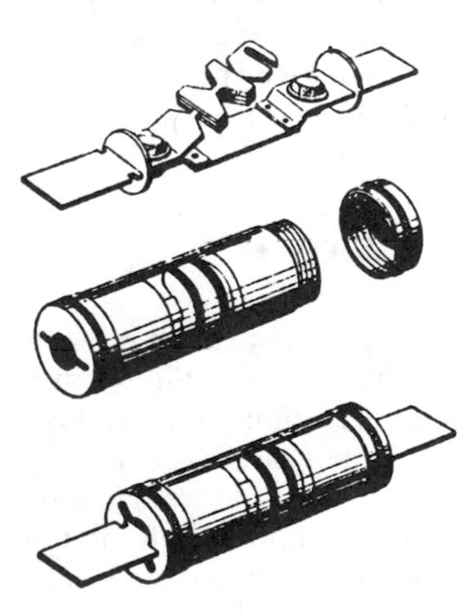

Figure 11-6 Assembly of enclosed fuse.

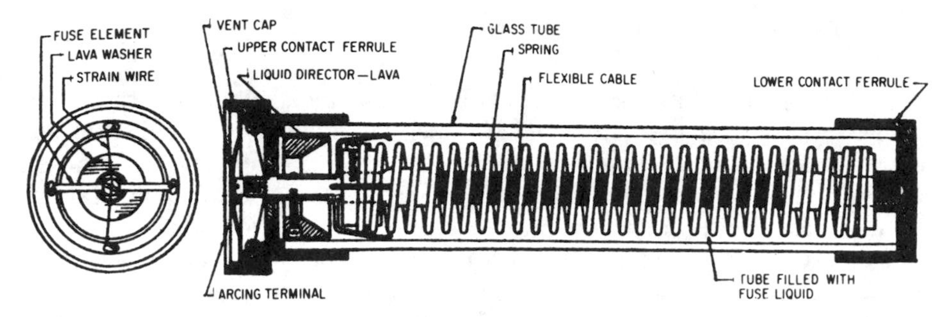

Figure 11-7 Liquid fuse with separate expulsion chamber.

Liquid-Filled Fuses

Larger-capacity, high-voltage fuses of the enclosed type may be filled with liquid (Fig. 11-7). One form of such a fuse employs a glass tube cemented into heavy metal end caps and filled with a fire extinguishing fluid such as carbon tetrachloride (Pyrene). The fuse element is held in tension by means of a spring that helps to break the circuit by quickly lengthening the gap and forcing the quenching liquid into the arc. The cap on this end of the tube is usually made to blow off if the pressure becomes excessive, but the gas produced is inert.

Sand and Boric Acid Fuses

In another type of large-capacity, high-voltage fuse, the fuse element is surrounded by sand. The sand tends to absorb the heat and gases generated when the fuse blows and also to squelch the arc. In still another type, powdered boric acid rather than sand surrounds the fusible element (Fig. 11-8). Exposed to the heat produced when the fuse melts, the boric acid decomposes and forms steam. The steam, being under pressure, acts to snuff out the resultant arc.

Oil Fuses

If severe conditions require an unusually reliable fuse with large interrupting capacity, the oil fuse and associated cut-out is used (Fig. 11-9). Its special features are that it is able to break full load current without expelling arcs or hot gases. The tank is made of steel. Stationary contacts inside the tank are brought out through pothead-type fittings. Moving contacts are of the clip or jaw type, to which the fuse link is connected. Here, the fuse link is fitted and fastened about an insulated assembly that fits into the tank.

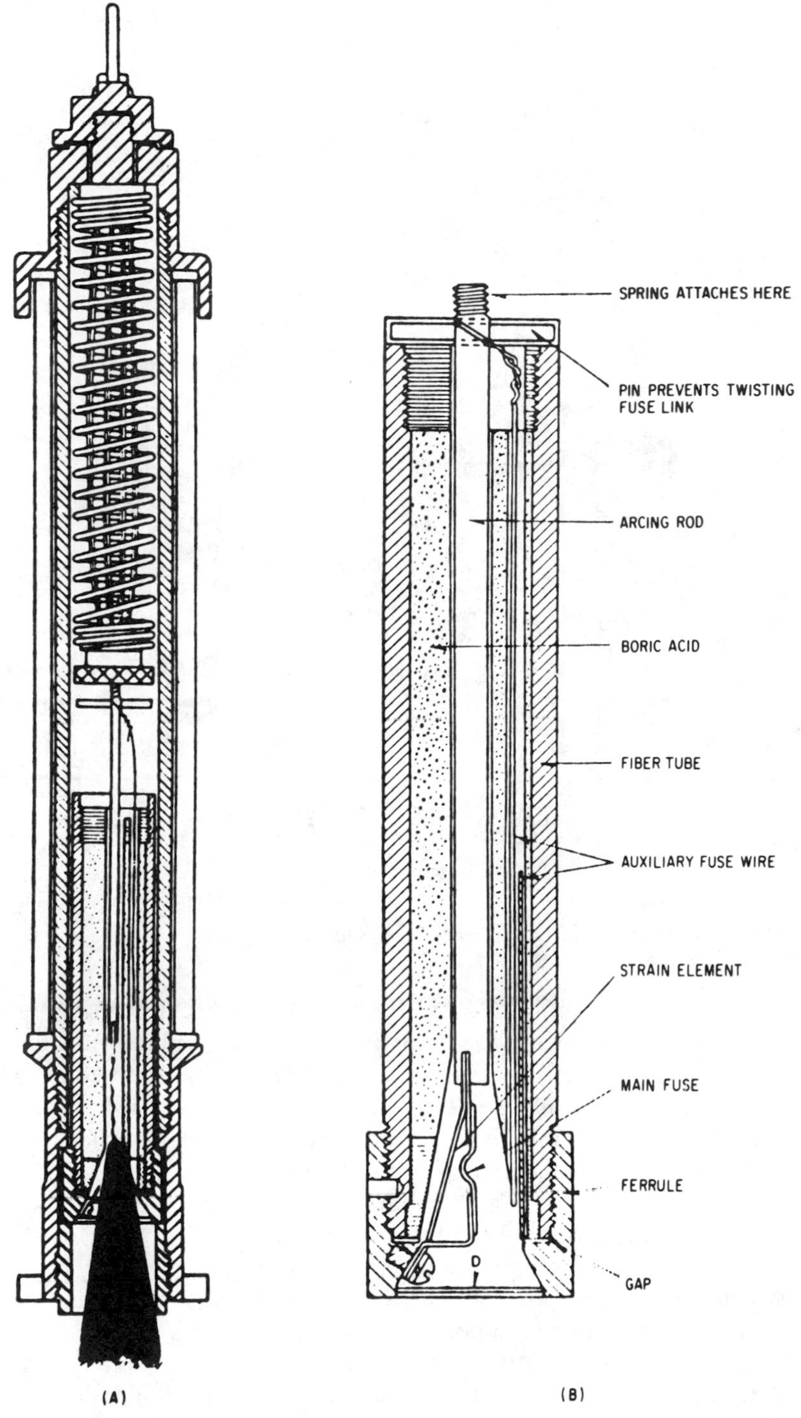

Figure 11-8 Boric acid fuse: (A) Cross section at instant the fuse link melts, and (B) cross section of refill unit.

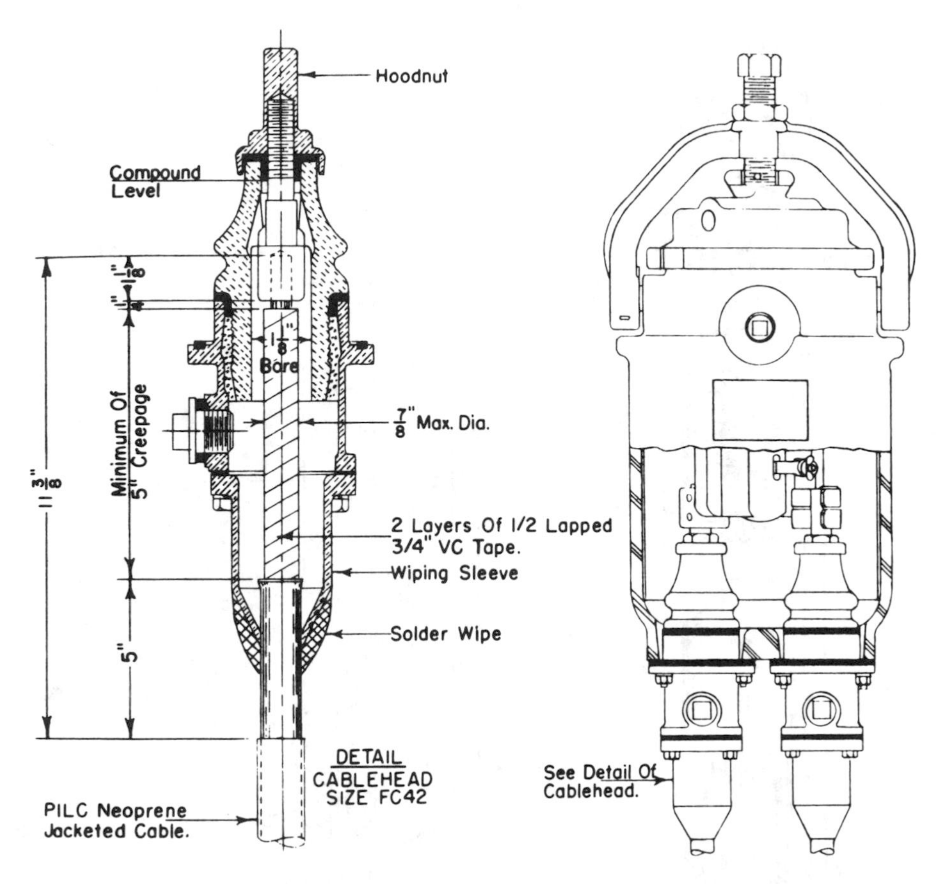

Figure 11-9 Oil fuse.

The heavy magnetic field produced by the current being interrupted induces electric currents in the tank, which, in turn, produce magnetic fields that tend to snuff out any arc that forms. Moreover, the confined chamber in which the fuse operates also acts as an "explosion chamber" to help extinguish the arc. The inside of the tank is lined with insulating material to prevent arcing to the tank, which is usually grounded. Gases that form are allowed to expand into sealed expansion chambers at the top of the cut-out, while the contacts and the remnants of the fuse link remain under cool oil at all times. The carrier assembly can only be inserted in the open position, providing a safety feature that prevents premature blowing of the fuse before the assembly is locked closed should a fault persist in the circuit.

Because of their relatively high cost, oil fuses are ordinarily limited in use to transformer vaults and underground lines where voltages do not exceed 15,000 V. As a rule, they are constructed as single-pole units.

EXPULSION FUSES

In expulsion fuses, open fuse wires are placed in a holder so designed that the expulsion of the gases formed by the vaporizing of the fuse and the expansion of the surrounding air is sufficient to blow out the arc (Fig. 11-10). Strong, heavily insulated tubes are used, sometimes reinforced by a metal end cap where the arcing first occurs. In some types, a fiber cylinder surrounding the fuse element is deliberately placed there so that additional gases formed when the arc attacks the fiber help snuff out the arc. In other types, a spring is attached to the fuse element as a means of drawing the arc apart faster when the fuse blows. Other types employ both of these measures.

One of the most popular fuses is the so-called *fuse link* associated with cut-outs in circuits operating at 15,000 V or below (Fig. 11-11 on page 210). This fuse is made up of three parts: a stranded flexible cable, a fusible link, and a metal button or cap. The button is attached to the top of the fuse link and the cable at the bottom. The fuse fits into an insulating tube made of fiber or Micarta. A cap screwed over the button seals the top end, and the flexible cable is held taut by a spring at the bottom. The arc is blown out of the bottom of the tube, and the spring pulls the fuse out of the tube when it melts.

For voltages under 7,500 V, the fuse element is often situated in an enclosed cut-out made of porcelain. The fuse is inserted in contacts fastened to the door. The hinge of the door is at the bottom, and, when the door is closed, the fuse contacts fit into the stationary contacts contained within the porcelain receptacle. For voltages between 7,500 and 15,000 V, the fuse and

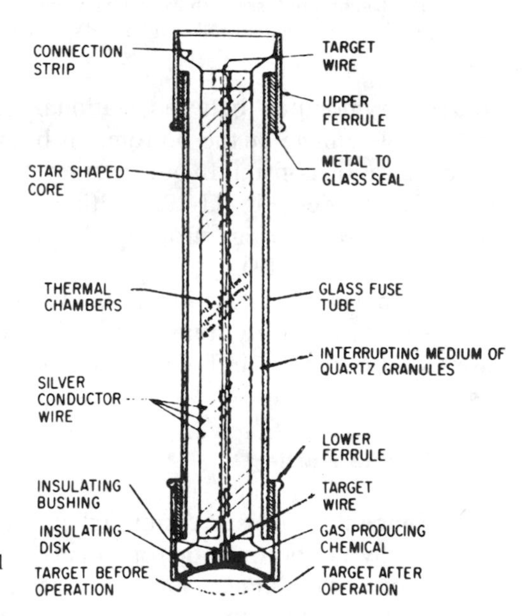

Figure 11-10 Cross section of typical expulsion fuse.

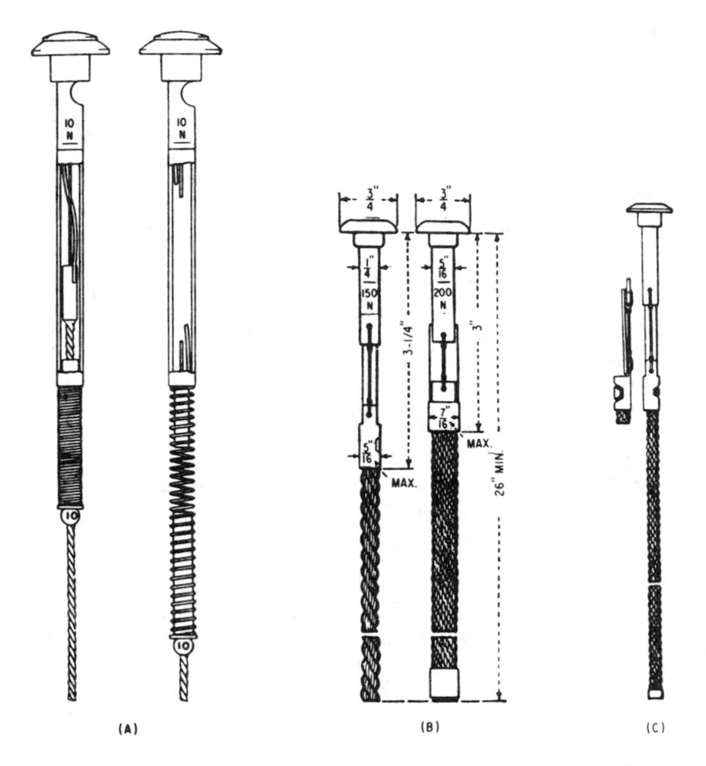

Figure 11-11 Link expulsion fuse: (A) 25-ampere fuse before and after blowing, (B) 200-ampere fuses with extra heavy pigtail for carrying large currents, and (C) 75-ampere fuse without fiber tube and spring. (*Courtesy*, James R. Kearney Corp.)

holder are mounted between stationary contacts in open air; such open cut-outs also hinge at the bottom. In both instances, when the fuse blows, some indication is given. In one instance, the door of the cut-out (Fig. 11-12A) or the fuse holder (Fig. 11-12B) is flipped to the open position. In the other, the blown fuse cartridge drops so that it can be seen from the ground (Fig. 11-13 on pages 212-13).

A similar fuse is used for some voltage ratings above 15,000 V, usually associated with electric transmission lines (up to about 69 kV), although the fuse element may be different; it may be a liquid, sand, boric acid, or de-ion fuse.

Double Venting

Unlike the enclosed cut-out where the expulsion tube vents only at one end, usually the bottom, the open cut-out may have an expulsion tube that vents at one or both ends, sometimes referred to as "double venting." In the latter case, the tube can vent accumulated gases more rapidly and can have a

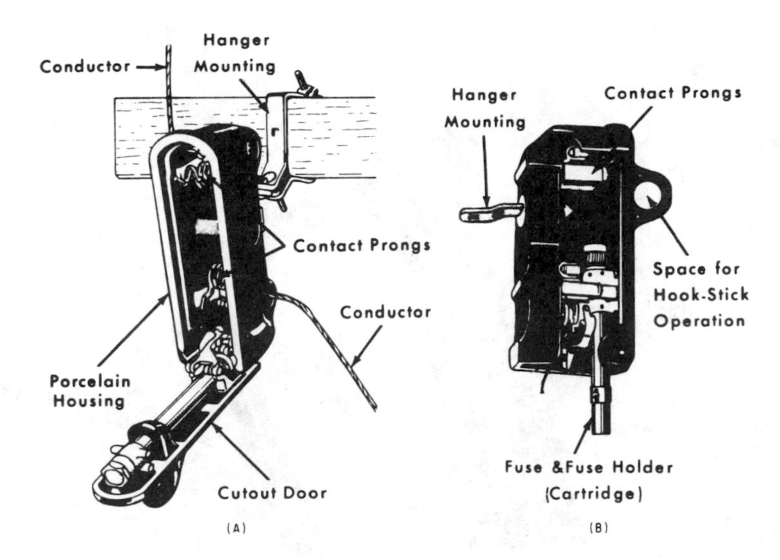

Figure 11-12 Door fuse cut-out: (A) Open door indicating fuse has blown, and (B) dropped cartridge indicating fuse has blown.

higher interrupting rating than a single-venting tube. In this type, one end (usually the top) remains completely closed during low pressures caused by low fault currents. When high fault currents produce large gas pressures, the top cap is blown off and the expulsion tube vents at both ends. Thus, on low fault currents, the tube is single-venting, and on high fault currents, it is double-venting.

De-ion Fuse

In this type of fuse, the fusible element is attached to an internal arcing rod, and both are held under tension by a spring (Fig. 11-14 on page 213). When the fuse melts, the arcing rod is pulled up by the spring, drawing the arc through a sectionalized chamber and breaking it into a series of small arcs. The section of this fuse are also filled with powdered boric acid; as previously mentioned, boric acid forms steam that tends to quench the arc. Since the end of the fuse is open, the pressure of the steam and gases developed during the operation is vented safely.

FUSES AS DISCONNECTS

Fuses often form a part of the disconnecting switch equipment, sometimes arranged as the blade of the switch and sometimes attached to auxiliary clips mounted on the same base (see Chap. 10).

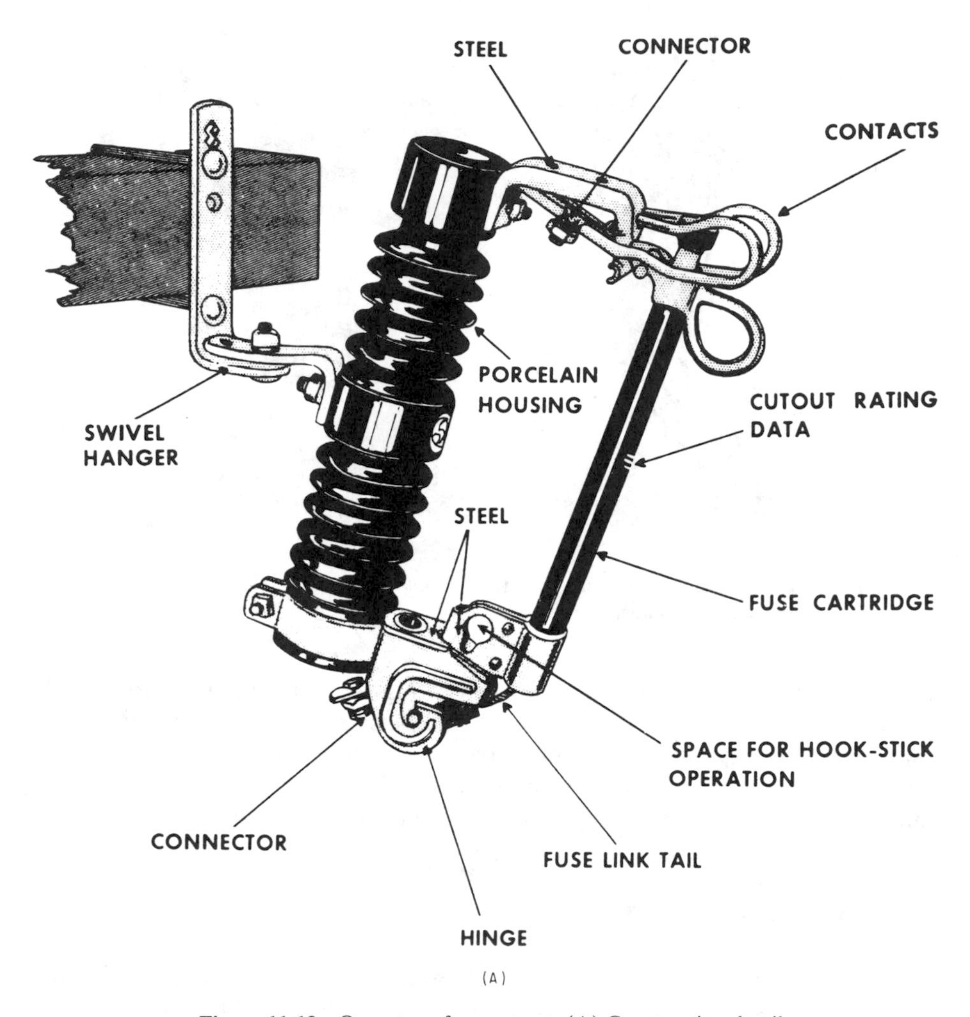

Figure 11-13 Open-type fuse cut-out: (A) Construction details.

THREE-SHOT FUSES

At remote locations or under certain special circumstances, it may be desirable to have a circuit back automatically after a temporary fault. If the line is fuse-protected, a three-shot fuse holder is used (Fig. 11-15 on page 214). Three fuses per pole or phase are mounted in parallel, but only the first fuse is connected to the load. When a fault or overload occurs, the first fuse will blow. As it swings open (in the drop-out type of fuse holders), it causes a latch to operate that places the second fuse in service. If this blows, a similar action places a third fuse in service. If the third fuse blows, the circuit remains disconnected. While all three fuses are similar as a rule, on special occasions

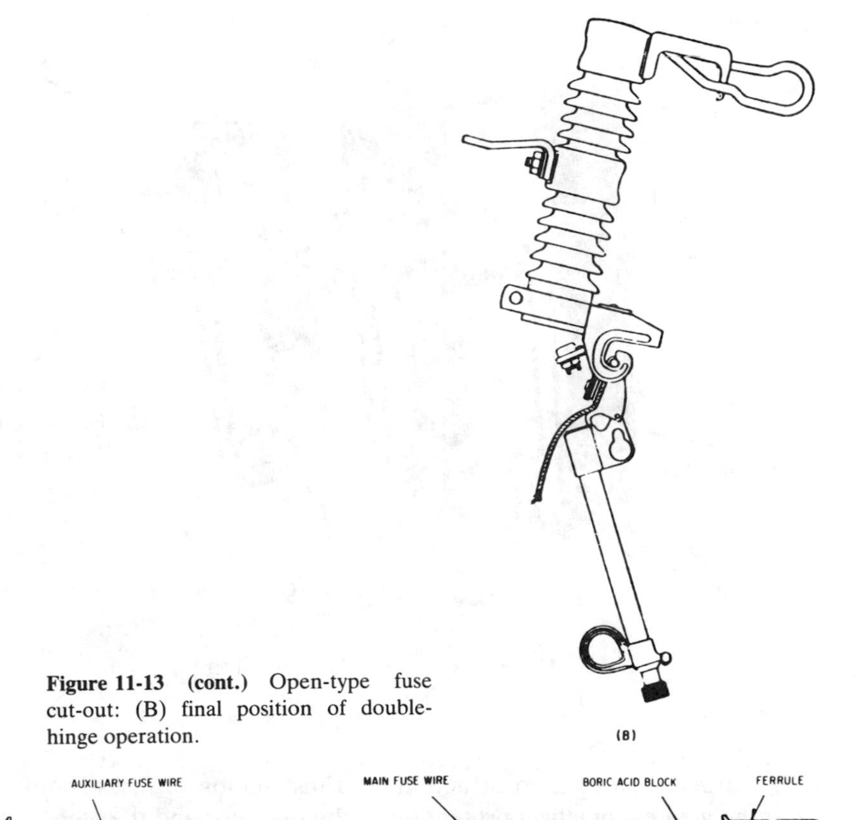

Figure 11-13 (cont.) Open-type fuse cut-out: (B) final position of double-hinge operation.

Figure 11-14 Cross section of de-ion type of boric acid expulsion fuse.

the second and third fuses may be of a different or larger rating so as to introduce a time-delay before blowing—an operation similar to the operation of a recloser.

SELECTION OF FUSES

In addition to the current and voltage rating, there are other considerations in the selection of fuses and their mountings that should be taken into account, as follows:

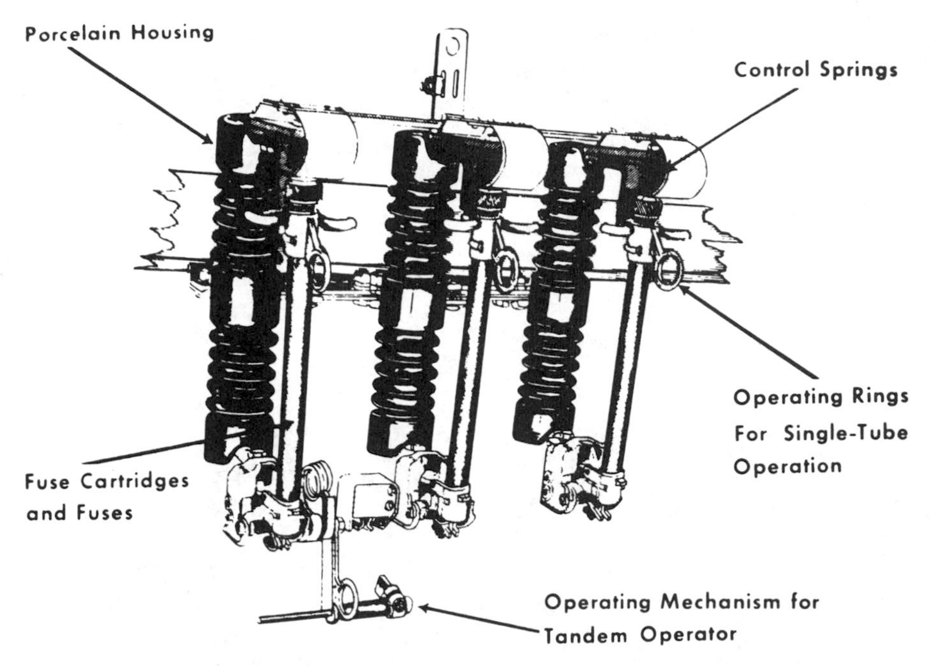

Figure 11-15 Three-shot fuse holder.

1. **Safety.** Design, construction, and installation of fuses should afford the greatest practical protection to the operator and the general public. Not only should they be installed out of reach of the public, but the cut-outs and other appurtenances should be operated with auxiliary devices such as insulated switch sticks to ensure that an operator will not get too close to parts that may not be de-energized. They should also be located so that the flame and loud report associated with their operation do not frighten or harm the public.
2. **Interrupting capacity.** The type and capacity of the fuse and mounting should be such as to interrupt the maximum short-circuit current that is likely to occur at the point of installation. A fuse and cut-out near a generating plant or source of electric power might have to handle a very large short-circuit current, whereas a fuse and cut-out in an outlying location might handle only relatively small currents. Because of economic considerations in purchasing and storing and because circuits are often rearranged, it may be prudent to select only a single mounting, one that will withstand the maximum short-circuit current. The correct fuse can then be installed in this mounting for each particular situation or application.

3. **Insulation.** All live parts should be sufficiently insulated so that there will be no danger of breakdown either from line to ground or from contact to contact after a fuse has blown. The cut-out or mounting should be able to withstand a dielectric test of two to three times the rated voltage or any tests specified by regulatory bodies.
4. **Durability.** The mounting or cut-out should be able to withstand severe climatic conditions (both heat and cold, rain, ice, etc.) and should be protected against decay and insects, as well as from birds, squirrels, snakes, and other likely animals. Metal parts should either be noncorrosive or protected against corrosion (including salt-spray atmospheres) and rust.
5. **Contacts.** Contacts should be sufficiently large to carry a normal rated current without heating. Radiating surfaces of the cut-out or mounting should be sufficient to dissipate heat generated by the melting fuse link and any arc that may form.
6. **Grounding.** This is an added safety consideration. All nonenergized parts of the fuse mounting, including metallic parts, should be bonded together and connected to one or more grounds, or to a common ground mesh where one exists. Furthermore, the ground resistance should be sufficiently low to drain off high voltages should the mounting or cut-out fail as well as to operate other protective devices that will de-energize the portion of the circuit on which the failure takes place. Caution should be taken in using water pipes as grounds. Although they formerly provided excellent grounding sources, the use of nonmetallic pipes for water supplies make water systems questionable. Connecting wires should provide some disconnecting point at which ground resistance may be periodically checked. Such disconnecting points and the wire itself should be protected from possible tampering by the general public or any unauthorized personnel.

MAINTENANCE

Little can be done about the maintenance of fuses and their mountings or cut-outs other than to inspect them visually. Since the life of practically all fuses is indeterminate (they do not weaken with age), no periodic replacement of them is usually undertaken. In a multipole or polyphase fuse installation containing two or three fuses, it is good practice to replace all fuses after one has blown. At some critical or sensitive installations, it is also good practice to keep a set of spare fuses for each different rating or type of fuse in service (see Table 11-2).

TABLE 11-2 Troubleshooting Fuses And Cut-Outs

Trouble	Cause	Suggested Remedy
Fuse blowing with no apparent cause	Fuse weakened as a result of being partially melted when another fuse blew at same installation	Adapt practice of always replacing all fuses at an installation where any one fuse blows.
Fuse blows during in-rush currents; motor starts; lightning	Fuse melts because of too much current for time applied	Improper fuse with too small a rating or wrong time-current characteristic must be replaced by an appropriate fuse. Also causes of high in-rushes should be checked.
Overheating of cut-out or mounting	Loose or dirty contacts in fuse carrier	Clean and repair contacts and cut-out of mounting.
	Fuse carrier not properly aligned	Realign fuse carrier.
	Current may exceed rating of cut-out or mounting; overload	Check and reduce load or replace with larger rated cut-out or mounting.
Oil leaking (oil fuse cut-out)	Oil level too high or cut-out mounted at angle that leaves bushing below oil level	Remove extra oil; straighten cut-out; check gaskets and replace if necessary.

REVIEW

1. Fuses serve to protect electric circuits and equipment, serving essentially as a one-shot circuit breaker. They are designed to melt at predetermined abnormal values of current flowing in a circuit. They are perhaps the most common protective device in use.
2. All fuses consist of a metallic link that melts when more than rated current flows through it. The composition and dimensions determine the magnitude of the current and the minimum time to melt it; its mounting and enclosure also affect the melting time, and the arc-clearing time as well. The sum of these two values is known as the *clearing time* of the fuse, (Fig. 11-1).
3. Power systems are divided into sections, each of which may be protected by a fuse, whose locations are known as *coordinating points*. If two or more are installed in a circuit, they must be coordinated so that only one section (faulted) becomes de-energized. Fuses at the coordinating points farthest from the source will have the lowest rating, while fuses at other coordinating points along the main line will increase in rating as they approach the source (Fig. 11-5).
4. Fuses may be mounted in the open, or enclosed within a container that

may be filled with solids, such as sand or boric acid (Fig. 11-8), or with liquids (Fig. 11-7), such as fire extinguishing liquids (carbon tetrachloride, for example); high duty fuses may be filled with oil (Fig. 11-9).

5. Expulsion-type fuses are designed so that the gases formed by the vaporizing of the fuse and the expansion of the surrounding air and filler material act to blow out the arc (Fig. 11-10). In another de-ion type, the arc is drawn over a grid so that it breaks up into smaller arcs more readily extinguished (Fig. 11-14).
6. Fuses may act singly (and also serve as disconnects), and may be mounted together to furnish a *three-shot fuse* (Fig. 11-15), arranged so that if the first fuse blows, the second automatically takes its place, and should this blow, a third takes its place. When all three have blown, the line or equipment is de-energized.

STUDY QUESTIONS

1. What is the function of a fuse?
2. Name several types of fuses and where they are used.
3. Of what are fuses made?
4. What is meant by the time-clearing characteristic of a fuse?
5. What is meant by fuse coordination?
6. What is an expulsion fuse? Where is it generally used? What is meant by double venting?
7. What is a de-ion fuse? How does it function?
8. What is a fuse cut-out?
9. What is a three-shot fuse? Where is it used?
10. What considerations should be taken into account in the selection of fuses and their mountings?

12

LIGHTNING OR SURGE ARRESTERS

FUNCTION

Lightning or surge arresters serve the same purpose on an electric circuit or line as a safety valve on a boiler. A safety valve on a boiler relieves a high pressure by allowing steam to blow off until the high pressure is reduced to normal, at which time the safety valve closes again and is ready for the next abnormal condition. So also is the operation of a lightning or surge arrester. In an electric system, this abnormal pressure takes the form of high-voltage surges in the conductors that may be brought about by the static electricity created by lightning (hence the name, lightning arrester). High-voltage surges may also result from improper switching and other abnormal operations in which the circuit may be involved. The function of the surge arrester is to allow the discharge of any dangerous overvoltage before it can do damage and then to restore the line to normal operation after the discharge.

To accomplish this protective function satisfactorily, arresters must:

1. Not allow current to flow to ground so long as the circuit voltage remains normal
2. Provide a path to ground, when the voltage rises to a predetermined value above normal, to dissipate the energy from the surge without raising the voltage at which the circuit is operating
3. Stop the flow of current to ground, as soon as the voltage drops below the predetermined value, and restore the insulating qualities between the conductor and ground

4. Not be damaged by the discharge and be capable of automatically repeating the discharge action as soon and as frequently as called upon to do so.

PRINCIPLE OF OPERATION

The principle upon which lightning or surge arresters usually operate is that when a voltage surge traveling along the conductor reaches the point at which an arrester is installed (usually at some equipment or line to be protected), it finds a lower resistance path to ground than that presented by the equipment or line. The voltage surge usually breaks down the insulation of the arrester momentarily, allowing the voltage surge to travel to ground and dissipate itself. The insulation of the arrester then recovers its properties and prevents further current from flowing to ground.

ARRESTER TYPES

The elementary arrester consists of an air gap in series with a resistive element. The high-voltage surge causes a spark to jump across the air gap and pass through the resistive element to ground (Fig. 12-1). One resistive element, sometimes referred to as a *valve* is usually made of a material that provides a low resistance path for the high-voltage surge but presents a high resistance to the flow of line energy at normal voltages. Arresters of this kind are referred to as *valve arresters*.

Another resistive element consists of an explusion chamber in which the surge, after traveling through the series air gap, permits current to flow through a tube in which the gases formed blow out the arc, thereby restoring the high resistance to the flow of current from normal voltage. The tube or chamber is made of, or lined with, a fiber material. When this material comes in contact with the arc, it generates gases that increase the pressure in the chamber and help blow out the arc, thereby restoring the arrester to its initial condition—ready for another operation.

There are many different types of arresters, all of which have a resistive element and may or may not have a series gap. They differ only in mechanical construction and in the type of resistive element employed.

VALVE ARRESTERS

Pellet Arresters

In the pellet arrester (Fig. 12-2), the series gap is contained within a porcelain receptacle, and the valve element is made up of a column of pellets some 2 inches in diameter with a height varying according to the voltage rating

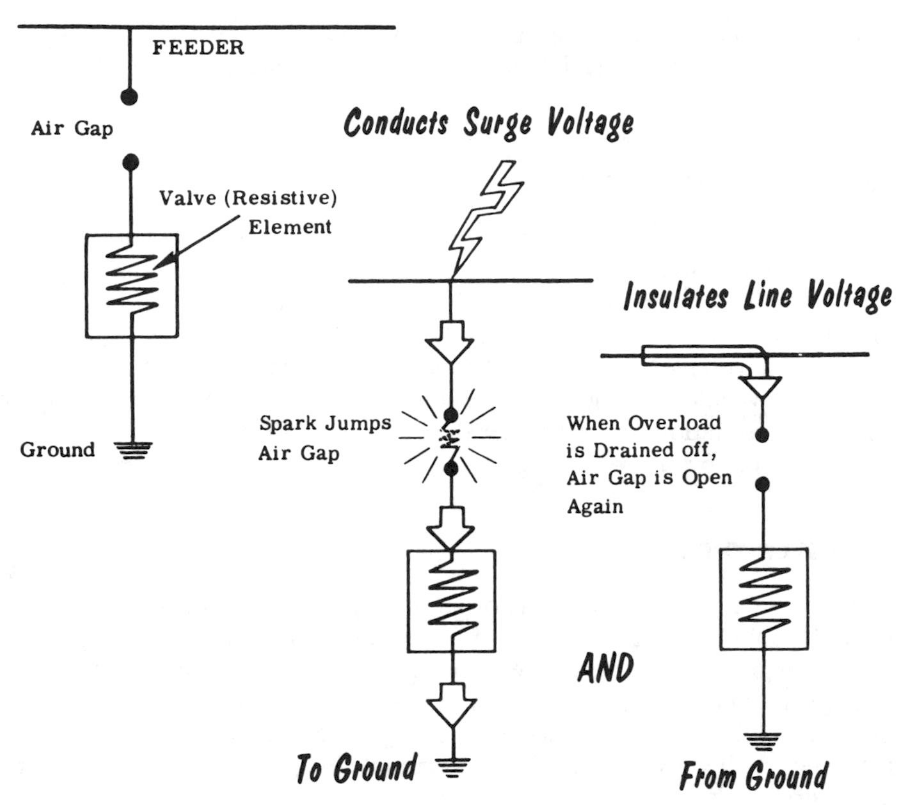

Figure 12-1 Elementary lightning or surge arrester.

(approximately 2 inches for each 1,000 V of arrester rating). The pellets, from ⅛ to ¼ inch in diameter are made of lead peroxide (PbO_2) coated with litharge (or lead oxide, PbO). The series gap assembly is usually located at the top end, where it is subject to spring pressure; one end of the gap assembly is connected to the line and the other to the pellets, with its connection to ground coming from the bottom of the pellet column. Though rapidly becoming obsolete, many arresters of this type still exist and will be around for a long time.

When there is an excessive voltage surge on the line, the spark jumps across the gap or gaps and flows through the pellets through multiple paths to ground, heating the pellets along the way. Under normal temperatures, lead peroxide is a relatively good conductor of electricity. When heated, it turns to lead oxide or litharge, which is a very poor conductor. As the surge voltage is drained to ground, a sealing action takes place at the contact surfaces of the pellets in the path of the discharge current, the lead peroxide changing to lead oxide or litharge. This chemical reaction restores the pellet column to its nonconducting state for normal line voltages.

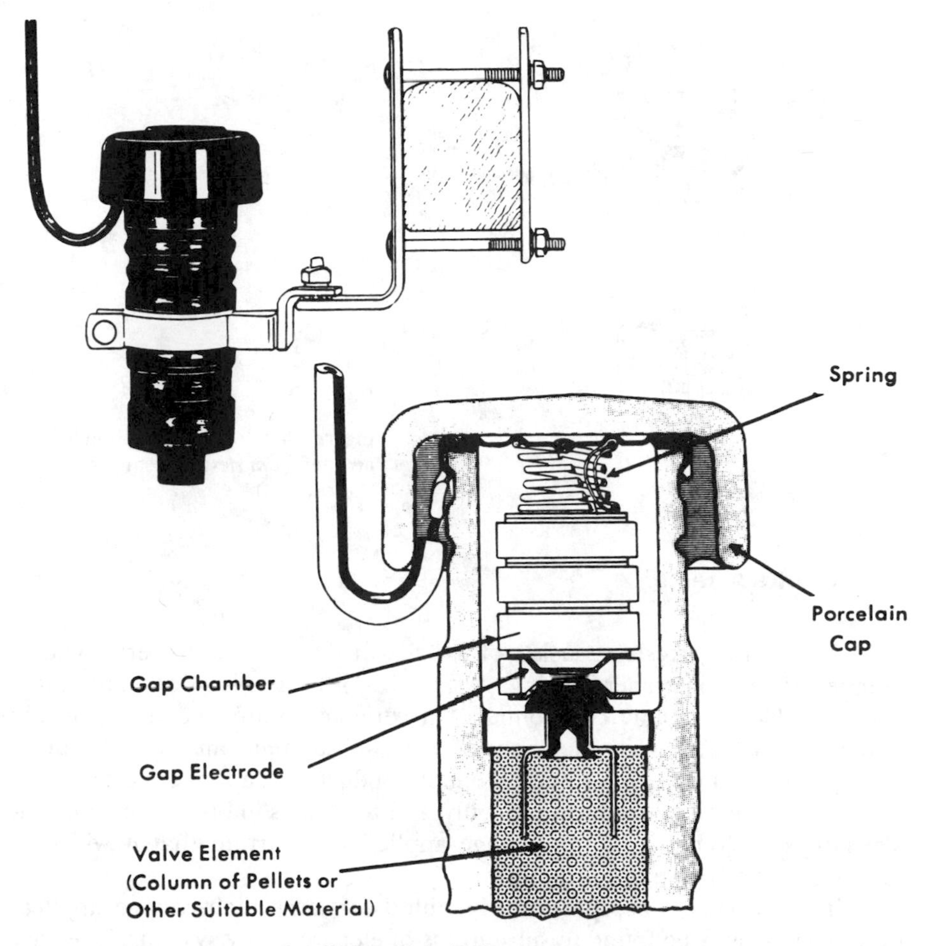

Figure 12-2 Pellet arrester.

Autovalve Arrester

In the autovalve arrester (Fig. 12-3), the valve element is made up of ceramic material and conducting particles. The arrester consists of a stack or column of flat discs or electrodes made of this material and separated by thin insulating mica washers; the column is connected through a series gap to the line and ground. At voltages above the critical value, the short gap between the discs breaks down and current flows. Since the discs are thin and of relatively large area, the resistance through which the discharge current flows is low, even though the resistivity of the discs is comparatively high. When the voltage falls below the critical value, no current will flow and the arrester is returned to its nonconducting state.

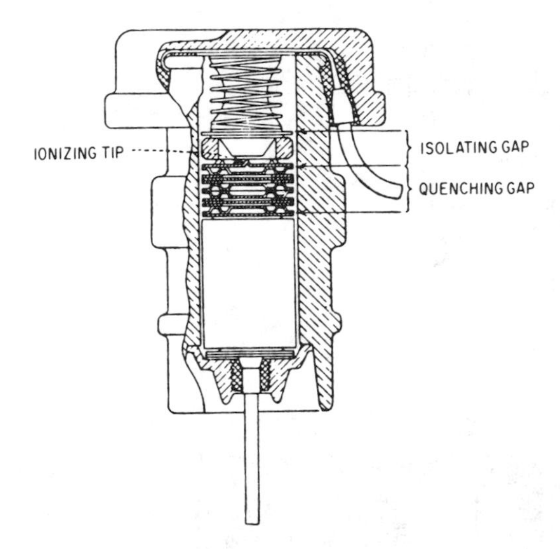

Figure 12-3 Autovalve lightning arrester. (*Courtesy*, Westinghouse Electric Co.)

Thyrite Arrester

The thyrite arrester consists of a stack of thyrite discs in series with a number of small, interspersed air gaps (Fig. 12-4). Thyrite is a dense, homogeneous, stable, inorganic compound of a ceramic nature and mechanically strong. It possesses the characteristic of being substantially an insulator at one voltage and then changing to an excellent conductor at a higher voltage; the transition is due to voltage changes only, not to heat as in other valve materials. For each doubling of the voltage applied, the current that it will pass increases some twelve or more times.

These arresters are particularly suited for large high-voltage applications, such as may be found in substations of electric utility systems. They are built in modules or units, each rated at approximately 11,500 V; the modules are stacked in series to obtain the desired voltage ratings in 11,500-V steps above that rating.

Granulon (or Isolator) Arrester

The granulon differs from other valve arresters in that the valve element (also of a ceramic type) and the series gap are enclosed in clear pyrex glass, making them visible for inspection. The action of this arrester under surge voltage conditions is the same as for other valve element arresters. High-voltage surges spark across the air gap, and discharge current flows through the valve to ground. Since the valve has a low resistance under high voltage and a high resistance at normal voltage, the arc is extinguished and the arrester returns to its original state.

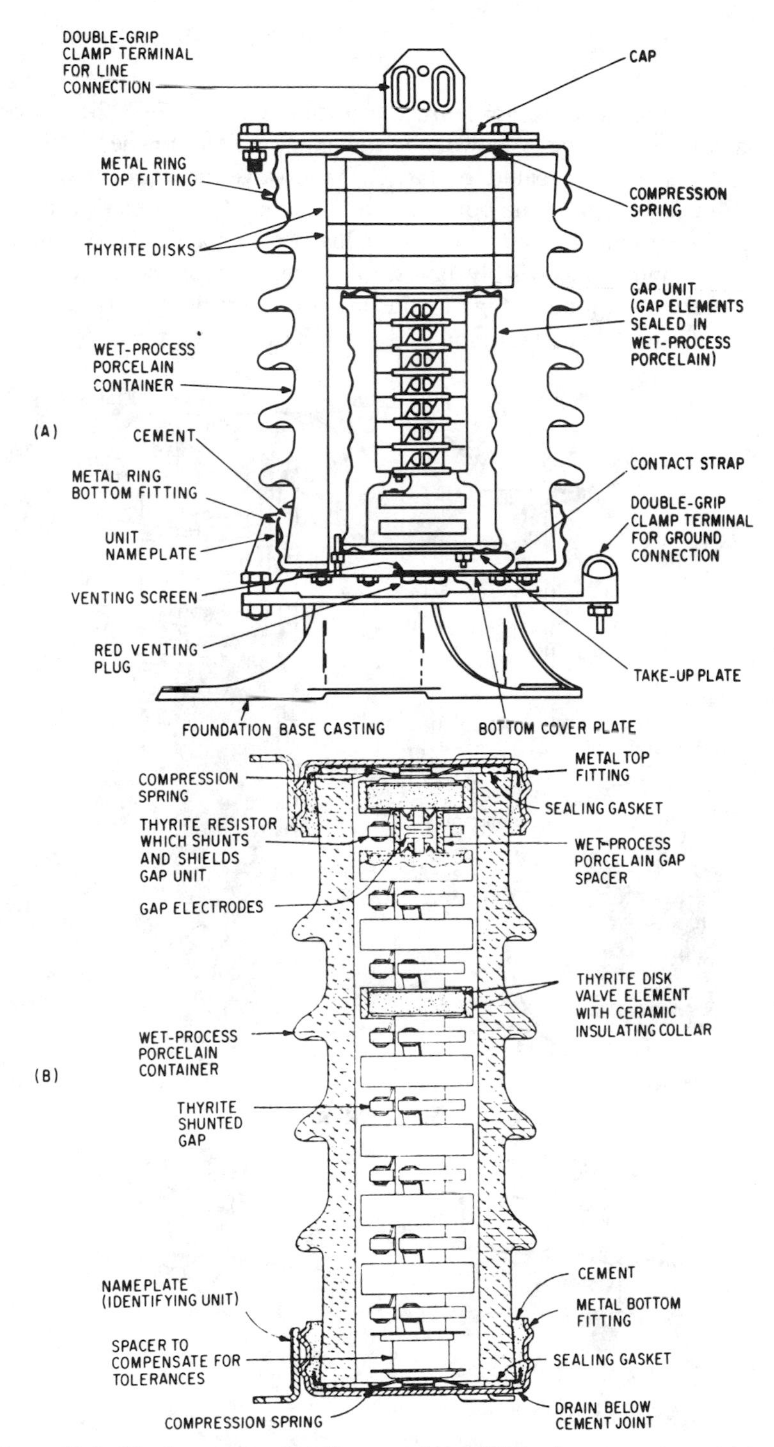

Figure 12-4 Thyrite arrester: (A) Cutaway of 11.5-5kV station-type unit, and (B) cross section of high-voltage unit. (*Courtesy*, General Electric Co.)

This arrester has another feature called an *isolator* (Fig. 12-5). It consists of two auxiliary electrodes between which combustible powder is placed, the whole apparatus being sealed in plastic. If the valve element is damaged, the heat produced will ignite the powder and the expanding gases will rupture the housing. The ground lead attached to the lower electrode is disconnected from the arrester and hangs visibly below it. Thus disconnected, the damaged arrester prevents a long outage to the line or equipment to which it is connected. Although obsolete, many will remain in service for a long time.

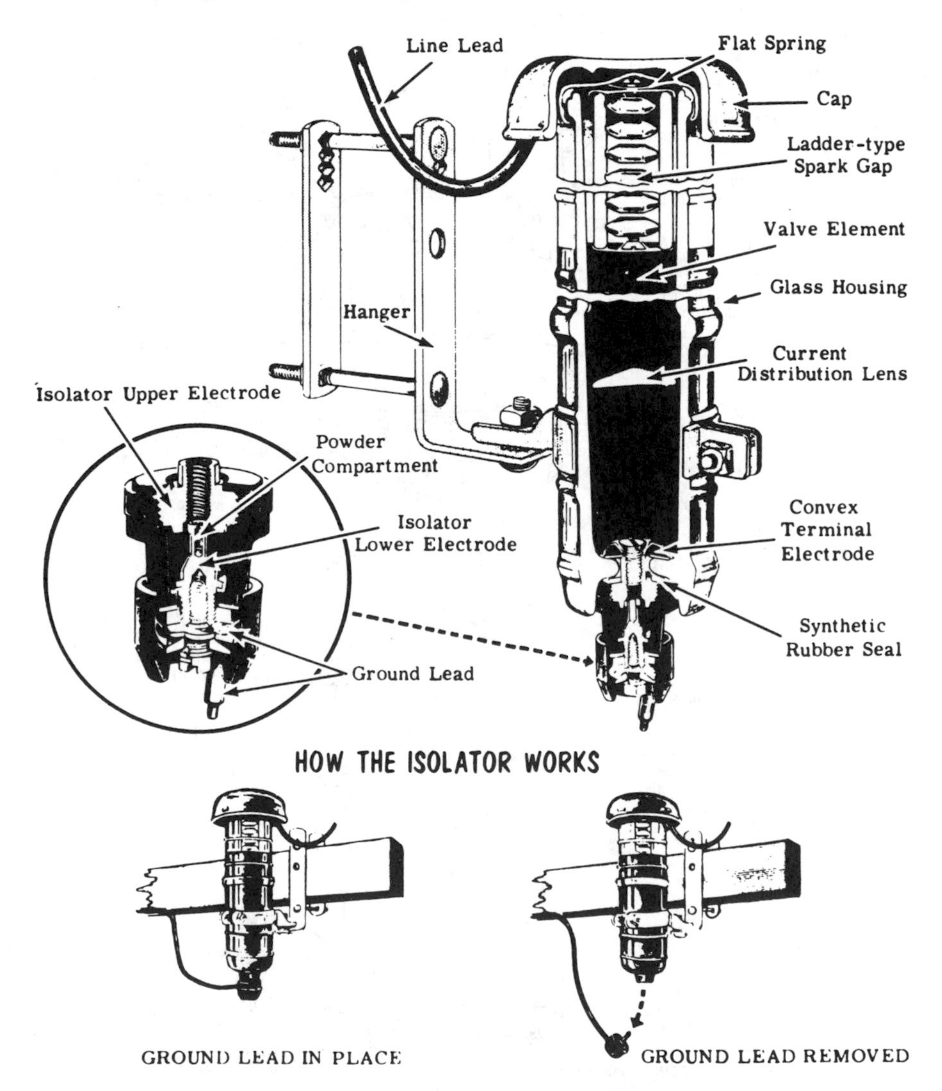

Figure 12-5 Granulon (isolator) arrester.

Expulsion Arrester

In the expulsion arrester, there are two series gaps, one of them being the usual gap in series with the valve element (Fig. 12-6). The valve element is essentially another gap in series with the first, although it is contained inside a fiber tube, with an electrode placed at each end of the tube. It is so designed that the voltage breakdown through the tube will be less than that in the insulation of the line or equipment to be protected. When a voltage surge occurs that is sufficient to spark over the series gap and the gap in the fiber tube, discharge current flows to ground. The arc in the tube attacks the fiber, releasing large volumes of a relatively cool, nonconducting gas. The gas produced acts not only to extinguish the arc but also builds up pressure that blows

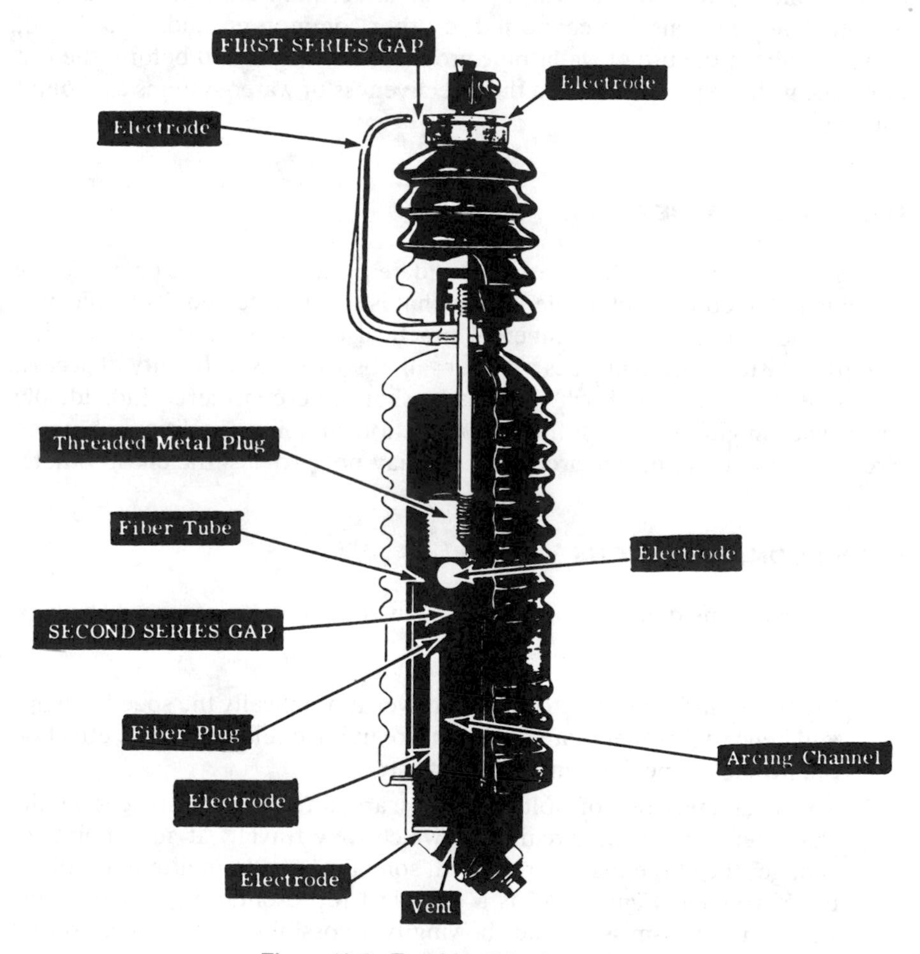

Figure 12-6 Expulsion lightning arrester.

the arc out of the tube, returning the arrester to its original condition—ready for the next operation. In later types, the external gap is sometimes omitted.

GROUND CONNECTION

The effectiveness of any arrester is dependent on a good connection to ground, and such connections must be tested frequently. This cannot be emphasized too strongly as arresters will not function without a proper ground; they are totally useless. Even worse, if they are not known to be in this state, supposedly protected equipment may be severely damaged or completely destroyed. If possible, arresters should have their own separate ground connection, usually the shortest and most direct path from the arrester to ground. They may also be connected to other common grounds. Take care, however, about the use of water pipe grounds. As mentioned before, the use of plastic water pipe has reduced the effectiveness of water systems as ground sources.

LOCATION OF ARRESTERS

As a good general rule, the arrester should be placed as close as conveniently possible to the equipment, cable, or line that is to be protected. This rule may be modified for any of a number of practical considerations: construction hazards, existence of high trees or other tall obstructions, difficulty of access, and other local factors. Each installation should be considered individually since what might provide sufficient protection in some localities, on some circuits, and under certain circumstances may not prove sufficient in others.

OTHER CONSIDERATIONS

In addition to considerations previously discussed, other points should be kept in mind:

1. Lightning and other electric surges travel at practically the speed of light, and hence a few feet more or less of circuit have relatively little effect on the protective performance of an arrester.
2. The characteristics of voltage surges are affected by changes in the characteristics of the circuit over which they travel. At such points of change, the surge may be reflected, sometimes with an attendant piling up of voltage. Figure 12-7 is a graphical representation of a traveling wave on a transmission line showing two possibilities of voltage surges occurring at points of discontinuity, such as an open switch, a transformer bank, or change of overhead to underground. In Fig. 12-7A, the

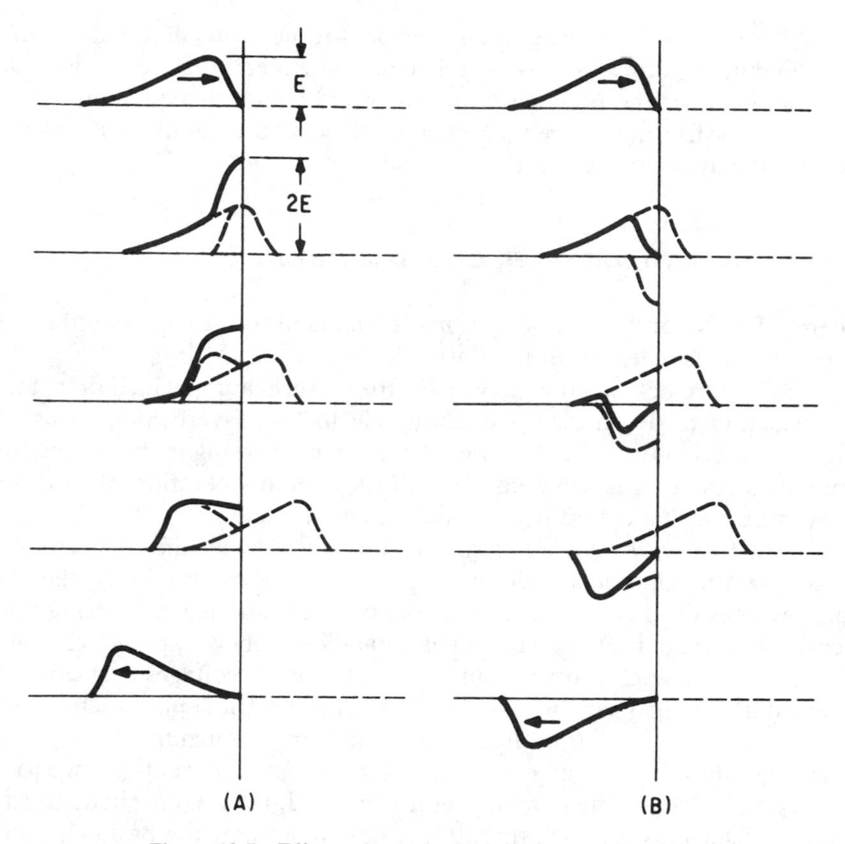

Figure 12-7 Effects of voltage surges on transmission line.

characteristics of the point of discontinuity are such that the reflected wave is superimposed on the original surge wave and the crest voltage is double the original value at the point of discontinuity. In Fig. 12-7B, the characteristics of the point of discontinuity are such that the reflected wave is subtracted from the original wave.

3. An arrester will not necessarily prevent a surge from striking or entering the equipment to be protected. Its function is to discharge fast enough after the surge has been applied to prevent the surge voltage from having enough time to damage the insulation of the equipment to be protected.
4. The effectiveness of an arrester depends on, in addition to its internal characteristics, the ability of the ground connection to discharge the abnormal current to the ground.
5. The fulfillment of the arrester's primary function, which is to prevent excessive voltage stress on protected equipment, requires the coordination of the equipment's insulation characteristics with the arrester's protective characteristics.

6. An arrester is only one of many devices for the protection of a circuit or equipment; other devices include circuit breakers, reclosers, fuses, and the like. For the proper protection of the circuit or equipment, all of these devices must have their characteristics coordinated so that each of them may operate correctly and positively.

INSULATION AND ARRESTER COORDINATION

Schemes for the protection of lines and equipment are usually based on the protection of insulation against failure.

The failure of insulation may result from overheating, which first causes the insulation to deteriorate and ultimately to fail. Overheating is usually allied to the magnitude and duration of the current flowing in the conductors. Protection against it is achieved through the speedy operation of fuses and circuit breakers that cut off the damaging current.

Insulation may also fail from punctures caused by the application of voltages of a magnitude and duration greater than those for which the insulation was designed (including the factors of safety and manufacturing tolerances). The level of insulation is not based on the supply voltage, which it must withstand under normal operations, but on the peak voltages that might be expected from surges capable of occurring on the electric supply circuits. Such surges usually are caused by lightning or improper switching. Protection is achieved by draining the surge to ground fast enough to prevent damage to the insulation of the equipment or line being protected. Hence, the characteristics of the arrester must be such that it operates well below the basic insulation level that has to withstand any surges that may be imposed on the equipment or line being protected.

Basic Insulation Level (BIL)

The coordination of insulation requires that the insulation of all components of a system be above a minimum level and that a selected protective device operate satisfactorily below that minimum level, known as the *basic insulation level,* or BIL (Fig. 12-8).

The minimum level of insulation must withstand not only the normal operating voltage applied continuously, but also the surge voltages that may exist for a comparatively very short time but can nevertheless cause damage. To select the value of this surge voltage, it is necessary to define its rise to its peak value and its fall back to lower values in terms of time, that is, its duration, for, although the peak voltage may rise considerably above the normal voltage, the insulation need sustain its stress only for an extremely short period of time. Put another way, the insulation need not be such as to

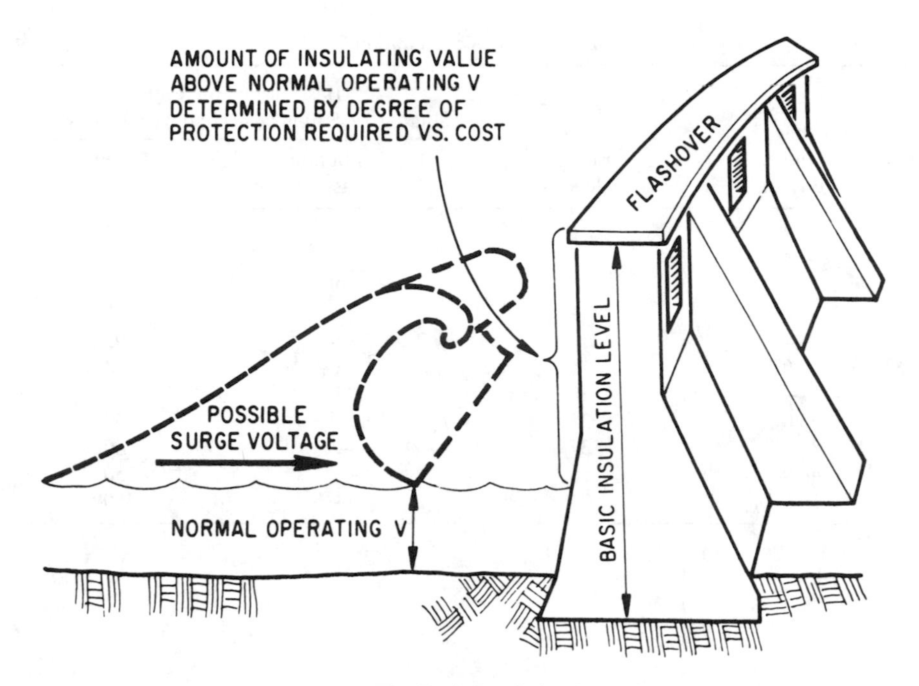

Figure 12-8 The Basic Insulation Level (BIL).

withstand a continuous stress from the high peak voltage. A committee representing engineers, utilities, and manufacturers, has therefore, established standards for the characteristics of a voltage surge and recommended insulation levels for equipment of various voltage classes.

The surge voltage is defined as one that rises to its maximum or crest value in 1.5 microseconds and falls to one-half that value in 40 microseconds (a microsecond is one one-thousandth—1/1,000—of a second). This wave is illustrated in Fig. 12-9. For ease of reference, it is designated as a 1.5/40 wave. The steep rising portion of the wave is called the *wave front* and the receding portion, the *wave tail.*

Recommended insulation levels for a number of voltage classes are listed in Table 12-1. Note that the ratio of the BIL (capability to withstand peak

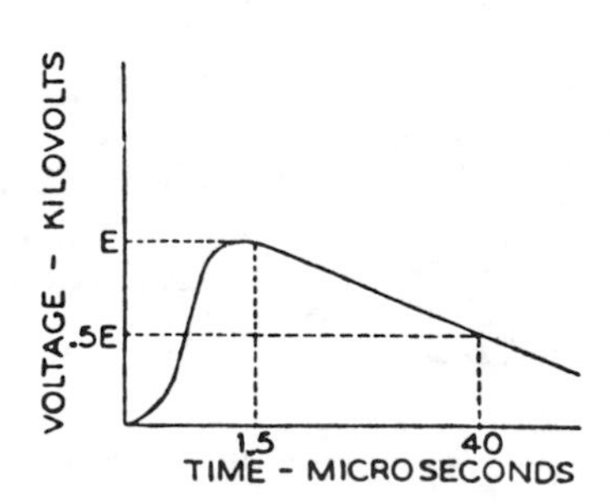

Figure 12-9 1.5/40-microsecond voltage wave.

TABLE 12-1 Basic Insulation Levels

Insulation voltage class, kilovolts	Arrester voltage rating, kilovolts	Basic insulation level (Standard 1.5/40-microsecond wave), kilovolts	
		Distribution class	Power class
1.2	1	30	45
2.5	3	45	60
5.0	6	60	75
8.66	9	75	95
—	12	85	102.5
15.0	15	95	110
—	18	125	—
23.0	25	—	150
34.5	37	—	200
46.0	50	—	250
69.0	73	—	350

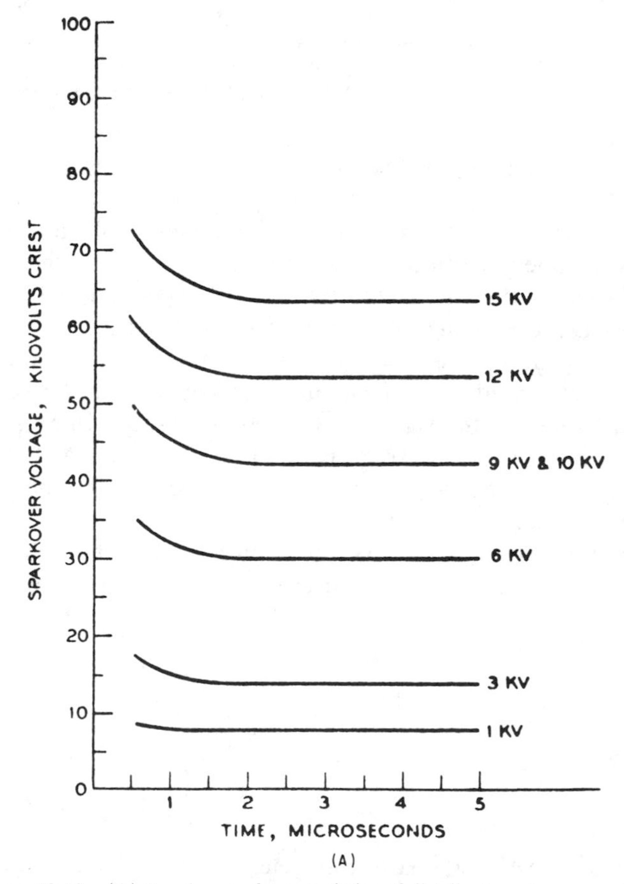

Figure 12-10 (A) Sparkover characteristics of distribution class arresters.

surge voltage) to the voltage class (capability to withstand this normal voltage continuously) decreases as the latter increases. The reason is that as the operating voltages become higher and higher, the effect of a surge voltage becomes less and less. Note also that the BIL for "distribution class" insulation is less than that for "power class," that is, the BIL for utility equipment is less than for consumer equipment. This difference takes into account that a failure from a surge should preferably occur on the facilities of the utility geared to handle them.

Distribution class arresters are usually restricted to electric distribution circuits (Fig. 12-10A). Power class arresters find application at electric stations and transmission lines as well as consumer equipment (Fig. 12-10B).

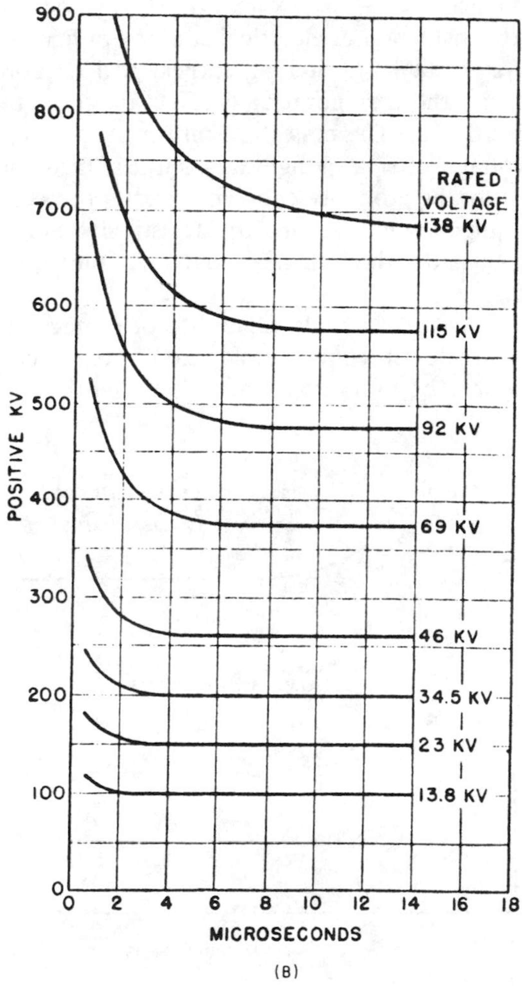

Figure 12-10 (cont.) (B) impulse characteristics of power class arresters. (*Courtesy*, McGraw Edison Co.)

The insulation characteristics of the equipment (its impulse level) are represented by the voltage-time curve that the equipment insulation can withstand. The characteristic of the protecting arrester is also a voltage-time curve that indicates the voltage and time at which the series gaps of the arrester spark over and begin to pass the surge to ground. The insulation characteristics of the equipment must always be at a higher voltage level than that of the sparkover characteristic of the protecting arrester, and a sufficient voltage difference must exist between the two curves. These characteristics are shown in Fig. 12-11.

The impulse level of equipment must be high enough for the arrester to provide adequate protection and low enough to make insulation costs economically practical.

Although the insulation of electrical equipment may be subject to overvoltages from system faults, switching surges, and occasional higher-than-normal line voltages, these do not usually affect the insulation but may damage the protective device. The major reason for overvoltage protection is to limit the high-voltage effects of lightning surges. Insulation coordination, therefore, consists of the proper selection of arrester protective characteristics for a particular equipment insulation characteristic so that lightning surges will be discharged to ground without damage to the equipment insulation or to the protective device.

Since electric systems include a number of protective devices, each of which has characteristics of its own, in addition to the circuit conductors and the equipment connected to them, it is essential that the characteristics of all

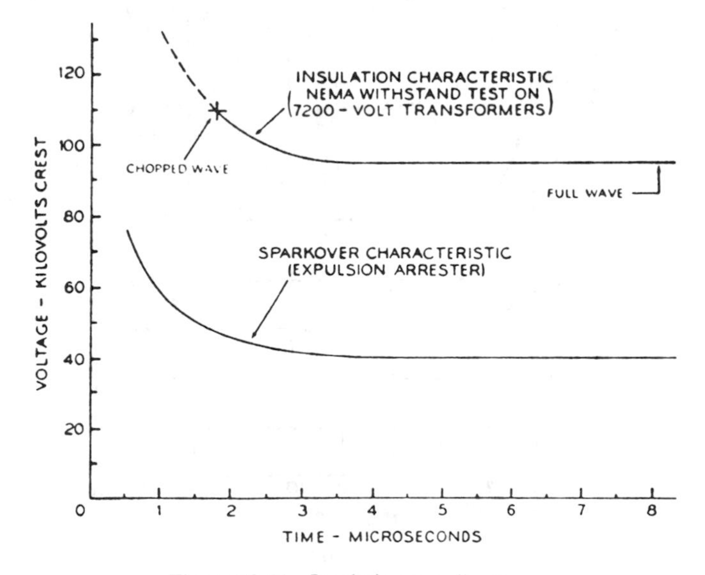

Figure 12-11 Insulation coordination.

TABLE 12-2 Troubleshooting Lightning or Surge Arresters

Trouble	Cause	Suggested remedy
Flashover of porcelain	Dirty or polluted surfaces	Clean or replace if necessary
	Insufficient line to ground or to line clearances	Replace with unit having greater clearances
	Unusual severity of discharge; incorrect arrester rating	Replace arrester with one having correct rating
Lightning damage to protected equipment	Arrester rating too high for circuit voltage	Replace arrester with one having correct rating
	Improper arrester location or connections; poor ground	Install arrester closer to equipment and obtain a reliable low-resistance ground. Interconnect all equipment, arresters, lines, cable sheaths, structures, etc., and grounds. Use shorter ground connections.
	Equipment insulation of inadequate impulse strength	Make sure equipment has minimum standard insulation level
	Air gap set too wide	Adjust gap according to specifications
Arcing or corona discharge from gap assembly; too frequent sparkovers	Loose or broken connection	Tighten or repair connections
	Incorrect gap setting	Set gap according to specifications
Excessive operating temperature; discoloration of metal parts	Excessive ambient temperature	Relocate to better location
	High internal losses; incorrect arrester rating	Replace arrester with one having correct rating
	Loose connections	Tighten or repair connections

these elements be coordinated for their proper operation and protection. Coordination of this kind will be discussed in connection with relays.

For causes and remedies when troubleshooting lightning or surge arresters, see Table 12-2.

REVIEW

1. Lightning or surge arresters serve the same purpose on an electric circuit or line as a safety valve on a boiler. When a voltage surge induced by a nearby lightning stroke, or caused by switching, reaches an arrester, it finds a lower resistance path to ground than that presented by the equip-

ment or line. The voltage momentarily breaks down the insulation or the arrester, allowing the surge to travel to ground and dissipate itself.

2. The insulation which breaks down momentarily and then restores itself is sometimes called a *valve,* but sometimes may consist of an air gap and resistor (Fig. 12-1).
3. The valve element may consist of lead oxide pellets (Fig. 12-2), ceramic material containing conducting particles, or such patented materials as thyrite (Fig. 12-4) and granulon (Fig. 12-5).
4. In an expulsion arrester, the arc is formed within a fiber tube which is attacked by the arc releasing large volumes of relatively cool non-conducting gas that flows out the arc (Fig. 12-6).
5. The effectiveness of any arrester is dependent on a good connection to ground, hence such connections must be tested frequently as arresters will not function and are totally useless without a proper ground.
6. Arrester characteristics must be coordinated with basic characteristics of the insulation of the lines and equipment they are to protect so that the arrester operates before the insulation is damaged, and before fuses or circuit breakers operate.
7. The coordination of insulation requires that the insulation of all components of a system be above minimum level and that a selected protective device operate satisfactorily below the minimum level, known as the *basic insulation level,* or BIL (Fig. 12-8).

STUDY QUESTIONS

1. What is the function of a lightning or surge arrester?
2. Describe the principle of operation of an arrester.
3. What types of arresters are commonly used?
4. What is the function of the fiber lining in an expulsion fuse? Where is this type of arrester commonly used?
5. Why is a good ground of the greatest importance?
6. Where should an arrester be located?
7. What conditions can produce electrical surges in an electrical circuit?
8. What is meant by the Basic Insulation Level?
9. What is meant by a standard surge voltage wave?
10. What other considerations should be taken into account in selecting and locating an arrester?

13

PROTECTIVE RELAYS: DESIGN AND CONSTRUCTION

FUNCTION

In order for an electrical system to function properly and with a minimum of disturbance under abnormal conditions, it is essential that faulty circuits and equipment be quickly disconnected from sources of energy. Protective relays are used as a means to accomplish this end by electrically tripping the proper circuit breaker.

The protective relay is an electrical device so placed between the circuit or equipment and the associated circuit breaker that any abnormal condition in the circuit or equipment acts upon it. The relay, in turn, after proper discrimination as to the magnitude and character of the fault or abnormal conditions, causes the circuit breaker, or breakers, to operate and relieve or protect the circuit or equipment.

A relay, then, is a low-powered device used to activate a high-powered device (Fig. 13-1). It may therefore, be viewed as a sensing device that triggers an action on or by the equipment with which it is associated. It may be likened to the human nervous system, including the senses. In this regard, it receives certain data, examines it, and then initiates action or actions on the part of the equipment (organs) that it controls. The actions may be positive (starting), negative (stopping), balancing (a combination of the two), or informative (reporting the status of specified conditions).

For electric power systems or equipment, such sensing commonly takes note of current, voltage, and frequency as well as temperature, pressure, and speed; also included may be sensitivity to sound, light, and other indicators. It

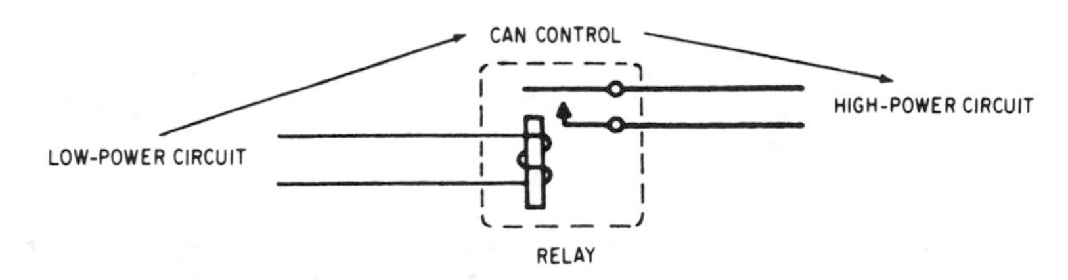

Figure 13-1 Features of the relay.

has the function of detecting actual or incipient faults or undesirable conditions and to initiate action that will result in switching or in giving adequate warning.

PRINCIPLES OF OPERATION

Modern protective systems are based on three fundamental requirements: accuracy, sensitivity, and selectivity. A protective system must *accurately* distinguish between normal and abnormal conditions. It may be required to operate at a current, voltage, or other quantity that is very small when compared with the capacity of the circuit or equipment to be protected; that is, it must be *sensitive* to small quantities or changes in value of the actuating forces. Under abnormal conditions, it must be able to distinguish in a positive fashion, between a fault in the circuit or equipment that is being protected and a fault in the rest of the system; that is, it must be *selective.*

Abnormal conditions include not only those involving the upper (or lower) tolerance limits of allowable continuous overcurrent or overvoltage, but also those transient conditions that, even if they involve values of current or voltage in excess of normal, are not of sufficiently large magnitude and long duration to have harmful effects. Such transient conditions include motor starting currents, switching of transformer or capacitor banks, synchronizing of machines, and so forth.

The several elements of an electrical system may include various circuits or lines and pieces of equipment connected together by circuit breakers. The opening of one or more breakers isolates an element from the rest of the system. Sometimes, however, two or more elements may be joined together as a unit by means of fuses or switches—all supplied through one circuit breaker. In such cases, protection may be achieved by blowing a fuse or by the operation of a switch. Relays, therefore, when called upon to act jointly with other forms of protection, must distinguish between conditions requiring the operation of a circuit breaker and conditions where other protective devices should function (Fig. 13-2).

Protective relays vary widely in their operating principles and design. They must be responsive to electrical quantities that are different during normal and abnormal conditions. They must be so arranged that those con-

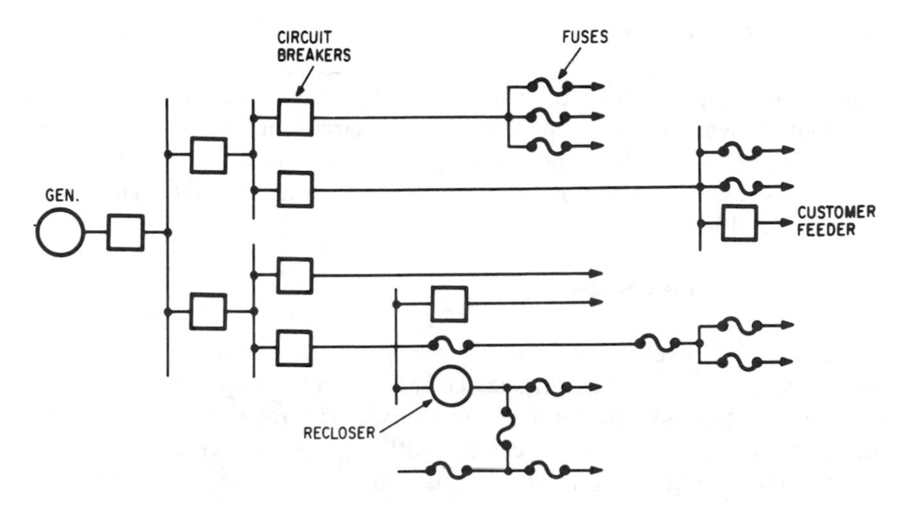

Figure 13-2 Protective devices on a circuit to be coordinated.

trolling the breaker or breakers in a faulted circuit or piece of equipment operate before, after, or simultaneously with other protective devices.

Since some electrical quantities may change during fault conditions—for example, current, voltage, their relationship or phase-angle (in an ac system), and frequency—it is usually necessary to provide relays that are responsive to more than one of these conditions. In practically all ac circuits, the relays receive their operating energy through current transformers or potential transformers, or both.

In the design and application of protective relays, the principal considerations for achieving positive operation and selectivity are the magnitude, duration, and direction of the relevant electric currents. All electromagnetic relays operate on the basis of the currents flowing through them since these currents set up magnetic fields that actuate mechanisms to close or open contacts and thus cause a circuit breaker or other device to operate. A relay, therefore, is essentially a switch.

All relays are usually composed of three elements: an actuating element (relay coil), a movable element (core or armature, disc, cylinder), and one or more sets of contacts.

Any protective system employs one or more protective principles, depending upon the nature of the protection desired. Relays of many kinds and forms are used for the protection of electrical systems. Often, two or more relays—each sensitive to a different condition—are employed together. In this chapter, discussion will include overcurrent, directional, and differential relays, which are among the most common types used and, together with reclosing relays, are almost standard equipment. The principles and practices associated with these relays apply for the most part to the many other types of relays employed in specialized applications.

OVERCURRENT RELAYS

Overcurrent relays, perhaps the simplest of the many protective relays, close their contacts when the current flowing in the circuit or equipment reaches a predetermined value. The principle of operation of each of the several types is described below; the variations in their characteristics are illustrated by the curves in Fig. 13-3.

Instantaneous Principle

In the basic overcurrent relay, the circuit is designed to open instantaneously, that is, as quickly as the device can be made to operate with no time delay deliberately added. As a normal result, the relay closes its contacts within one-half cycle to twenty cycles. Although great accuracy can be obtained in the setting of the relay, the instantaneous feature often causes the settings to be made at values that are higher than desirable in order to prevent too frequent operation of the relay as a result of transient, nonpersistent conditions. The time-current characteristic of this type of relay is shown by curve (a) in Fig. 13-3.

Inverse-Time Principle

The inverse-time principle was developed as a refinement of the instantaneous (overcurrent) principle. With it, the time of operation of the protective device is made to vary approximately inversely with the magnitude of the current; that is, the greater the current, the shorter the time for operation. Current settings may be made lower and sensitivity increased because

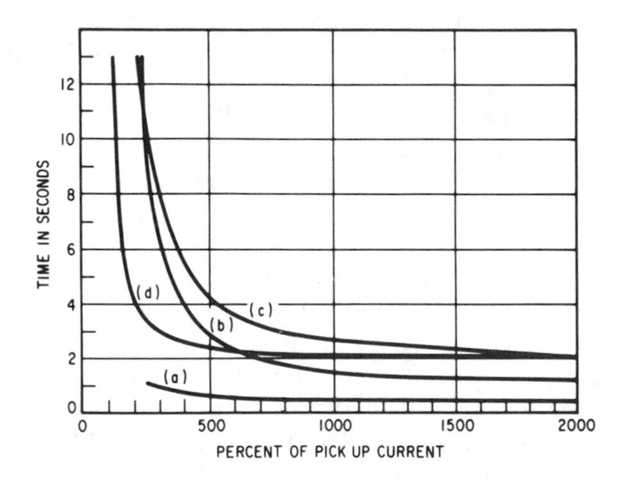

Figure 13-3 Representative time curves for overcurrent relays. (*Courtesy*, Westinghouse Electric Co.)

the normal transient currents (small in comparison to fault currents) that may start the operation of the relay will not last long enough to complete it. Furthermore, by varying the current setting and the restraint on the movable element, the time-current characteristics of this type of relay may be varied, as indicated by curves (b) and (c) in Fig. 13-3. Greater selectivity between relays and fuses in the circuit may thus be obtained, as will be discussed subsequently.

Definite-Time Principle

For still greater selectivity, the principle of definite time was introduced. It prevents the abnormal currents of any value from operating the relay until a definite time has elapsed after the relay is set in operation. This principle is often combined with the inverse-time principle to give inverse-time characteristics to a certain value of current, beyond which operation can be completed only after the definite time for which the device is set has elapsed. This combination is called the *inverse definite-minimum time principle.* The selection of a minimum time common to all such relays in series simplifies their coordination. The large flat portion of its characteristic, shown as curve (d) in Fig. 13-3 results in only a small increase in the relay time for smaller values of fault current. Relays of this type are used in the majority of overcurrent relay applications.

OVERCURRENT RELAY DESIGN AND CONSTRUCTION

The design and construction of overcurrent relays may be divided into two general classes based on their method of operation: (1) the electromechanical or plunger type, and (2) the induction or disc type. Since both methods, particularly the second, are employed in many other types of relays, a description here will prove useful in discussing the latter.

Electromechanical or Plunger Type

The plunger-type (overcurrent) relay consists of a core or plunger, sometimes also called an *armature,* that is movable within a solenoid or electromagnetic coil. When a sufficient amount of current is passed through the coil, the core is pulled up, thus causing the cone-shaped disc—the movable contact—to bridge the gap between the stationary contacts. The bridging of the contacts completes the tripping circuit, energizing the associated circuit breaker trip coil that will open the breaker. The main parts of this relay are shown in Fig. 13-4A. Single-phase and three-phase circuits are shown in Figs. 13-4B and 13-4C, respectively.

By adjusting the position of the plunger with respect to the solenoid, the

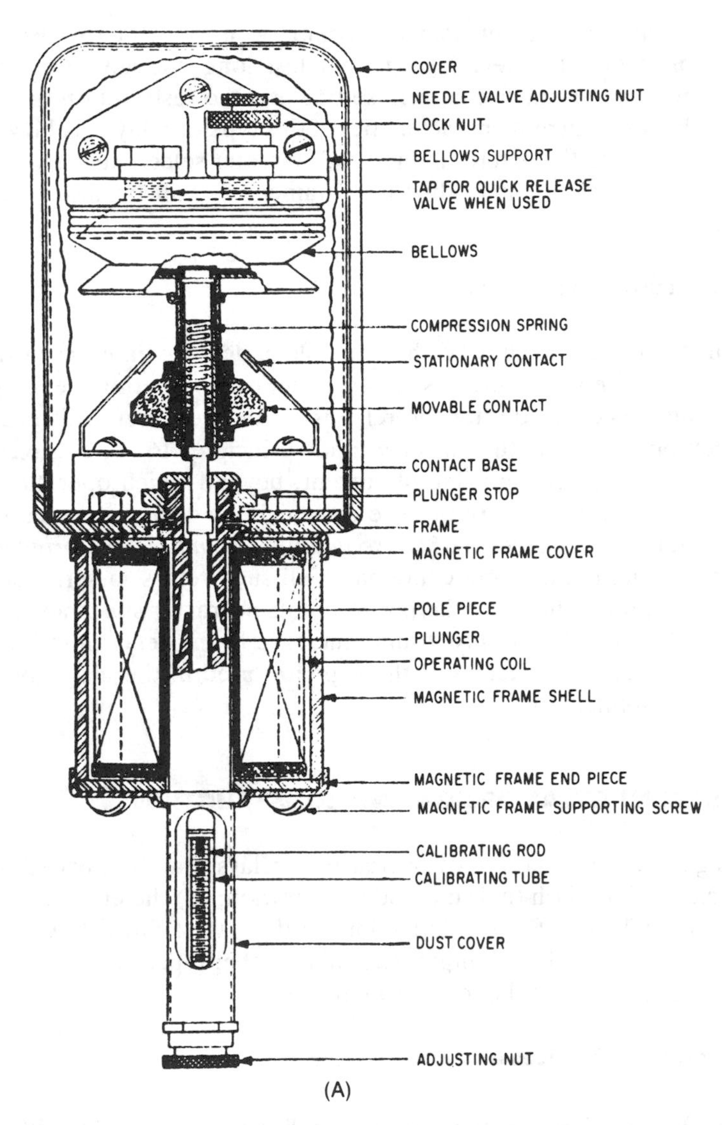

Figure 13-4 Overcurrent relay: (A) Plunger type.

relay may be set to raise the plunger at any predetermined value within the range of the solenoid. The lower the position of the plunger, the more current that is required to pull it into the closing position. Current settings may also be varied by taps in the coil.

Inverse time delay in the operation of this relay is obtained by means of a bellows (leather or rubber) arranged to retard the motion of the movable contacts. Time is varied by changing the setting of the needle valve that governs the escape of air from the bellows. An oil-filled dashpot is sometimes

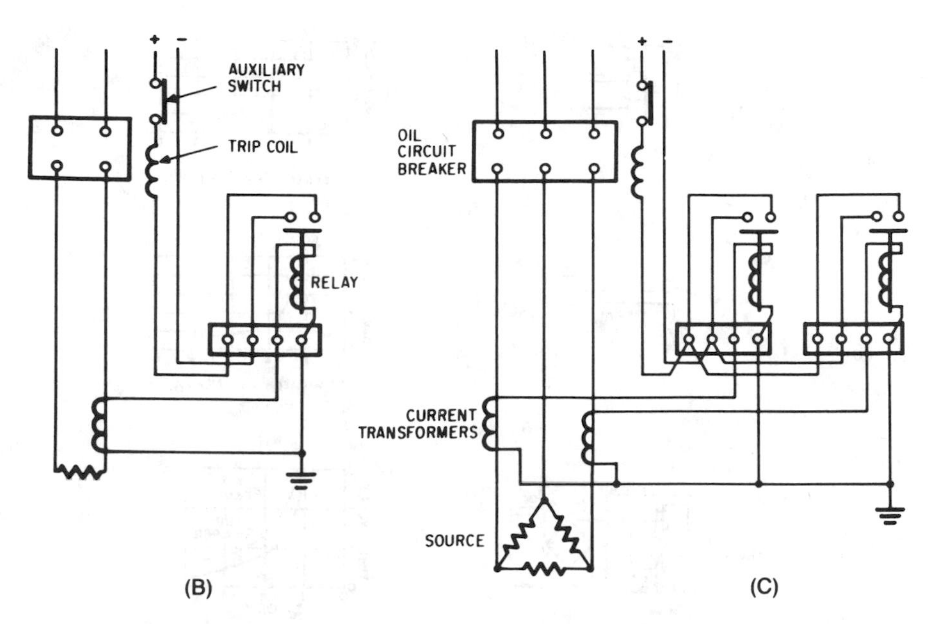

Figure 13-4 (cont.) Overcurrent relay: (B) singlephase, and (C) three-phase.

used in place of the bellows. The rate of flow of oil interchange determines the time variations. The dashpot is a mechanical retarding device consisting of a piston and a cylinder that is usually filled with oil. The piston is attached to the plunger to introduce a time delay. This delay is governed by an opening in the piston that permits the oil to flow from one side of the piston to the other as it moves in the cylinder.

Although the accuracy of this type of relay is adequate, it is not as good as that available in the induction type. Many of them are presently installed, particularly in 2300- and 4000-V distribution circuits. Although apparently vanishing, this type of relay will continue to exist for many years to come, if only for economic reasons.

Other such relays employing magnetic attraction include balanced beam and hinged armature or clapper types, which are shown schematically in Fig. 13-5. Also shown are their operating characteristics, indicating typical orders of magnitude only. A diagram of the plunger type (or instantaneous type, defined as having no intentional time delay) is included for comparison.

Induction or Disc Type

These relays operate on the induction principle, being essentially small induction motors, the contacts of which are closed by the rotation of an aluminum or copper disc (Fig. 13-6A). When sufficient current flows through the electromagnet to produce the necessary torque (or tendency for rotation), the disc rotates until, at the end of its travel, it touches a stationary contact and

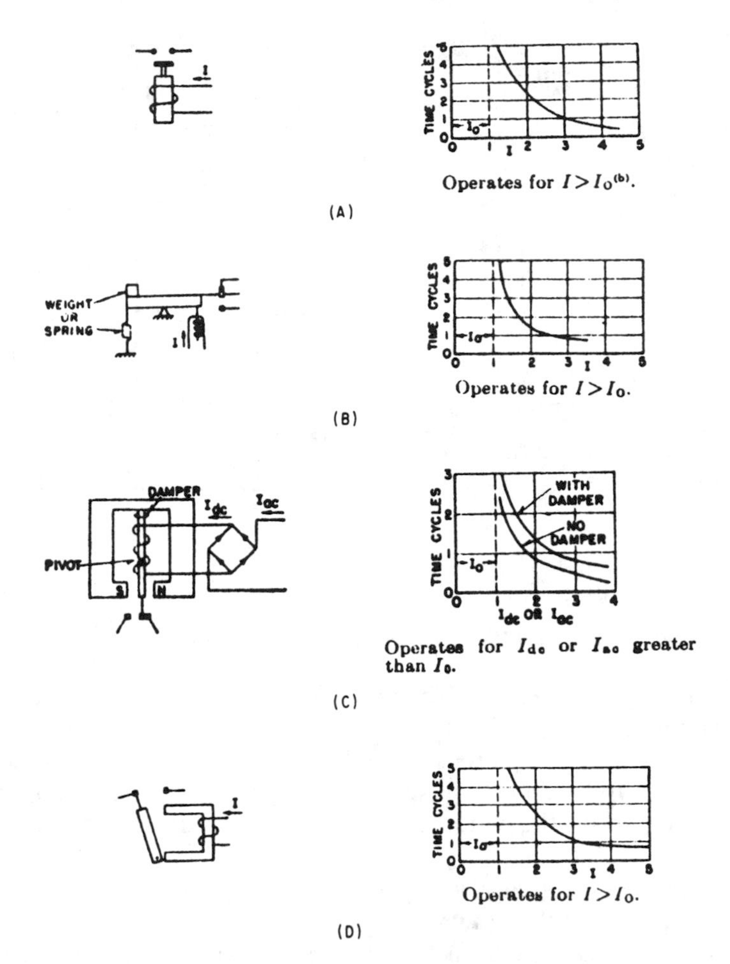

Figure 13-5 Electromechanical relays: (A) Solenoid type-instantaneous, a-c or d-c; (B) balanced-beam type--instantaneous, a-c or d-c; (C) polar type--instantaneous, d-c (a-c used with rectifier); (D) armature or clapper type--instantaneous, a-c or d-c. (*Courtesy*, Westinghouse Electric Co.)

completes the tripping circuit that operates one or more circuit breakers. Since the disc is balanced against the pull of a spring, it is necessary to have a certain quantity of current flowing before the disc will begin to rotate. Once it has started, the speed of rotation will be proportional to the current flowing in the relay coil until the latter reaches an exceedingly high value. This relationship gives the relay an inverse-time characteristic; that is, the greater the current flowing in the coil, the sooner the relay contacts will close. In order to prevent the instantaneous closing of the relay contacts under heavy short circuits, a small "torque compensator," or reactor, is placed in the circuit of the relay coil (Fig. 13-6B). This compensator acts upon the relay in such a way that a definite minimum time must elapse before the contacts will close, regardless of the value of the current flowing in the relay coil. Typical connections for overcurrent relays are shown in Fig. 13-6C.

The time-current characteristics of this type of relay may be altered to change the time for closing the contacts by varying the distance the relay disc has to travel to close them. The current characteristic is made to vary by taps on the relay coil, which, with a fixed current flowing, will cause the rotation of the disc to speed up or slow down.

Combination Types

Although rapid clearing of short circuits is important, it is often necessary to provide an intentional time delay in relay operation in order to ensure selectivity between individual relays or between relays and fuses. To overcome this difficulty, two overcurrent relays can be placed together in the same case: a time-relay induction element and a small, instantaneous, plunger type of overcurrent element, known as the *instantaneous attachment* (Fig. 13-7). The instantaneous element can be set to be much higher in current than the tap setting of the time-delay element. The coils of the two elements are in series and their contacts in parallel.

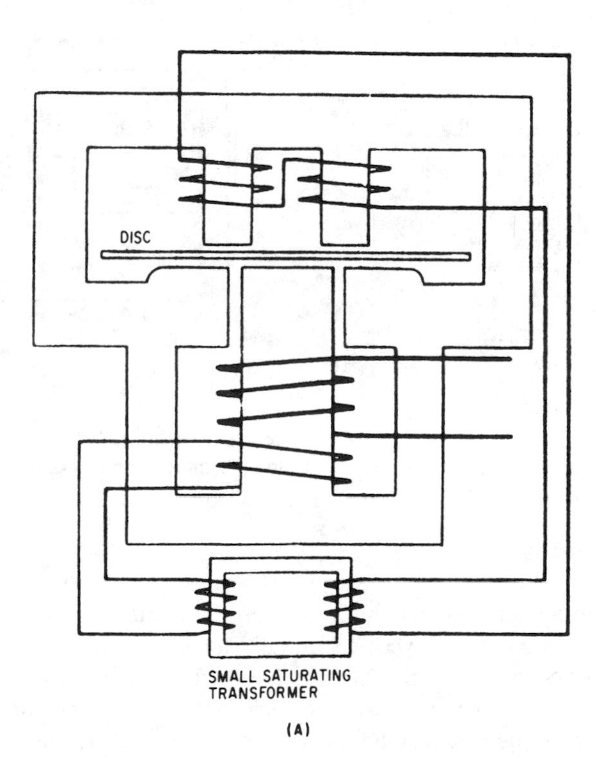

Figure 13-6 Induction type of overcurrent relay: (A) Elementary type.

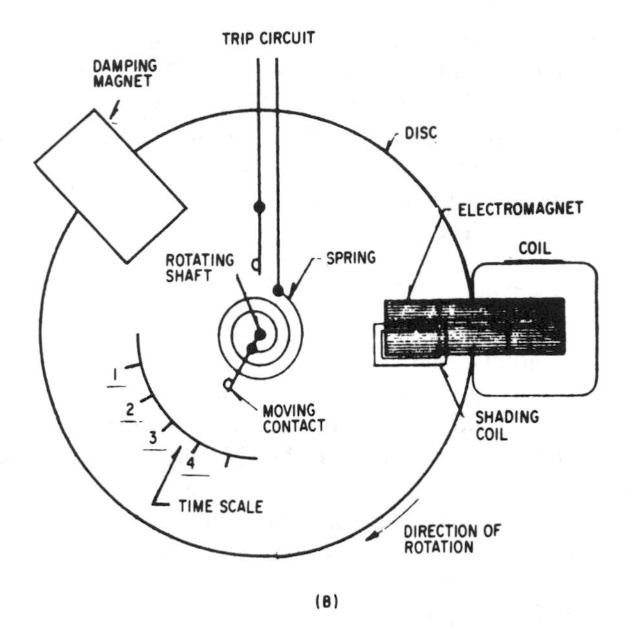

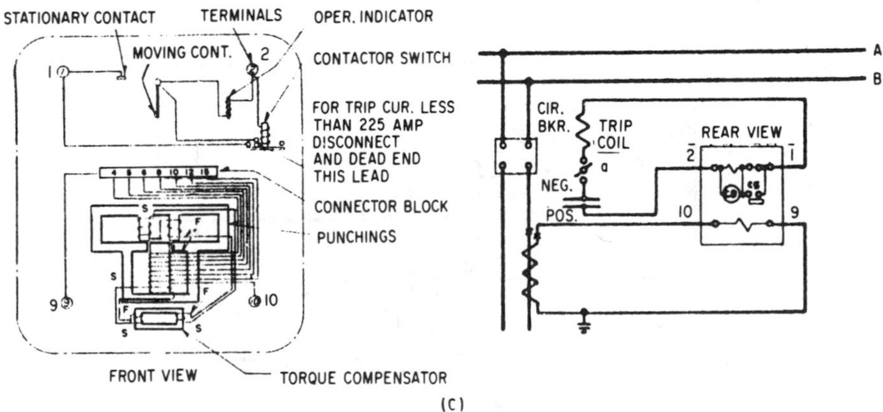

Figure 13-6 (cont.) Induction type of overcurrent relay: (B) induction overcurrent electromagnet with torque compensator, and (C) internal connections.

Other Types

Other relays employing the induction principle use more than one disc and more than one actuating coil. Some use a cylinder in place of one or more discs. Such relays are shown schematically in Fig. 13-8, also shown are their operating characteristics, indicating typical orders of magnitude only. A diagram of the disc type of overcurrent relay previously described is included for comparison.

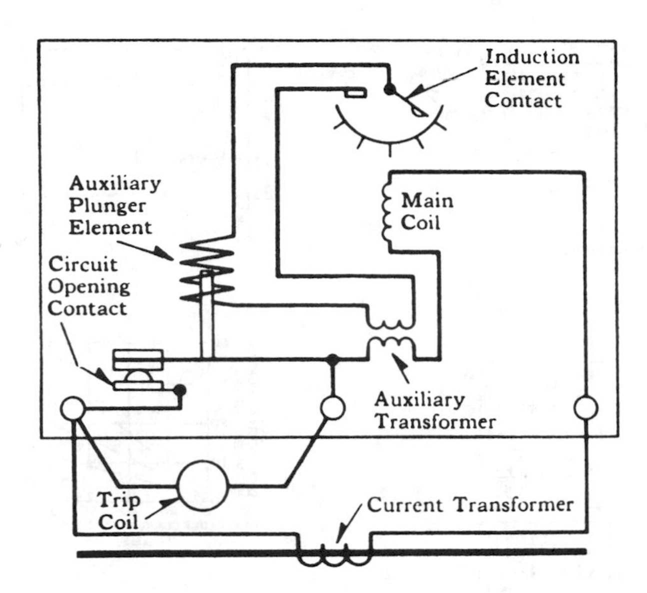

Figure 13-7 Arrangement of auxiliary circuit-opening contacts of induction relay.

DIRECTIONAL RELAYS

This type of protective relay operates on the principle of an abnormal direction of current flow. Both sensitivity and selectivity may be obtained by using the directional element, since a relatively low-current setting of the relay may be made in the direction opposite to normal. To guard against operation caused by transients that result in a momentary reversal of power, a sensitive instantaneous directional element is combined with an overcurrent element having a time characteristic (Fig. 13-9 on page 248).

All relays dependent on current alone for their operation function when sufficient current flows in their coil, regardless of the direction of the power flow in the line. To make such relays sensitive to the direction of power flow, a second element must be introduced. This second element operates essentially as a wattmeter that tends to turn the relay disc or cylinder in one direction when power flows normally and in the other direction when power flows in the opposite direction. The contacts of the overcurrent element are connected in series with the contacts of the wattmeter element. Both sets of contacts, therefore, have to be closed and the power flowing in a given direction before the relay will operate as a unit.

Since this relay has all the characteristics of an overcurrent relay, three conditions must exist before the tripping circuit is completed:

1. Excess current must be flowing
2. Power must be flowing in the opposite direction to its normal flow
3. The excess current must be flowing for a sufficient length of time.

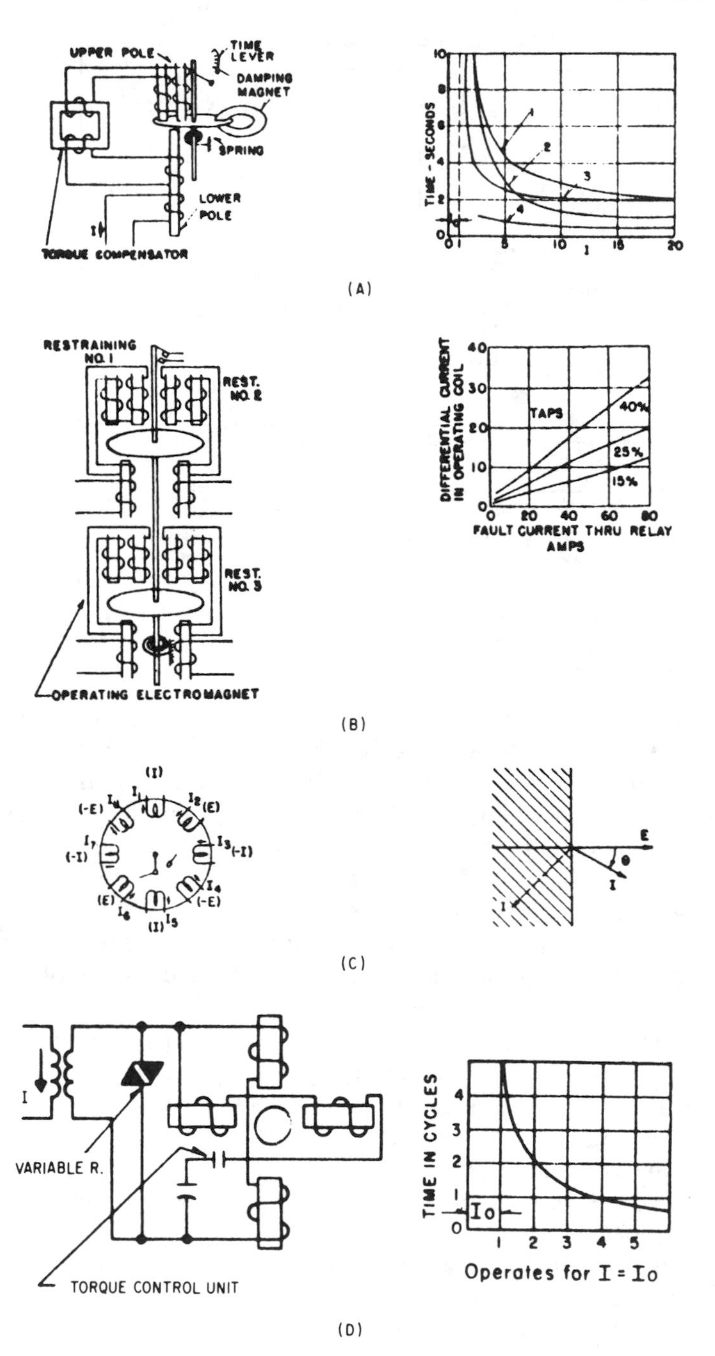

Figure 13-8 Induction type of relays: (A) Induction disk inverse-time unit, (B) multiple electromagnet or disk unit, (C) multiple-pole cylinder or disk unit, and (D) four-pole induction-cylinder unit. (*Courtesy*, Westinghouse Electric Co.)

Hence, even though the directional or wattmeter element may close its contacts upon the flowing of current in the reverse (from normal) direction, the tripping circuit will not be completed until both elements close their contacts.

In another type of directional overcurrent relay, only one set of contacts exist. Here, the two relay elements, the overcurrent and directional, are combined into one. In this induction type of relay, the magnetic field of the overcurrent coil is made to induce a current in the disc of the directional element. The tendency for the disc to rotate in one direction is balanced against a spring that also acts to restrain the action of the overcurrent coil. When a fault current in the opposite direction develops, the directional element turns the disc in the reverse direction; it is aided by the magnetic field of the overcurrent coil and the force of the spring, all joining to produce a rapid closing of contacts that actuates the associated breaker. This type of relay is often used in three-phase systems; only two discs are mounted on a single vertical shaft to which one set of contacts is mounted.

DIFFERENTIAL RELAYS

This type of protective relay operates on the difference between the currents entering and leaving the protected line or equipment. It utilizes the most positive selective principle.

With the differential principle, the current flowing into any part of an

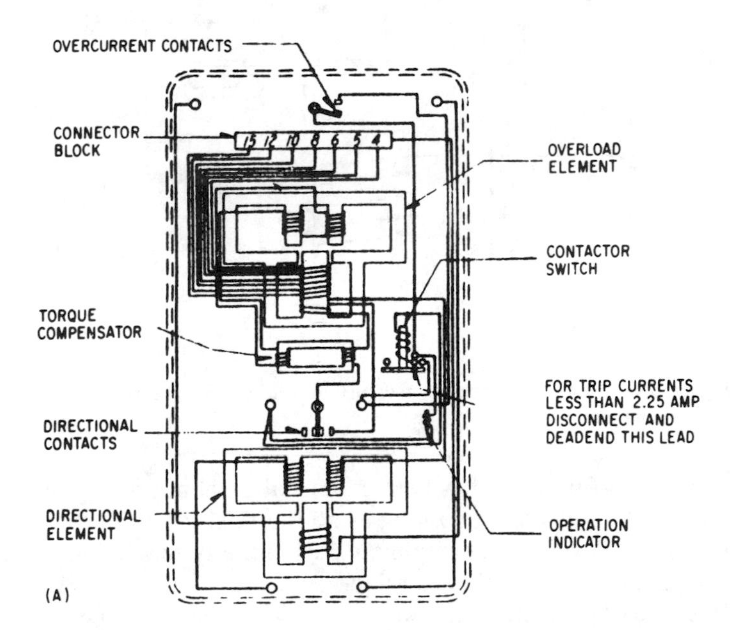

Figure 13-9 Directional overcurrent relays: (A) Rear view of internal wiring.

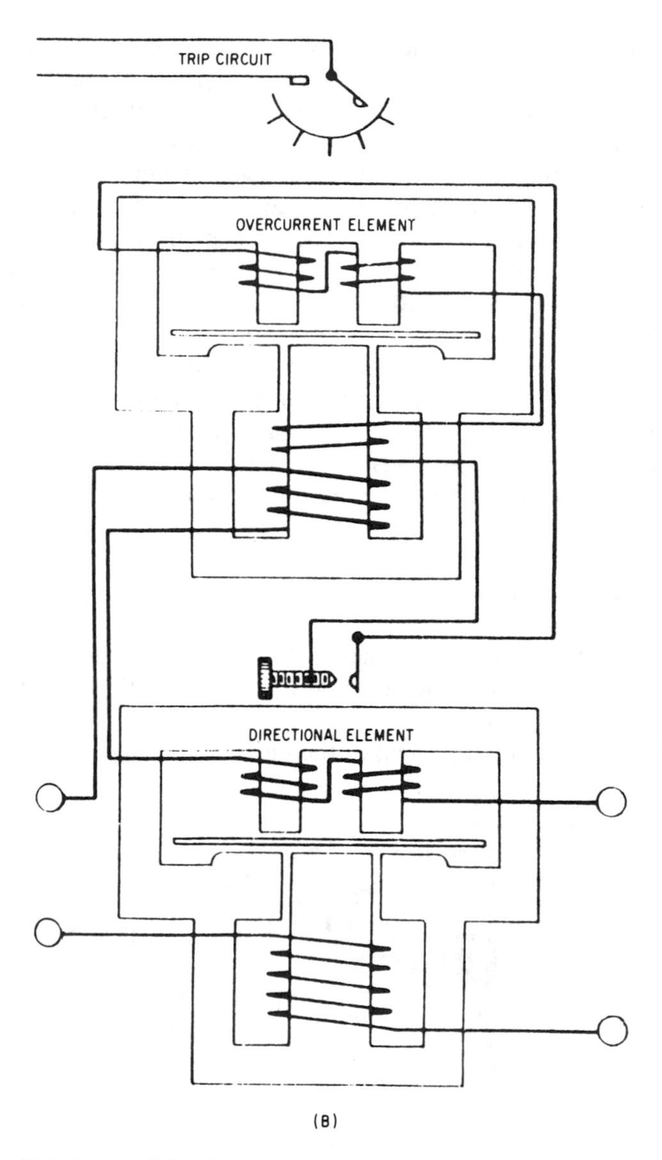

Figure 13-9 (cont.) Directional overcurrent relays: (B) overcurrent and directional elements.

electric system so protected, whether normal or abnormal, must equal the current flowing away from it so long as no fault exists within that part. Transient currents or power losses within the part of the system being protected are taken care of by the setting of the relay.

This type of protective relaying—usually used to protect equipment such as generators, buses, regulators, and motors—is perhaps most often used in disconnecting power transformers from an electrical system. Where there are internal faults, it operates by opening the circuit breakers on both the input and output sides of the transformer. Since this method of protection is obtained by connecting the elements of induction-type-overcurrent relays differentially, the relay operation depends on the difference in magnitude or direction of the currents supplied by current transformers on both sides of the power transformer (Fig. 13-10).

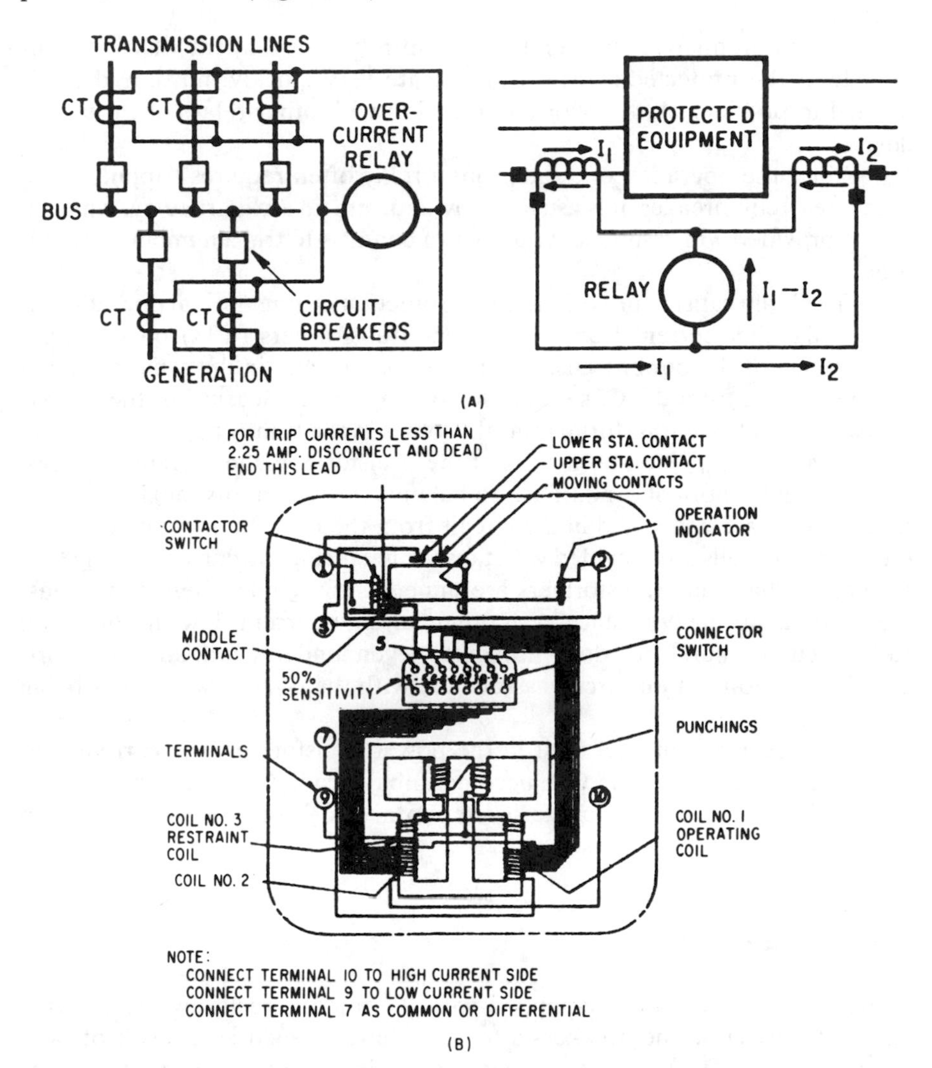

Figure 13-10 Differential relay protection: (A) Difference in current on either side of protected equipment is detected by current transformers and will not pass through the relay-actuating coil, $1_1 - 1_2$, if $1_1 = 1_2$; (B) wiring diagram showing internal connections.

Current transformers on each side of the piece of equipment being protected are connected to a differential relay. As long as the current flowing to the equipment through the one current transformer is equal to the current leaving the equipment through the other current transformer, the differential relay will not operate. A fault or short circuit between the two current transformers will result in more current flowing into the area of equipment being protected than out of that area, thereby causing the differential relay to operate.

As differential relay operation is entirely independent of conditions outside of the protected area, it can operate very quickly; no time delay is needed to provide selectivity or coordination with other relays or protective devices.

Since the operation of a differential relay often requires tripping more than one circuit breaker, it is usually provided with an auxiliary relay, which in turn is provided with a sufficient number of contacts to trip a number of circuit breakers.

To obtain satisfactory differential protection during normal operation, it is essential that current from the current transformers (CTs) on the high-voltage side of the power transformer be equal to, and in the same direction as, the current from the CTs on the low-voltage side. Because of the power-transformer ratio of transformation, the losses in the transformer, and the use of standard-ratio CTs on both sides of the transformer, the currents may not be equal under normal conditions. To balance these currents, auxiliary autotransformers are connected in the circuit from the CTs. In one type of relay, the operating coils are provided with taps for balancing the currents of the CTs so that auxiliary autotransformers are unnecessary. Under these conditions, the currents normally balance each other, and no current flows in the relay. Relay action, therefore, does not occur even under overload and short-circuit conditions on the circuits associated with the power transformer being protected.

If there is an internal fault in the power transformer, the currents between the circuits from the CTs become unbalanced. This difference of current flowing through the relay causes its contacts to close, completing the trip circuit of the circuit breakers.

RECLOSING RELAYS

Some circuit breakers are equipped with "automatic reclosing relays" that automatically close the breakers after they have opened as a result of fault conditions (Fig. 13-11). The reclosing operation takes place in accordance with a predetermined cycle. The first automatic reclosure usually takes place immediately after the fault causes automatic tripping. If the fault persists, the breaker opens automatically again and after a time interval recloses a second

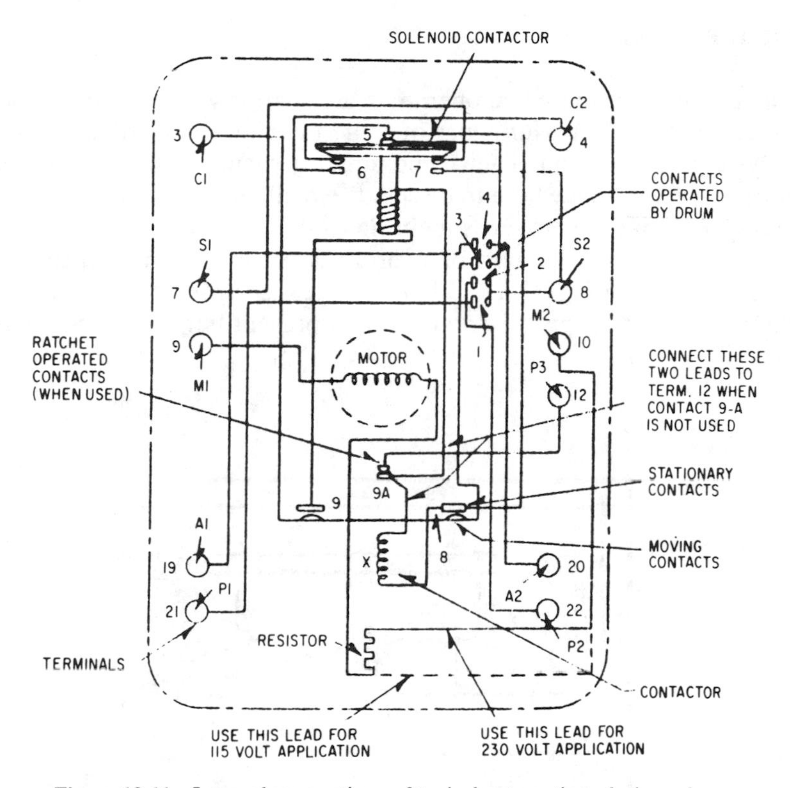

Figure 13-11 Internal connections of typical automatic reclosing relay.

time. The same cycle is repeated for a third reclosure after a time interval that is usually somewhat longer than that for the first two reclosures. If, upon the third reclosure, the fault persists, the breaker opens again automatically and remains open or "locked out." The purpose of this reclosing cycle of operations is to restore to service as quickly as possible any circuits that may be temporarily faulted because of lightning, wires temporarily crossed by tree limbs or wind, or other causes of a temporary or transient nature.

The reclosers consist of a small motor actuated by the initial tripping of the breaker. When the overcurrent relay contacts make contact, the tripping circuit of the breaker is completed and the breaker opens. At the same time, an auxiliary switch on the breaker starts a small motor that, through a train of gears, moves a disc or drum on which the contacts are mounted. As these contacts make contact, the reclosing circuits controlling the operation of the breaker are completed. If the fault persists, the overcurrent relay again trips the breaker open. After a cycle of such operations, a final contact is opened to de-energize the recloser motor.

Succeeding reclosures are usually spaced at increasing intervals of time; a duty cycle of 0–15–60-second intervals is common practice.

AUXILIARY RELAYS

Auxiliary relays are those that operate in association with equipment, devices, or other relays in the performance of some function or functions and that are usually actuated by them. They come in many shapes, forms, and types, and no attempt will be made here to discuss them all. Some of their functions will be shown schematically in subsequent figures.

An elementary diagram of a protective relay connected to a circuit breaker is shown in Fig. 13-12. When the breaker opens, the auxiliary switch of the breaker interrupts the trip circuit to prevent burning of the relay contacts.

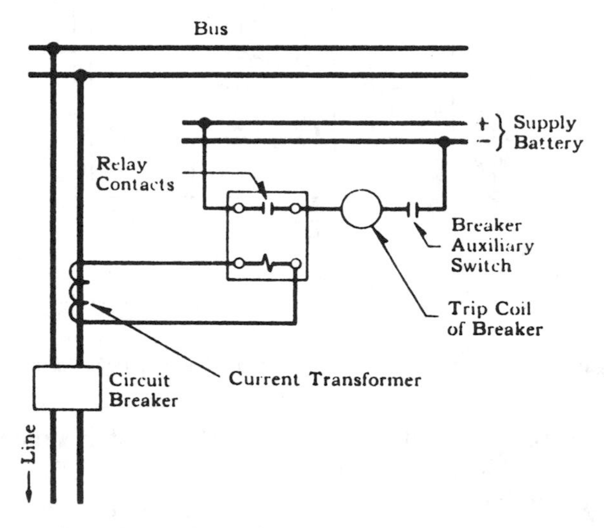

Figure 13-12 Protective relay and circuit breaker.

Large circuit breakers are usually electrically operated, being closed by a closing coil and opened by a trip coil that trips a latch. Breakers may also be operated manually by means of a control switch that closes the circuit both for opening and closing them. Together with indicating devices, all employ auxiliary switches of some kind, and the entire assembly is usually referred to as a control circuit. A typical dc control circuit, including a number of auxiliary switches, is shown in Fig. 13-13. The purposes of this assembly are as follows:

1. To open the trip circuit outside the relay since relay contacts usually cannot carry the heavy tripping current required.
2. To light the proper signal-indicating lights: green for open and red for closed.

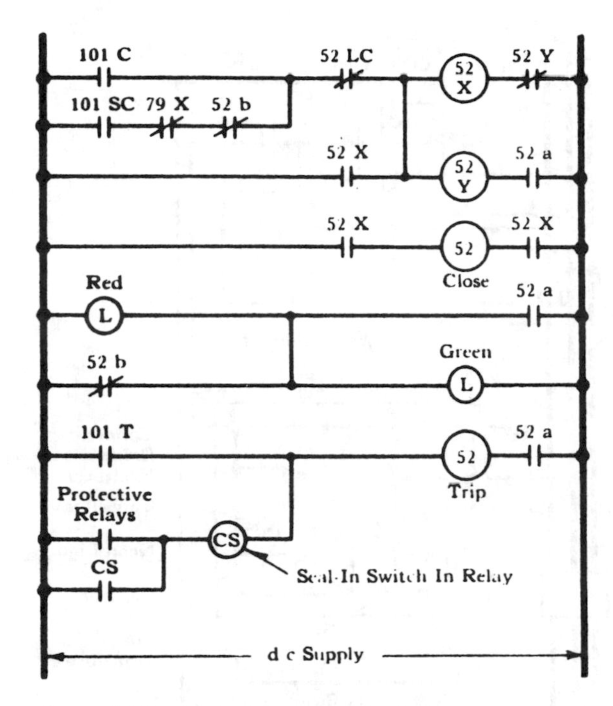

Figure 13-13 Typical circuit breaker in d-c control circuit.

3. To make the circuit trip-free so that it cannot be held closed on a fault; these sometimes energize electrical interlock circuits.

Since the supply of control power is usually dc from storage batteries, breakers may still be operated if the main ac supply sources are de-energized. These may be 6-, 12-, 24-, 48-, or 120-V systems, although the 24-, 48-, and 120-V systems are usually preferred because possible voltage drops across relays or from poor connections will still leave sufficient voltage to operate the relays, breakers, and auxiliaries.

If only small amounts of control power are required and a dc source is not economically justified, an ac source is sometimes used in conjunction with a capacitor, as shown in Fig. 13-14. A combination of a copper-oxide rectifier (to obtain a dc) in series with a capacitor is energized from a potential transformer. The charged capacitor furnishes current in place of the battery, but since it is quickly discharged, it must be recharged before it can be used to trip a breaker again. It will retain sufficient charge to trip a breaker for only some five or six seconds after the charging potential is removed.

In some circuits, control circuit switches are normally closed and initiate action when contacts are opened. More often, however, control circuit switches are normally de-energized and initiate action when contacts are closed.

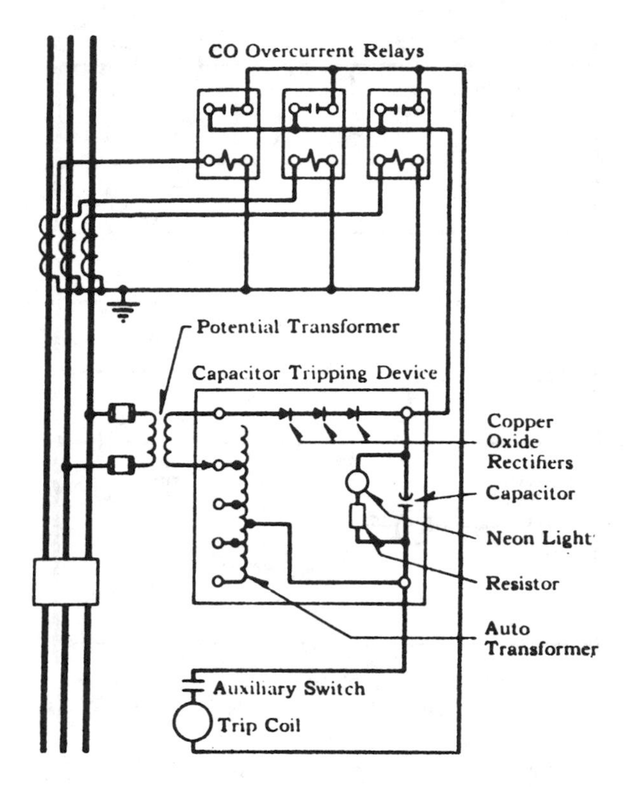

Figure 13-14 Capacitor as a source of tripping energy.

ELECTRONIC RELAYS

Electronically activated relays are essentially data processing systems programmed for specific purposes. As such, they perform the same functions in the electric power field as their counterparts of electromagnetic and induction type relays. As indicated earlier, relays are intermediate devices that perform the function of a switch controlling the operation of some other device, but does not itself open or close the main circuit that is to be controlled.

Electronic relays have many advantages over the electromagnetic and induction relays. They require no physical break in connections, they have no moving or wearing parts, they have a quicker response in the overall switching function, and they require less energy to operate. Further, greater accuracy in settings is obtainable than with the coil taps and spring adjustments of electromagnetic and induction relays, resulting in greater flexibility and selectivity in line and equipment protection systems.

Basically, electronic data processing systems work on a series of controlled electrical impulses. These impulses may be generated directly by moving contacts or initiated through magnetic impulses. These magnetic impulses are controlled through various schemes to deliver "on" and "off" signals that are coded for further initiation of a particular function. The signals need not be "on" and "off," but may be based on a difference in magnitude of the impulses, either large or small signals may be regarded as having "yes" or "no" connotations.

These signals are controlled at a tremendous rate of speed (billions per second) by means of electronically controlled "semiconductors", such as the silicon chips that may be found in radio and television sets, cellular telephones, and other similar applications. These devices have the faculty of allowing an electric current (or impulse) flow or stop flowing in extremely short periods of time; in short, they operate as very rapid switches in an electric circuit.

These impulses may be used singly or in combinations to initiate predetermined actions through the medium of a code, and to record (or store) data through a series of such impulse combinations, again through the medium of a code. This is done through a device known as a computer.

While the computer, strictly speaking, may be only one part of an entire data processing system, the term "computer" is often used to denote the entire system. The computer has five basic parts: input devices, storage or memory, processors, control and output devices; the relationship between them is shown in Figure 3-15A. These parts are interconnected so that intricate and involved operations may be performed as one continuous sequence.

In contrast, in the electromagnetic and induction relays (electromechanical devices) the control and processing mechanisms are designed to accomplish only one (or a limited number of) specialized functions (e.g., signal to operate a circuit breaker and the indicating lights, targets, and other associated safety and informational devices). A schematic diagram of this type processor, showing the relationship between the input, control and processor, and output, is shown in Figure-15B.

The processors in the processing system or computer in this case, the electronic relays, are sometimes referred to as microprocessors and are programmed to accomplish a particular function or functions. These programs include all of the functions described earlier for the electromagnetic and induction type relays.

(The marvelous powers of the computer do not stem from its ability to make the "yes" or no" decisions. Rather, it is man's ingenuity in transcrib-

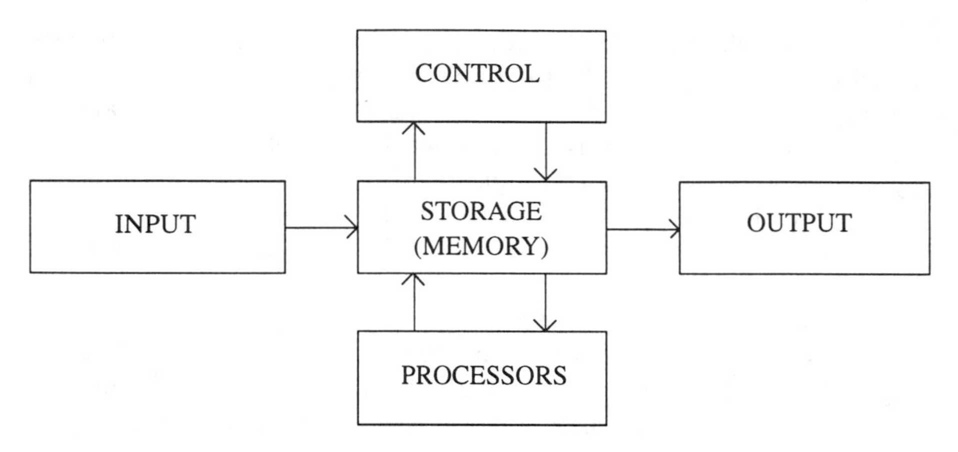

Figure 13-15A. Schematic diagram of a data processing system.

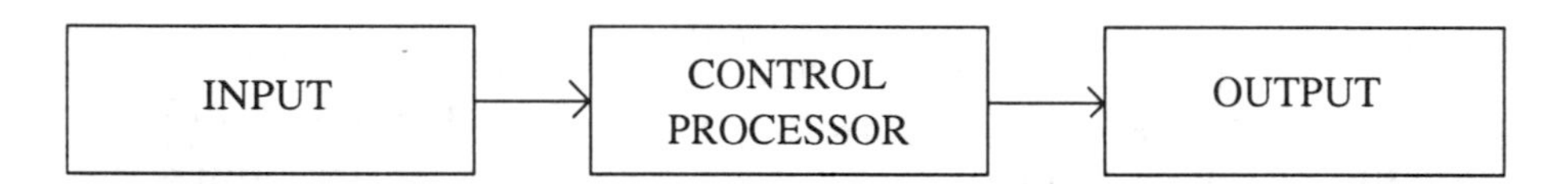

Figure 13-15B Schematic diagram of a single function data processing system.

ing his problems or functions into a series of such "yes" or "no" decisions. This transcription, broadly, is called a program. Typically, a program may encompass many thousands of such "yes" and "no" decisions, but the fantastic speed with which such decisions are made make the computer feasible. Further, once these programs are made up for a particular function, they are thenceforth available generally whenever such a function is again under consideration.)

As an example, an electronic relay having the same characteristics as its electromagnetic counterpart is outlined below. It will accommodate the same time-current selectivity, namely, instantaneous, inverse and definite time. A functional block diagram of the time-current electronic relay showing the relationship of its components is contained in Figure 3-16A. Time curves are selected by means of a rotary selection switch; the times and the codes for their selection are:

0 = Extremely Inverse
1 = Very Inverse
2 = Inverse
3 = Short time
4 = Definite Term
5 = Long Time, Extremely Inverse
6 = Long Time, Very Inverse
7 = Long Time, Inverse

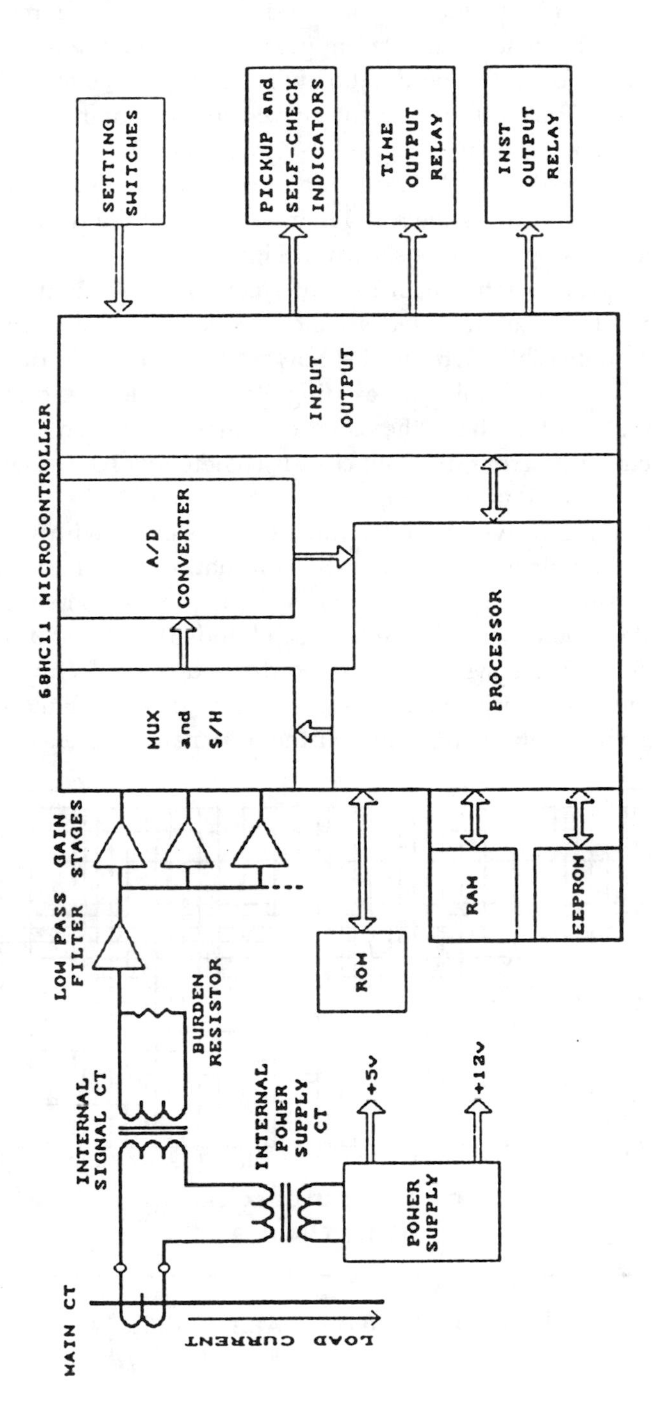

Figure 13-16A Block diagram of Micro-51 Time-Overcurrent Relay (*Courtesy* ABB Power T&D Co.)

Dial positions 8 and 9 are not used, and default to Extremely Inverse. Note the refinements available are greater than for the same overcurrent relay, making for better coordination with other protective devices that may be involved. Typical time-current curves for instantaneous, inverse, and definite time are shown in Figures 3-16B, C, and D.

A functional block diagram for a more complex illustration, a multiple shot reclosing relay, is shown in Figure 3-17A and the flow chart planning each step of its operation is shown in Figure 3-17B.

The great investment in present electromagnetic and induction type relays, and their excellent performance, make large scale replacement economically unfeasible. Retrofit kits, however, are available that permit electronic relays to be installed in existing electromagnetic and induction type relay facilities—they have the same terminal connections and contact arrangement, and permit drawout chassis (where involved) to be placed into existing cases without rewiring.

Electronic relays for other functions, together with miniaturization technology, have been utilized in the automation of elements of the generating, transmission and distribution systems. In their operation, they are often called upon to coordinate (and be included) with the protective relays associated with the system being coordinated. The electronic relays with their greater flexibility and available memory banks facilitate such automation while insuring the reliability of the automated systems.

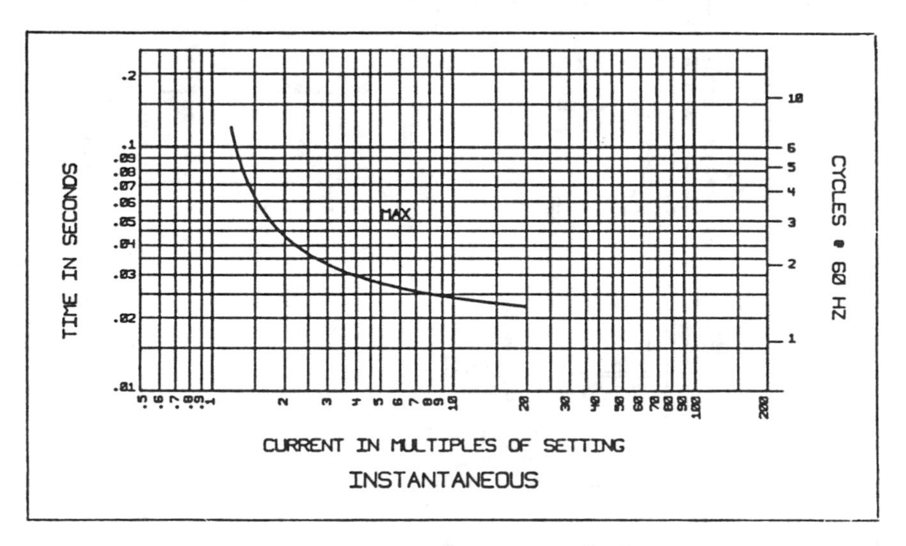

Figure 13-16B (*Courtesy* ABB Power T&D Co.)

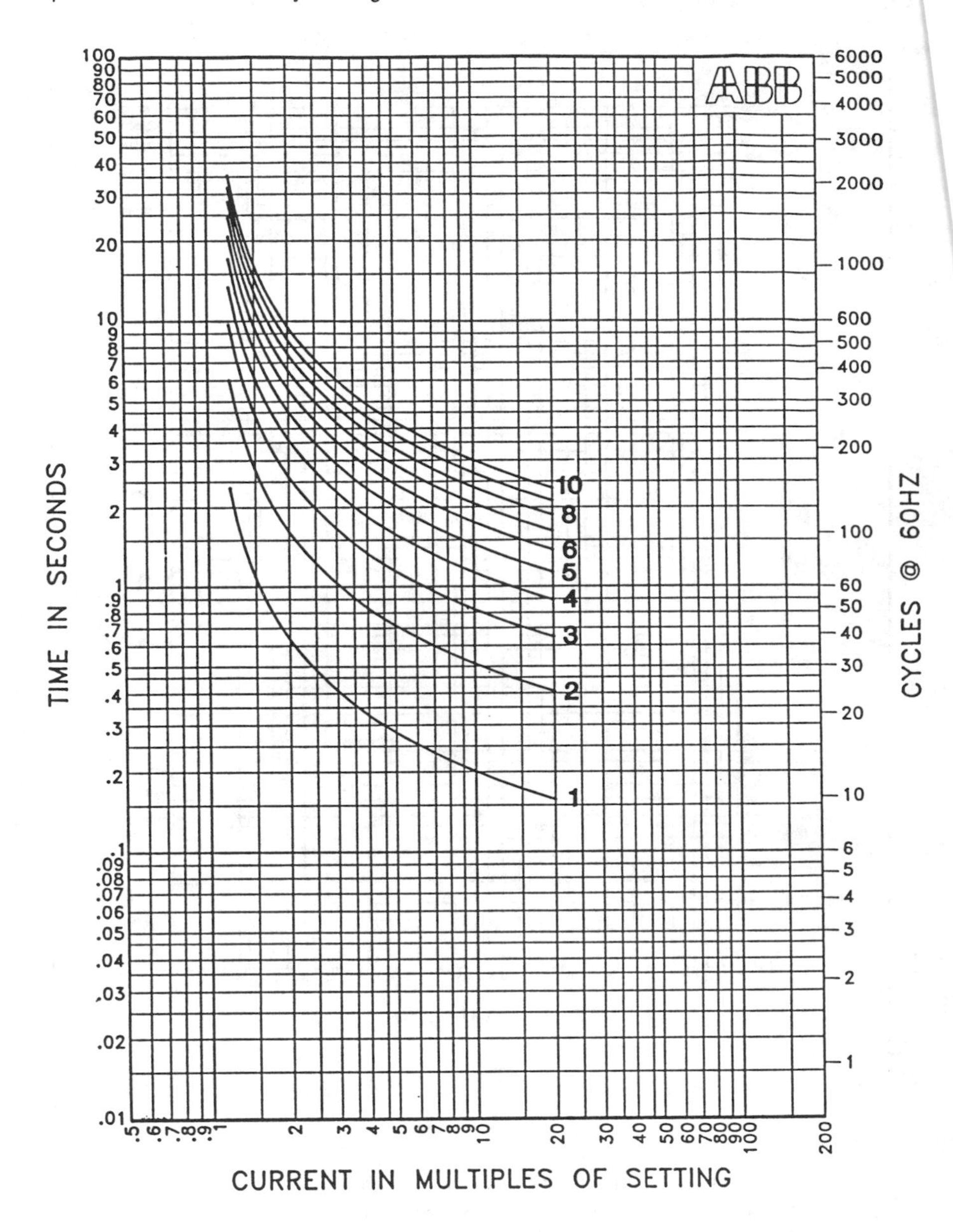

INVERSE

Figure 13-15C Inverse: time curve selector position #2. Time delay as shown. Long-time-inverse: time-curve selector position #7. Multiply time delay shown by 10. (*Courtesy* ABB Power T&D Co.)

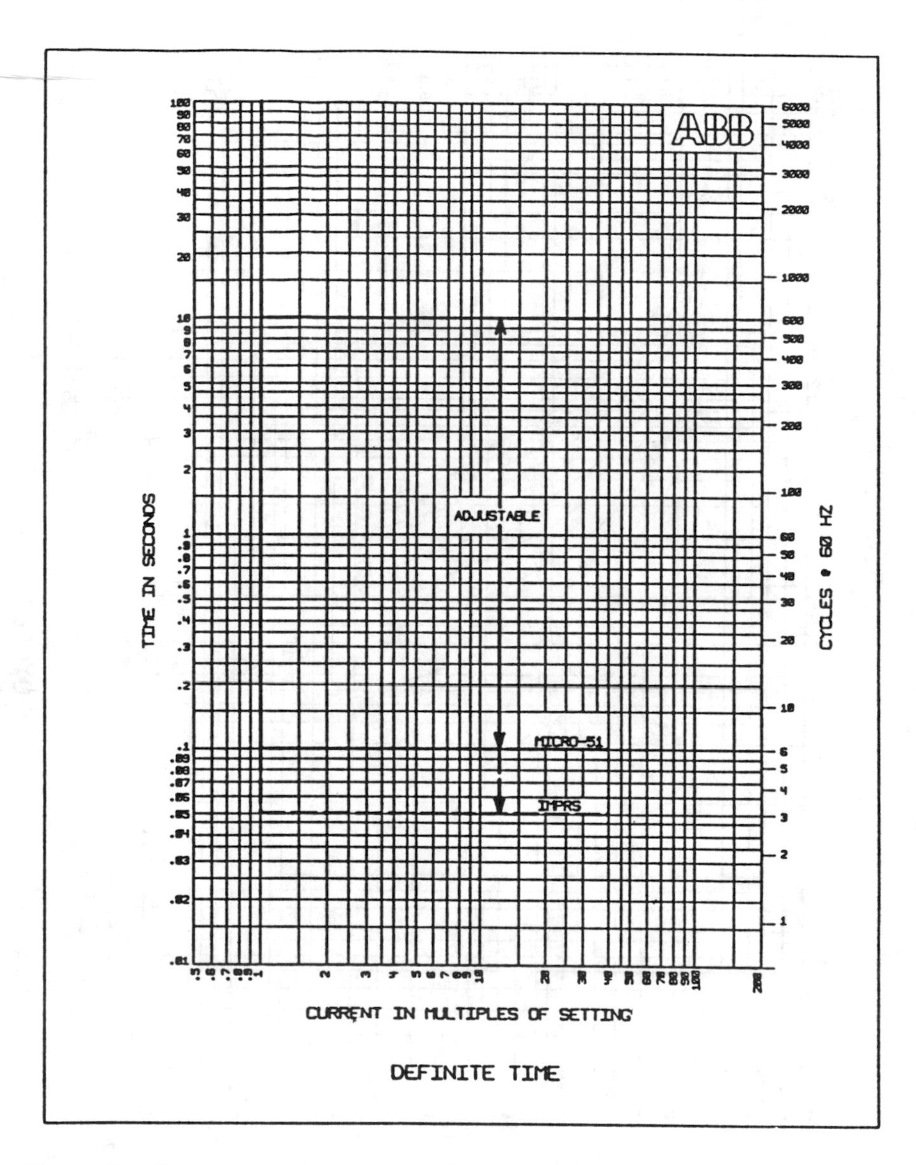

Figure 13-15D Definite time time-curve selector position #4. Set the delay directly in seconds on the time-dial selector switches. Range 0.1-10 seconds in 0.1-second increments. (A setting of 0 defaults to 0.1 seconds.) (*Courtesy* ABB Power T&D Co.)

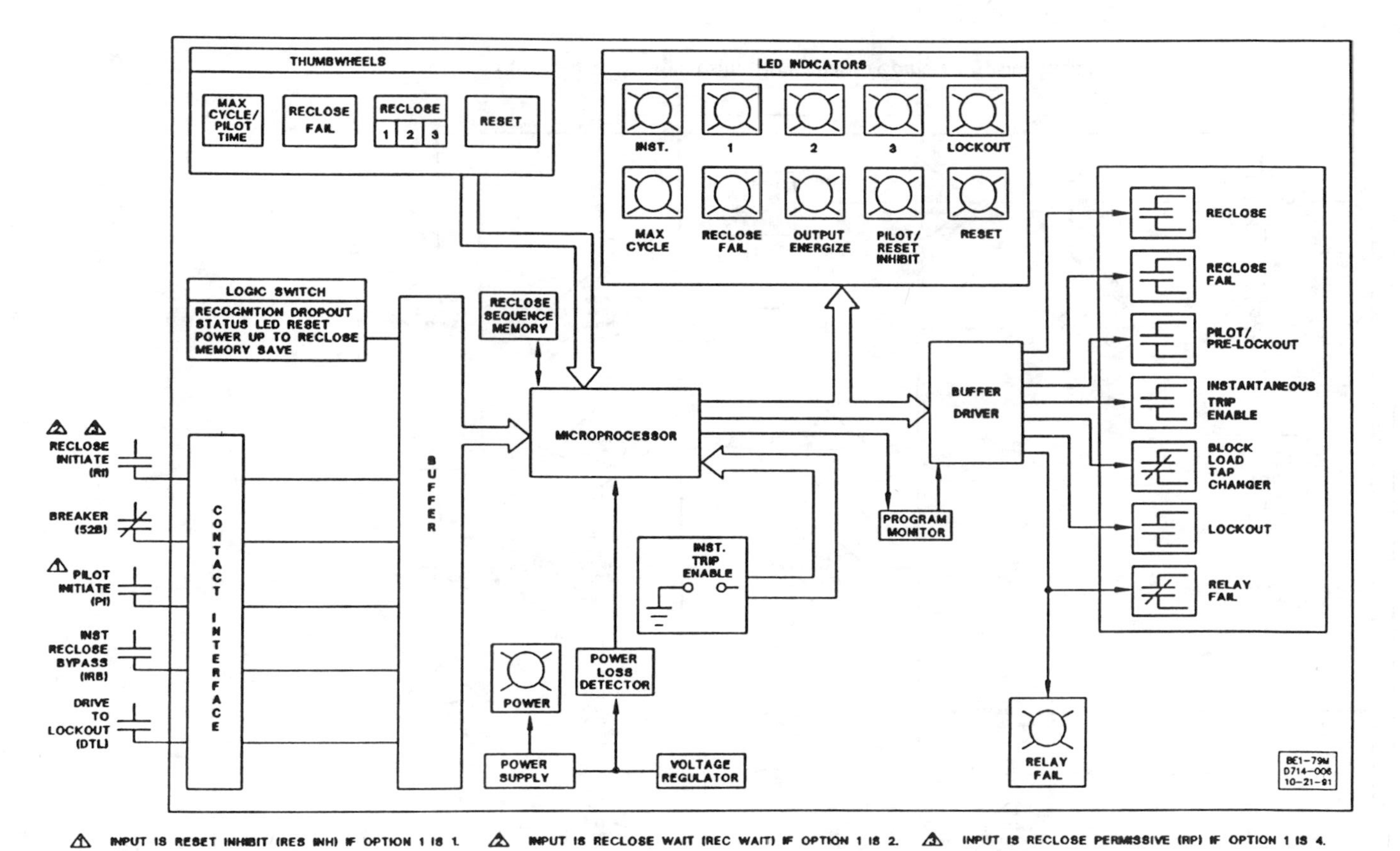

Figure 13-17A Functional block diagram. (*Courtesy* Basler Electric Co.)

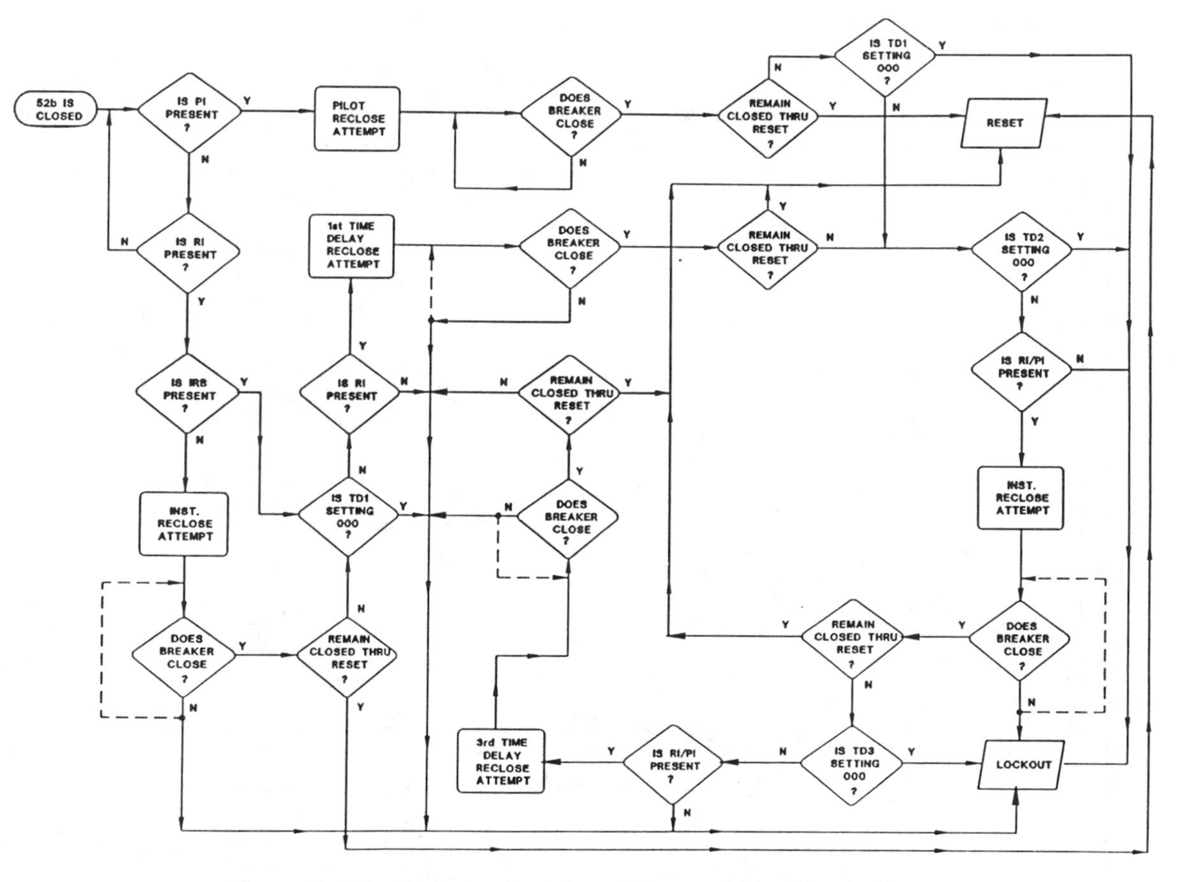

Figure 13-17B Reclosing flow chart. (*Courtesy* Basler Electric Co.)

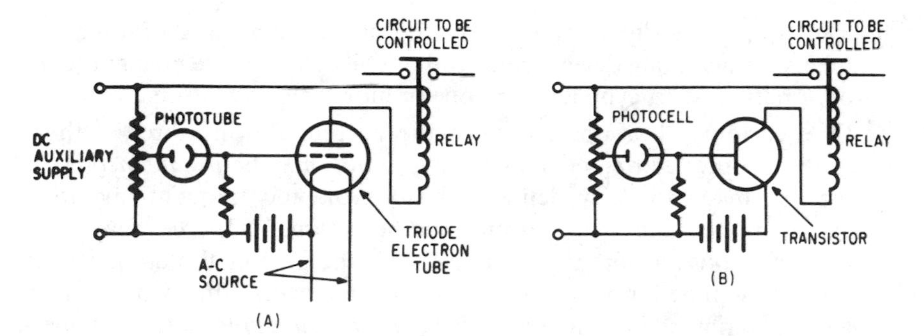

Figure 13-18 Basic photoelectric electronic relay circuit.

TRANSDUCERS

Transducers are devices that convert nonelectric quantities into electric ones, and vice versa. Among them are the thermostats and thermocouples that actuate relays to control heating or protect equipment from heat. Microphones, moreover, convert sound to electric current, which is then employed to make recordings as well as to actuate signals or to sound alarms. Photocells convert light to electricity to control lighting circuits (Fig. 13-18). Other devices convert pressure or pressure differences to actuate relays that open or close valves and give indications or alarms.

REVIEW

1. Protective relays are low-powered electric devices used to activate high-powered devices (Fig. 13-1). Placed between the circuit or equipment and associated circuit breakers, they reflect a small but definite proportion of the power in the circuit or equipment; they sense and analyze abnormal conditions, causing circuit breakers to operate to relieve or protect the circuit or equipment.
2. The relays may sense abnormal values of current, voltage, frequency, temperature, pressure, speed, as well as be sensitive to sound, light, and other indicators. Actions taken may be positive (starting), negative (stopping), balancing (a combination of the two), or informative (reporting the status of a specified condition).
3. Three fundamental requirements include accuracy, sensitivity and selectivity. In general, they do this by measuring the magnitude, duration and direction of the electric currents flowing through them. These currents are usually received from current transformers, potential transformers, or both.

4. Electromagnetically operated relays are usually composed of three elements: an actuating device (relay coil or coils), a movable element (core or armature, disc, cylinder), and one or more sets of contacts.
5. Over-current relays operate when the current flowing through them exceeds a predetermined value. Their action may be *instantaneous,* or may be made to *delay* deliberately to avoid too frequent operation caused by transient nonpersistent conditions (Fig. 13-3). They may operate on an *inverse-time* principle in which the greater the amount, the shorter the time for operation. The delay and inverse-time features may also be combined into an *inverse definite-minimum time* method of operation.
6. Directional type relays respond to abnormalities in the direction of the current flow. These are often combined with an overcurrent element to guard against momentary transient reversals of power (Fig. 13-9).
7. *Differential* relays operate on the difference between currents entering and leaving the protected line or equipment (Fig. 13-10).
8. Other type relays are used for automatically reclosing circuit breakers after abnormal conditions no longer exist (Fig. 13-11) and for miscellaneous auxiliary purposes.
9. Electronic or solid-state type relays may accomplish the same purposes as the electromagnetic or induction types, and are generally faster in their operation (Figs. 13-15 and 13-16).
10. *Transducers* are devices that convert nonelectric quantities into electric ones, and vice versa (Fig. 13-18).

STUDY QUESTIONS

1. What is the function of a protective relay?
2. What three things must a protective relay be capable of doing?
3. What is meant by normal conditions in a relaying system?
4. What three factors concerning electric currents are utilized in the design and application of protective relays?
5. What are three essential elements of a relay?
6. What are the more common types of protective relays? How do they operate?
7. What two general methods of operation are employed in the design and construction of protective relays?
8. Why and how are overcurrent relays modified to increase their sensitivity and selectivity?
9. What are reclosing relays? How do they function?
10. What are auxiliary relays? Where and how are they employed?

14

PROTECTIVE RELAYS: OPERATION AND MAINTENANCE

The operation of relays (and their associated circuit breakers) is meant to prevent or limit damage during faults or overloads and to minimize their effect on the remainder of an electric system. When the great variety of possible faults and types of relays is considered, it is soon seen that any attempt to coordinate different relays to encompass all possibilities becomes very impractical.

In the most popular three-phase ac systems, faults include single-phase-to-ground, phase-to-phase, phase-to-phase-to-ground, three-phase-to-ground, and fault resistance differences from very high to very low. There is also an almost infinite variation in distance of the fault from the relays. The system may be a grounded one or not, and it may be delta- or wye-connected. Other differences combine to produce a host of considerations that must be taken into account in the coordinating process.

When these circuit variations are set against the many variations in the types and characteristics of relays as well as those of other protective devices (fuses, reclosers, lightning arresters), it is soon realized that the practical coordination of all these factors must involve a series of compromises and approximations, and coordination becomes more of an art than a science.

ZONES OF PROTECTION

Nevertheless, electric systems can be adequately protected if small parts of the system can be disconnected. To do so, the system is divided into protective zones, which are then subdivided further as need or desirability dictates. As a first division, the electric system can be divided logically into three parts:

1. Sources of generation and their outgoing circuits (likened to a manufacturing plant)
2. A subsource or station handling relatively large amounts of power (likened to a wholesale warehouse)
3. Local sources utilizing equipment and circuits for distribution to the ultimate users (likened to retail stores serving local areas).

Each part, moreover, can be further subdivided into various zones of protection whose requirements can be considered more or less independently. In practice, however, coordination can best be achieved by analyzing conditions in reverse order. First the local distribution should be analyzed, then the intermediate subtransmission-substation area, and last the generation with its associated high-voltage transmission system.

DISTRIBUTION CIRCUITS

Distribution circuits are most frequently operated as *radial circuits,* that is, they are supplied from only one point. As shown in Fig. 14-1, each transformer is fused on the high-voltage side. The circuit is sectionalized with fuses and reclosers that provide points for clearing faults without interrupting the entire circuit.

The time-current characteristics of the overcurrent relays associated with the circuit breakers in this type of circuit must coordinate with those of the reclosers and fuses. Several characteristic curve shapes are available for different induction-type overcurrent relays, as previously indicated in Fig. 13-3. These provide some margin in the selection of the relay settings that coordinate best with the characteristic curves of the reclosers and fuses for the current values involved. The curves are shown in Fig. 14-2.

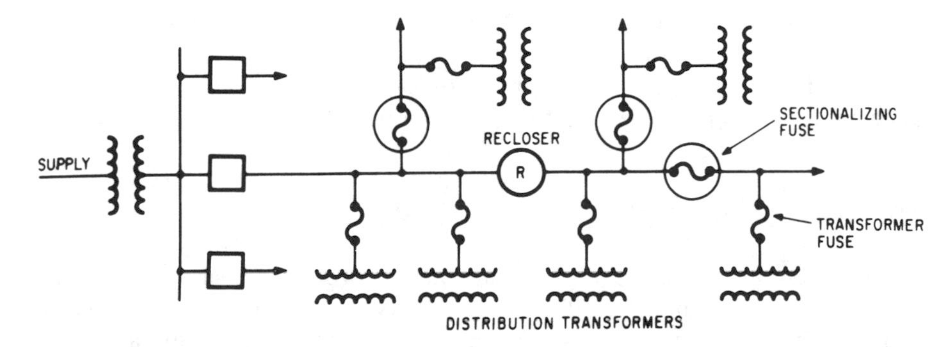

Figure 14-1 Radial distribution protection zone.

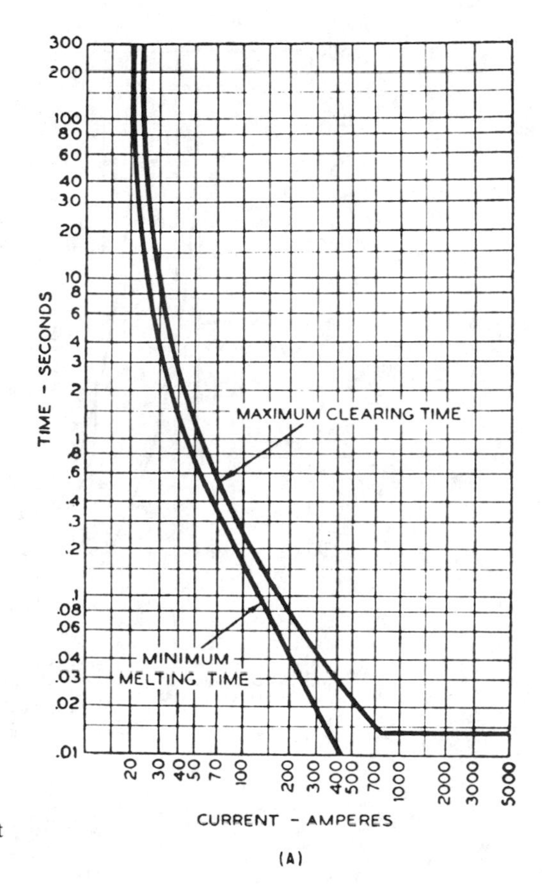

Figure 14-2 (A) Typical time-current curves for link fuse.

The general plan for coordinating relays on a local distribution system is shown in Fig. 14-3. Some typical time settings are given to illustrate the gradual increase in minimum settings for relays as they approach the source. The characteristics of various induction types of overcurrent relays, each of which may be further modified by imposing a definite minimum time setting, are shown in Fig. 14-4. By choosing the proper minimum time settings, it becomes possible for several of these relays (and their breakers) to be in series and still operate properly in event of a fault anywhere in the circuit. Several relays in series can be set for the same time to handle faults immediately beyond them. Thus set, they can still provide the requisite quarter second or more margin for faults beyond the next relay because of the smaller fault current that will flow for a fault in that direction.

Subtransmission–Substation

The intermediate system of electric supply—usually consisting of several incoming high-voltage transmission lines and moderate-voltage subtrans-

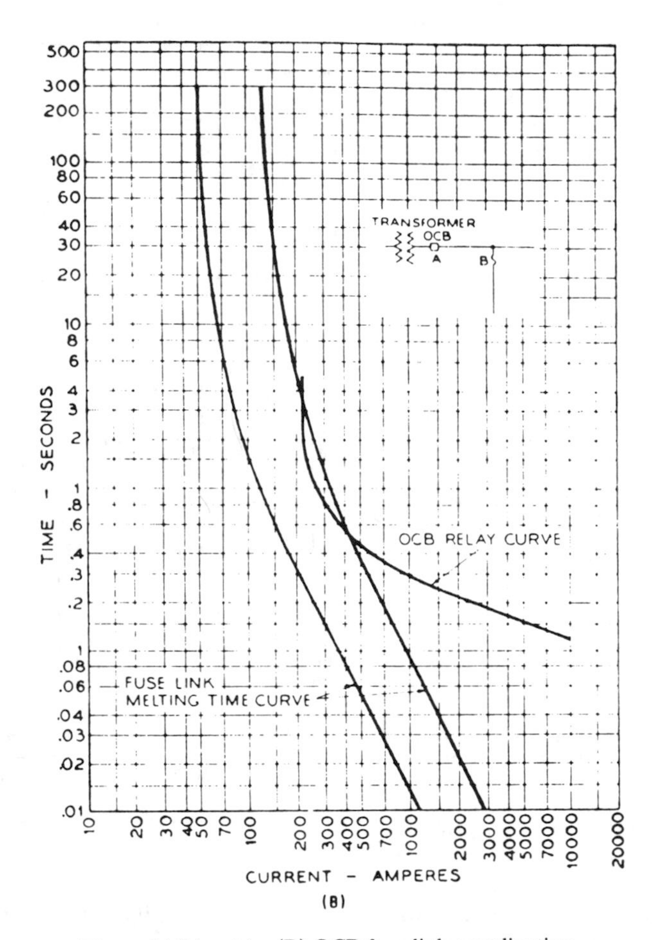

Figure 14-2 (cont.) (B) OCB fuse-link coordination.

mission lines—is shown in Fig. 14-5A, which also illustrates zones of protection. Differential relaying is used for the protection of incoming breakers and buses. This type of protection depends on differences between incoming and outgoing currents in the breakers and buses and is not affected by faults outside these zones.

The breakers, however, must operate to clear faults outside these zones and must therefore coordinate with other protective devices on the circuits connected to them. Coordination here becomes more difficult as fault currents may flow to the point of fault from two or more directions.

One way of attacking this problem is to assume that all but a single source are disconnected and that the circuit resembles a loop. Protection can be provided by a combination of overcurrent and directional-overcurrent relays, as shown in Fig. 14-5B. This process is repeated for each source and relay

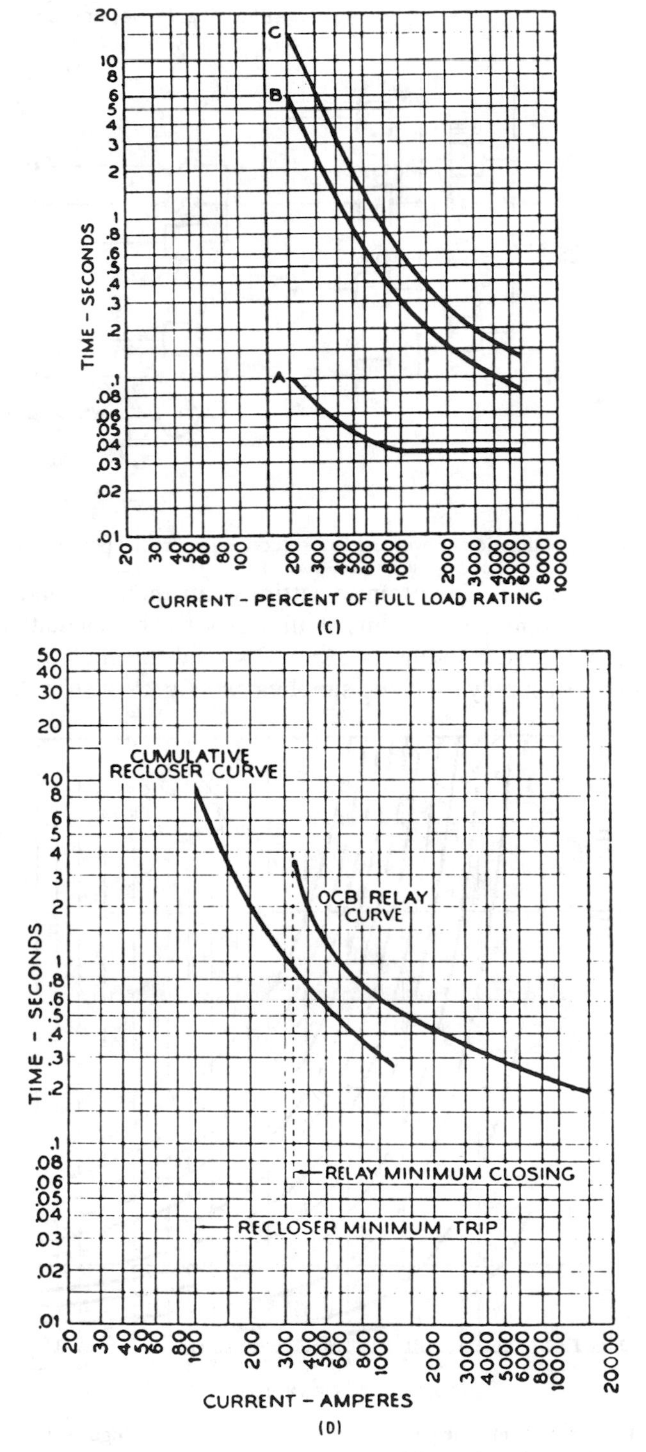

Figure 14-2 (cont.) (C) Series trip T-C curves of three-phase heavy-duty reclosers, and (D) coordination of OCB relay and recloser. (*Courtesy*, McGraw Edison Co.)

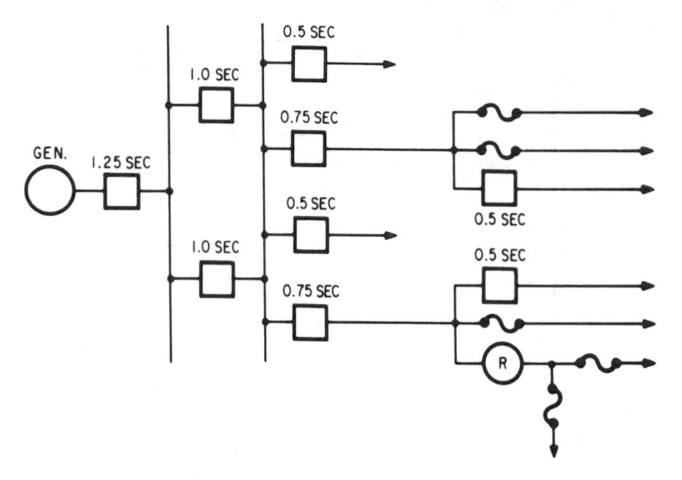

Figure 14-3 Coodination of overcurrent protection on a radial power system.

combination determined. Compromise settings can then be made after the maximum and minimum fault currents at all locations are calculated.

In the assumed system of Fig. 14-5B, a 23 kV subtransmission system has two 69 kV sources of supply itself and supplies two 7.6 kV distribution substa-

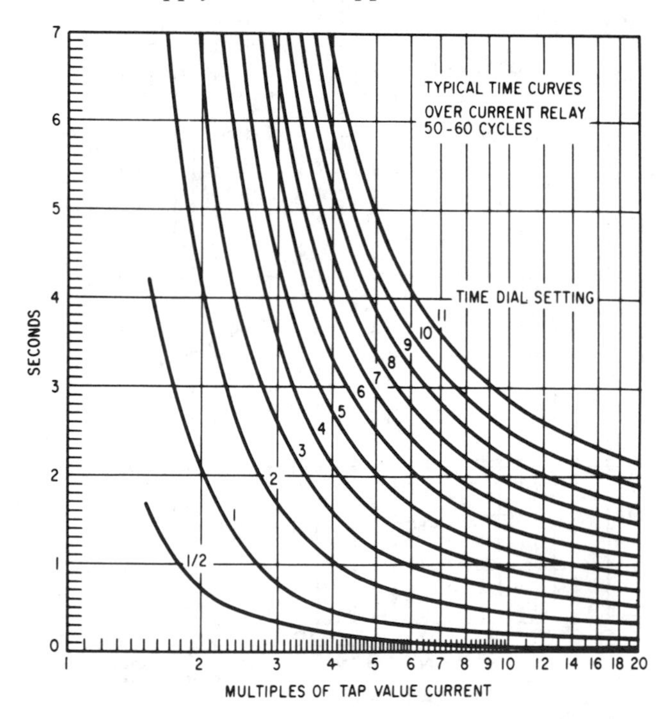

Figure 14-4 Overcurrent time-current curves. (*Courtesy*, Westinghouse Electric Co.)

tions. The upper and lower portions are assumed to be identical in electrical characteristics. Fault locations, indicated by an "x," are identified by a circled number. Alongside the number appear the maximum (on the left) and the minimum (on the right) fault currents; all currents are in amperes at 23 kV. The current distributions and directions for each fault location are also specified. For example, the current in line AB for fault number 2 is 1000 amperes maximum and 750 amperes minimum; there is no contribution of fault current from the 7.6-kV distribution system.

Where both overcurrent and directional-overcurrent relays (or elements in the relays) exist, the overcurrent setting is made high enough so that the directional element will normally close first, except for faults at or very near the breaker. In the latter case, the overcurrent relay will operate to open the breaker almost immediately. This, however, is a simplified explanation.

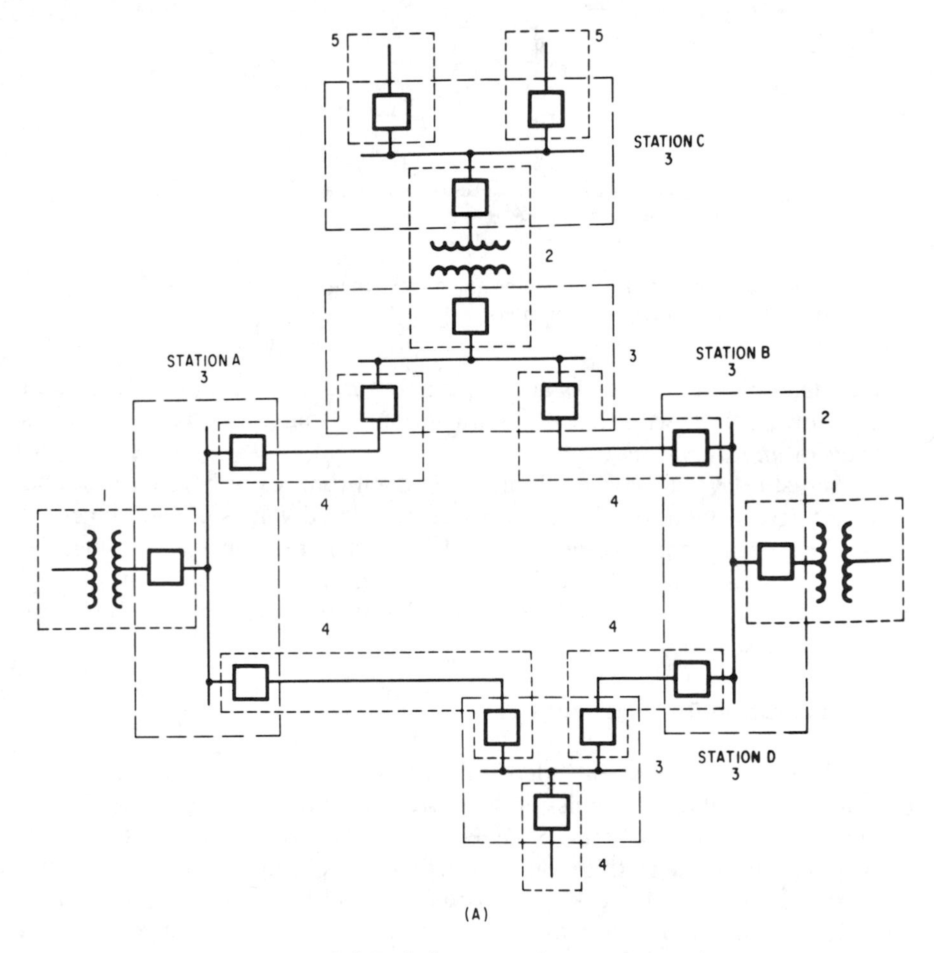

Figure 14-5 (A) Typical system and zones of protection.

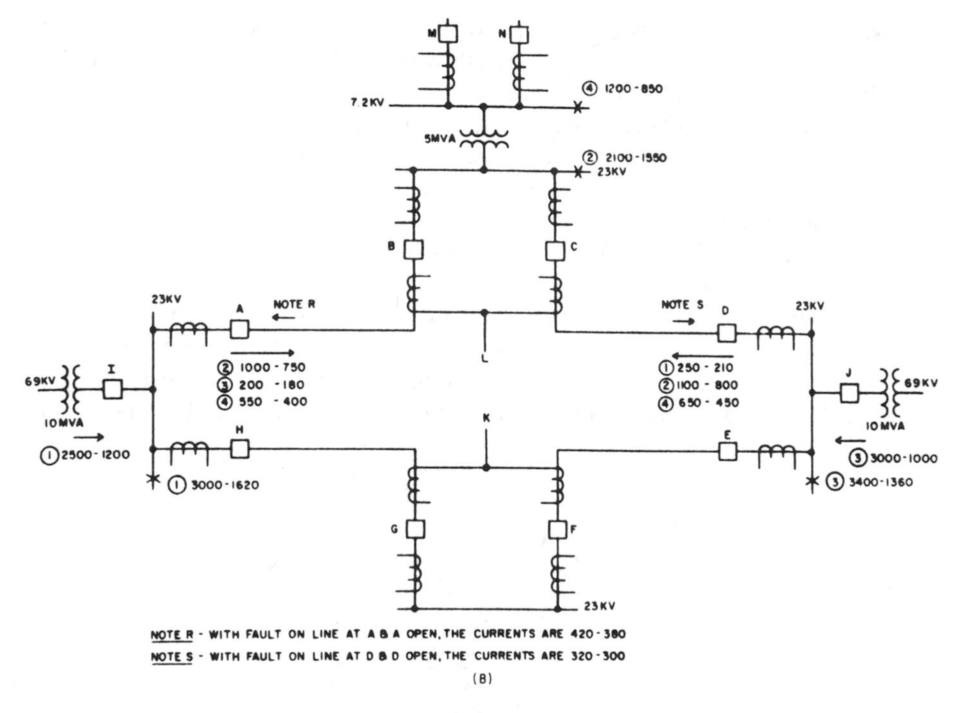

Figure 14-5 (cont.) (B) protection provided by overcurrent and directional overcurrent relays. (*Courtesy*, Westinghouse Electric Co.)

Actual determination of relays and settings is a more complex operation and may involve relays with special characteristics, sometimes referred to as *distance* or *impedance relays.*

Actual relay settings are obtained by compromises of both current and time settings, as mentioned earlier. Since calculated values of fault currents approximate relay coil settings specified in increments and time settings on inexact spring tensions, relay settings are often arrived at by cut-and-try methods. Several combinations should be tried out on paper or on computers before a final selection is made.

Generation–Transmission

The protective zones included in the relaying schemes for generators and outgoing high-voltage transmission lines are shown in Fig. 14-6. As indicated in the diagram, the protection of generators, transformers, and transmission lines is accomplished, in this case, by differential relaying associated with the generators, buses, and breakers. Coordination with the relaying associated with the transmission lines may be accomplished by directional-overcurrent

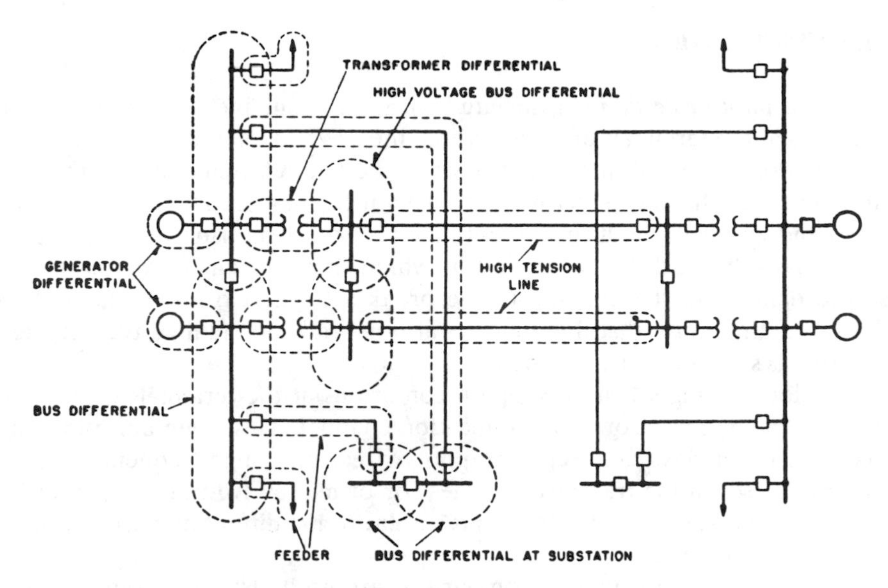

Figure 14-6 Protective zones for generators and outgoing high-voltage transmission lines. (*Courtesy*, Westinghouse Electric Co.)

relaying, somewhat similar to the coordination achieved by the subtransmission–substation schemes described above. Radial feeders supplied from these sources require only overcurrent relaying.

In schemes utilizing differential relaying where protective zones overlap, the protection of the overlapped equipment is accommodated by the cross-connecting of its associated current transformers (Fig. 14-7). In addition to their protective function in the schemes described above, the circuit breakers are also used for switching of circuits under normal conditions.

Whatever protective scheme is selected, the incorrect tripping of circuit breakers not directly affected by a fault can result in unforeseen, widespread interruption. Failure to trip, however, is also serious and more apt to occur, and backup protection has to be relied upon to clear a fault, usually resulting in interruptions to more than one section. When any relay fails to function, the next relay to operate should be the one that is immediately in series toward the source of supply.

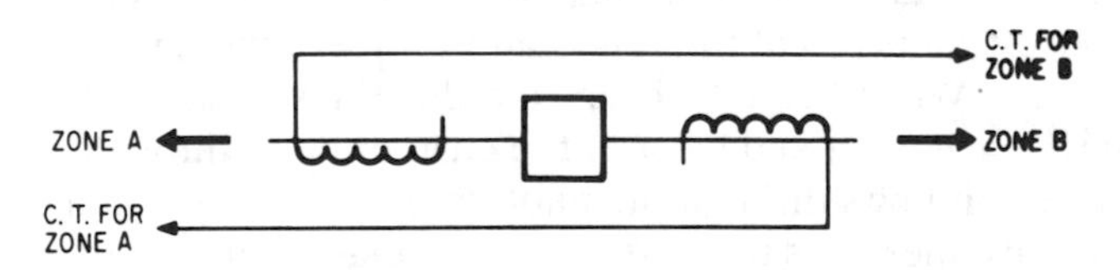

Figure 14-7 Principle of overlapping protection around a circuit breaker.

PILOT PROTECTION

So-called *pilot protection* is associated with the simultaneous tripping of the breakers at the terminals of transmission lines. When a fault occurs on such a line near one terminal, the greater part of the fault current will flow through the breaker of the nearest terminal, causing it to operate first and making the other end supply the fault current until the breaker there trips somewhat later. This relatively much longer period in which fault current flows may cause severe damage to the line and to the breaker furthest from the fault. It is desirable, therefore, that the breakers at both ends of a faulted transmission line trip as simultaneously as possible.

Pilot-relaying schemes compare conditions at the terminals of a transmission line by means of a communication channel. This communication link may consist of physically separate pilot wires, of a high-frequency carrier superimposed on the transmission lines, or of microwave radio communication. In many ways, pilot protection is similar to the differential protection of buses and equipment.

Two general methods of comparison are used. The first compares the magnitude and direction of current flow only; the second compares the direction of power flow.

Pilot Wire

There are several methods of employing pilot wires to trip the circuit breakers at each end of a transmission line simultaneously. Two schemes employing overcurrent relays at each end are shown in Figs. 14-8A and 14-8B. A scheme using differential relays at one end is shown in Fig. 14-8C. All of these employ from three to six pilot wires. Another scheme employing only two pilot wires, polyphase directional relays at each end, and a dc source is shown in Fig. 14-8D.

The most popular method, however, employs only two wires, an ac source, and special relays that combine the currents in each of the current transformers for any type of fault into a single-phase voltage that is compared to a similar quantity from the opposite end of the line. A simplified circuit is shown in Fig. 14-9.

For through-current conditions, the output voltages from the two special relays are in series and, by way of the pilot wires, circulate most of the current through the restraint coils of the relays and only a minimum of current in the operating coils. When a fault occurs on the transmission line, the output voltages buck each other and most of the current flows through the operating coils; the rest of it flows through the pilot wires and the restraint coils. Under these conditions, the relays trip and open the breakers at both ends of the line.

Beside the advantages of only two wires and an ac source, this scheme has the added advantage of tripping the breakers should the pilot wires be-

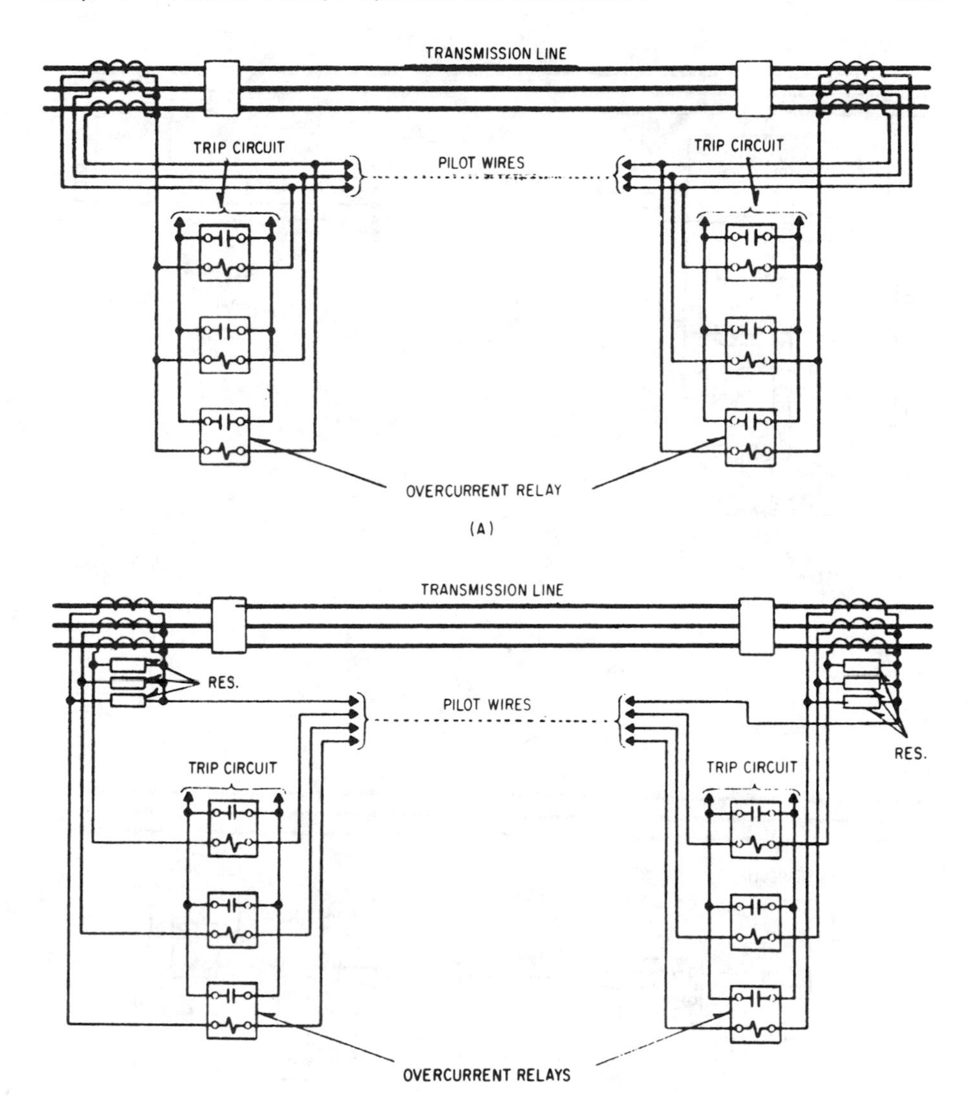

Figure 14-8 Pilot wire schemes: (A) Circulating-current scheme with load currents and through fault currents circulating over pilot wires; (B) balanced-voltage scheme with load currents and through fault currents producing equal opposing voltages at line terminals;

come faulted, that is, the system "fails safe." One of the disadvantages of pilot wires is that they may be adversely affected by the extraneous voltages and currents induced in them, especially when they are on the same right-of-way as the power circuit. To avoid this situation, pilot circuits sometimes make use of leased telephone lines and extend over routes far removed from the trans-

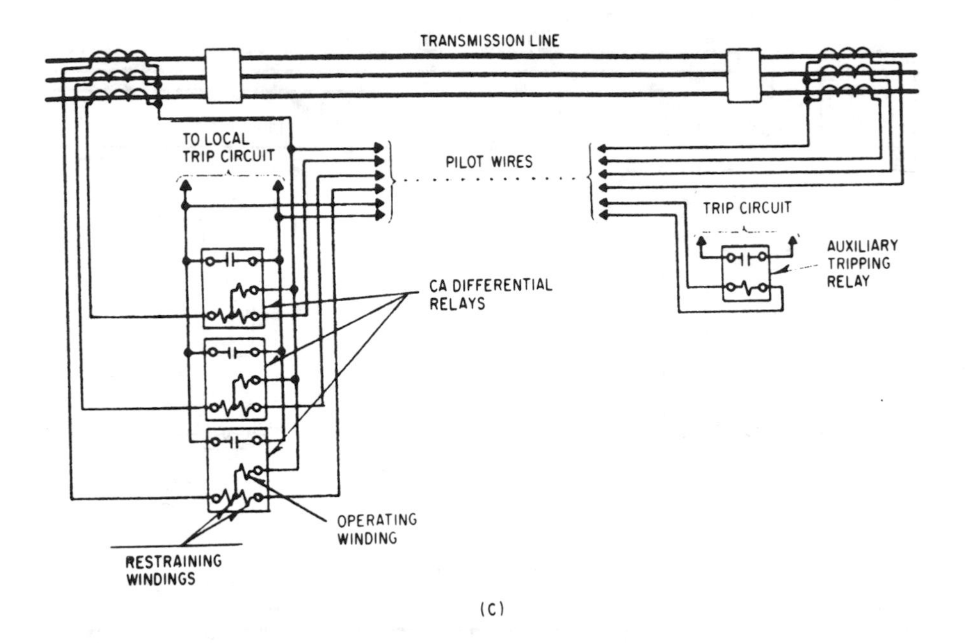

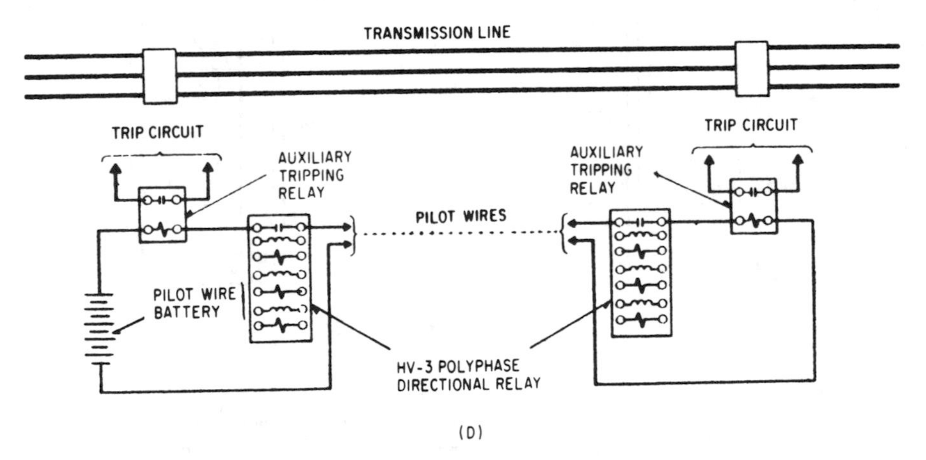

Figure 14-8 (cont.) (C) scheme using CA percentage differential relays; (D) directional comparison scheme using direct current over a pair of wires (a-c connection omitted for simplicity). (*Courtesy*, Westinghouse Electric Co.)

mission lines. The leasing adds to the cost of such systems. The greatest disadvantage, however, is the practical limit—some ten miles—for the positive and effective operation of such systems that is imposed by the burden limitations of the associated current transformers.

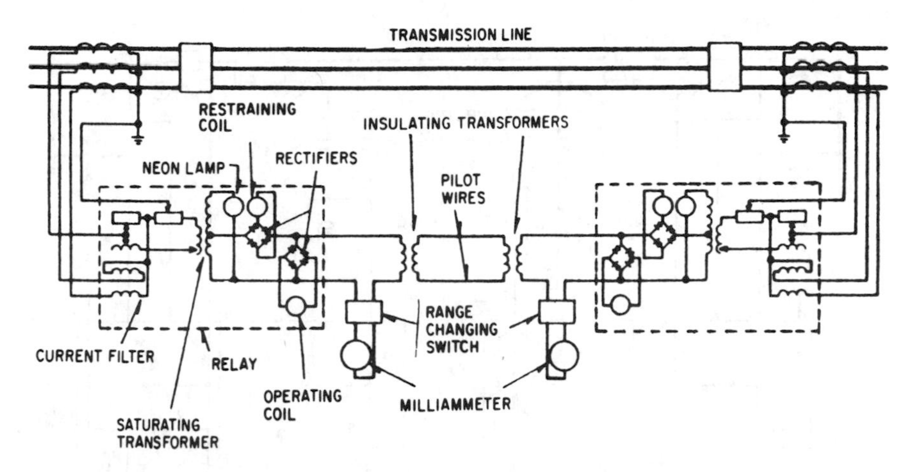

Figure 14-9 Alternating-current pilot wire scheme using HCB relays. (*Courtesy*, Westinghouse Electric Co.)

Carrier Pilot Relaying

In this type of pilot relaying, a high-frequency current is superimposed on a transmission line. The frequency employed, however, is confined to the lines and does not act as a radio broadcaster. A simplified diagram of an installation of one terminal is shown in Fig. 14-10. Power directional relays are installed at both ends. In a simplified explanation of operation, the carrier signal normally tends to block the operation of the relays. When the line is faulted, the carrier signal is interrupted and the relays operate to trip the breakers. Similarly, if the carrier unit fails for any reason, the breakers will open and the system fails safe.

The principal advantage of this type of pilot relaying is its ability to function positively and effectively over several hundred miles. Operation at frequencies from 50 to 150 kilocycles per second results in freedom from local inductive interference. Its first cost and operating expense are economically more favorable than the pilot wire schemes and are practically unaffected by the length of the lines. Moreover, the carrier channel can be used for other purposes, for example, for telemetering and supervisory controls that operate on coded impulses.

To transmit several functions simultaneously over the same carrier channel, the carrier is audio modulated with two or more frequencies, each of which is controlled by one of the functions (Fig. 14-11). Tone receivers (incorporating filters) are used at the receiving end to separate the incoming quantities. With this method, some ten separate quantities may be transmitted over a single-carrier channel. Although reception may not always be satisfactory, the carrier channel may also be used for telephonic communication for other line operations as well as for other nonrelated business.

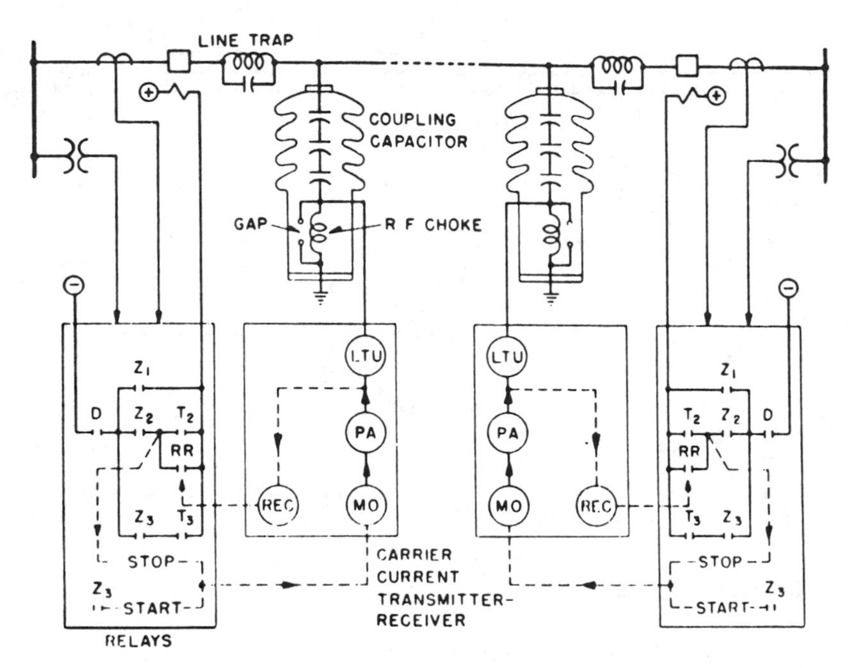

Figure 14-10 Carrier pilot relaying system (dotted lines indicate carrier controls; MO = master oscillator; REC = receiver; PA = power amplifier; LTU = line tuning unit). (*Courtesy*, Westinghouse Electric Co.)

Microwave Relaying

In microwave relaying, the pilot systems of relaying are transmitted over microwave radio channels without significant modification in the modes of operation. This system is not subjected to line faults. As in the carrier system, the microwave channel may accommodate several separate functions in addition to line protection. Because the transmission distances of the microwave

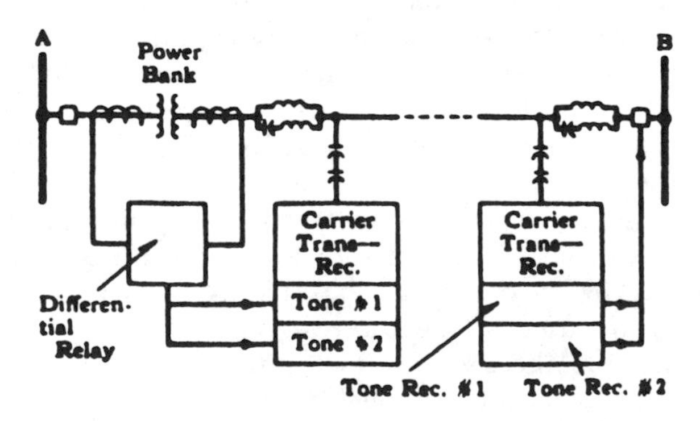

Figure 14-11 Basic circuit elements for remote tripping of a circuit breaker over a tone-modulated carrier pilot channel. (*Courtesy*, Westinghouse Electric Co.)

units are much affected by line-of-sight limitations, several units may be required for long transmission lines. The intermediate stations receive their signals and retransmit them to the next unit, referred to as *relay stations.* These relay the radio signals and are not to be confused with other relay functions connected with the power lines.

GROUND RELAYS

Another scheme measures the flow of so-called *ground current* and employs overcurrent relays. In a three-phase circuit or piece of equipment, the currents flowing in each of the conductors are usually fairly well balanced in magnitude. When wye-connected, the neutral or ground conductor carries little or no current. By measuring this current directly, or by measuring each conductor and determining the difference, this unbalance or ground current can be made to actuate relays whenever it exceeds certain predetermined values. Figure 14-12 shows a relay using ground-fault detection and illustrates

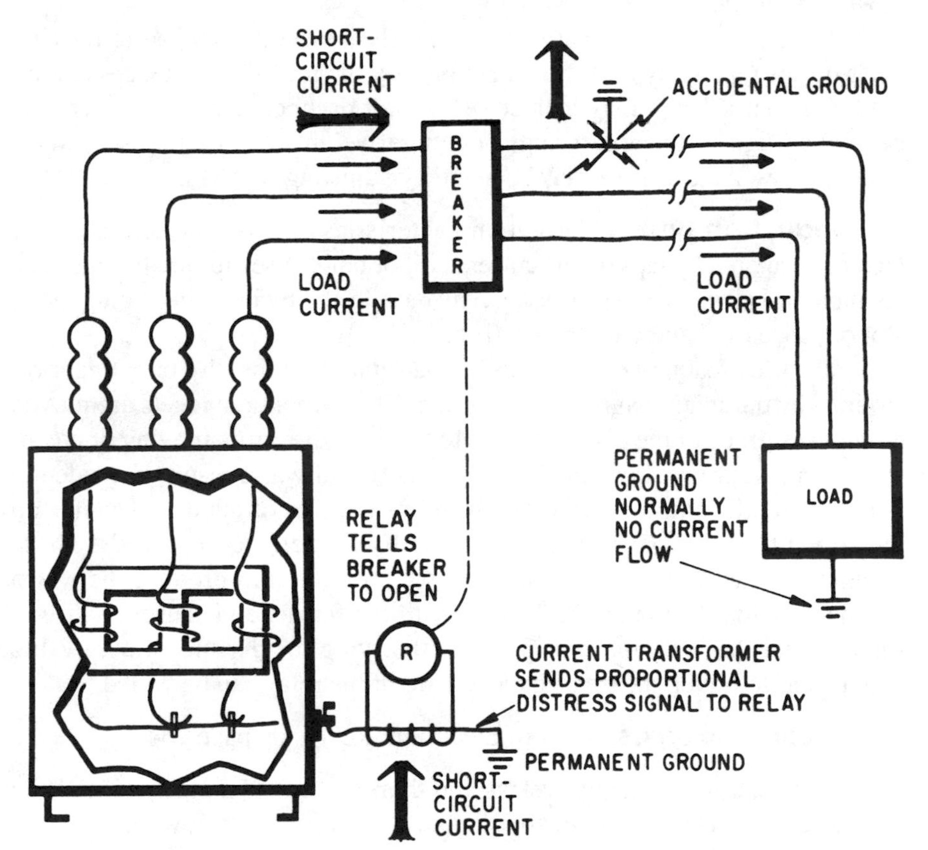

Figure 14-12 Ground relaying.

the ability of this protection scheme to discriminate between load current and fault current.

RELAY MAINTENANCE

Protective relays have two characteristics that determine their performance:

1 Current magnitude in the relay coils
2. Time to close the contacts.

Relay Setting

Electromechanical types. In these types, the current characteristic of the relay coil may be regulated by taps on the current coil or by the distance the plunger, clapper, or armature must move. In general, the greater the distance these must move, the greater the current required to pull them into closing position.

The time feature is controlled not only by the distances the moving parts of this type relay must travel but also by the restraining devices associated with them. These may include bellows or dashpots for the plunger type of relays (Fig. 14-13) and springs or counterweights for the clapper and armature types. Adjustments may be made by adjusting screws or nuts.

Induction types. Current characteristics of this type of relay may also be regulated by taps on the current coil or coils. The taps are brought to a terminal board where the proper tap may be selected by inserting a plug into a slot or jack associated with the tap (Fig. 14-14).

The time feature here depends on the rotation of the disc or cylinder restrained by damping magnets and a spring. The damping magnets are moved along the surface of the disc and serve to keep the disc from moving or creeping when no current flows in the relay coils (usually as a result of incidental magnetic fields in its vicinity). The speed of rotation of the disc depends on the strength of the magnetic fields set up by the relay coil or coils. These magnetic fields set up currents in the disc, which in turn produce their own magnetic fields. The latter fields react with the former and cause the disc to rotate under the restraint of a spring. By varying the distance the movable contact has to travel, the time characteristic of the relay can be varied.

Electronic types. Refer to Electronic Relays on page 254.

Inspection. Protective relays and their associated auxiliaries must be periodically checked for proper operation, chiefly for the following:

1. The assembly must be free of dirt, dust, and other foreign matter.
2. The contacts must not be burned, pitted, or corroded.

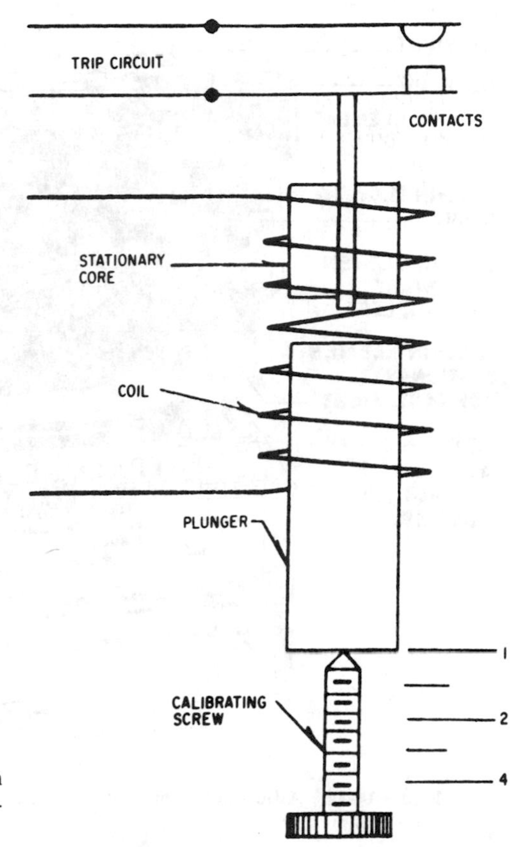

Figure 14-13 Plunger type of relay with coil wound for operation on either current or voltage.

3. The contacts must be properly aligned and correctly spaced.
4. The contact springs must be in good condition, flexible, and show no signs of corrosion.
5. The contact moving parts must travel freely and function in a satisfactory manner. The solenoids of relays of the plunger type must be free from obstruction. Bearings must be properly lubricated. Dashpot oil must be replaced if necessary.
6. The connections to the relay must be tight.
7. The insulation on wires and coils must not be frayed or worn.
8. The relay assembly must be securely mounted.
9. The coil must show no signs of overheating.
10. The relay must not be abnormally noisy.

Contacts. Contacts constitute the major element requiring maintenance. The mechanical action of the relays should be checked to make certain that moving and stationary contacts make positive contact and are directly in

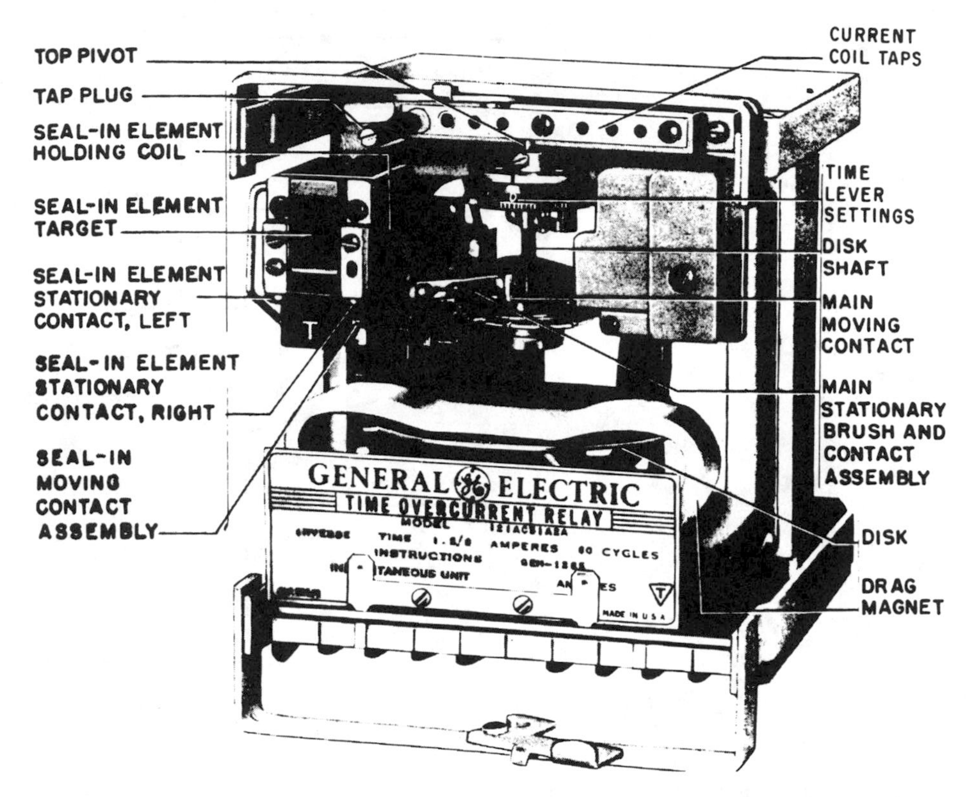

Figure 14-14 Adjustments for induction type of relays: Current coil taps and time lever setting.

line with one another. The contacts must have the required "wipe" and "gap" for the particular relay.

Relay contacts are usually of two kinds, hard-surface or soft-surface types. Hard-surface contacts are made of various alloys. Soft-surface contacts are either made of solid silver or plated. Improper cleaning of silver-plated contacts will soon remove the plating. All silver contacts need special attention since the soft metal will wear away at an excessive rate if carelessly cleaned.

Relay contacts have various shapes, depending on their size and application (Fig. 14-15). Sometimes, both contacts are flat; other times, one contact is convex and its mate flat. The original shape of a contact must be retained during its cleaning and dressing. If burning or pitting has distorted the contact, it must be reshaped and restored to its original shape or replaced.

Hard alloy contacts may be cleaned by drawing a strip of thin, clean cloth or paper between them while they are held together (Fig. 14-16). In some cases, the cloth may be moistened with a cleaning fluid. They may be polished with a dry cloth or paper strip. If hard contacts are corroded, burned, or

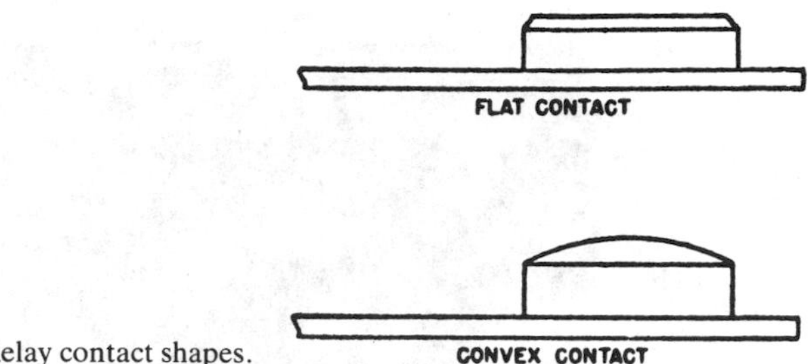

Figure 14-15 Relay contact shapes.

pitted, they must be cleaned with a burnishing tool, making sure that the original shape is retained.

Solid silver or silver-plated contacts may be cleaned with a cloth or brush dipped in cleaning fluid, then polished with a dry cloth. The dark discoloration found on silver and silver-plated relay contacts is silver oxide, which is a good conductor and should be left alone unless the contacts need to be cleaned or dressed for other reasons; it may be removed with a cloth moistened in cleaning fluid. Deeply pitted contacts may be reshaped very carefully with a fine file or fine sandpaper. Very badly burned or pitted contacts should be replaced.

Contact arms should be adjusted to the correct positions if they have been moved during cleaning, taking care that the settings of the relays are not changed.

Other items. If an ac relay is excessively noisy, it may be due to some loose connection or from a mechanical misalignment of the armature or

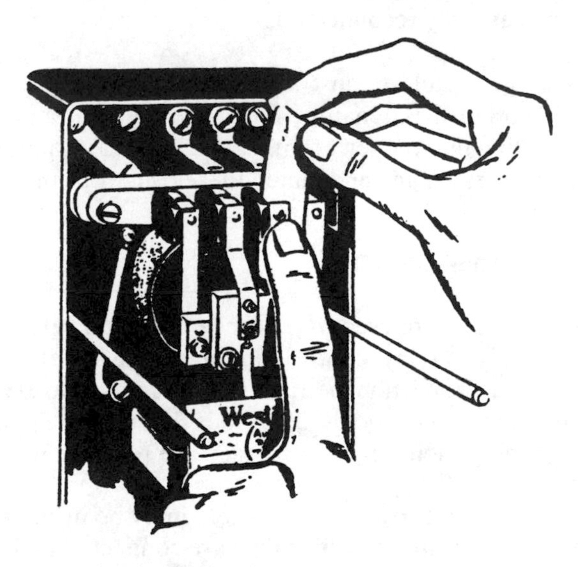

Figure 14-16 Cleaning relay contacts.

plunger. All loose connections and mounting screws, bolts, and nuts should be tightened. The armature or plunger mechanism should move freely without binding or dragging. If alignment cannot be made, the coil or armature or plunger should be replaced.

Some relays are replaced in the field and brought to the shop or laboratory for maintenance. When a relay is removed for any reason, extreme care should be taken in tagging and identifying both its terminals and the associated connecting leads. When the relay is reinstalled, corresponding leads and terminals should be carefully reconnected.

Relay testing. Relays must be checked periodically, and usually in conjunction with other maintenance, for proper current and time settings—usually by connecting them in series with a test kit that permits current flow to be adjusted to proper values and times the closing of contacts with a stop watch or timer (Fig. 14-17).

INSTRUMENT TRANSFORMERS

Instrument transformers are used to reduce voltage and current to values that can be handled safely in relay, control, and metering systems. This reduction by a predetermined proportion permits power circuits to be essentially reproduced in miniature scale models. The loads that they are permitted to carry while still maintaining their desired accuracy are referred to as their *burden,* rated in amperes.

Both potential and current transformers must be insulated to withstand the voltage of the power lines to which they are connected and must be rugged enough to carry short-duration fault currents.

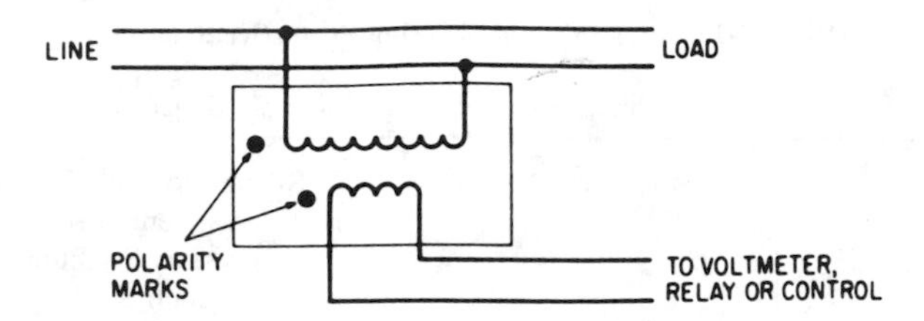

Figure 14-18 Potential transformer connection.

Potential transformers are connected across the line as shown in Fig. 14-18. Current transformers are connected in series with the line, as shown in Fig. 14-19. Because the turn ratio of current transformers can produce very high voltages, extreme care should be exercised when handling them; they should be short-circuited when being disconnected. See Chap. 1.

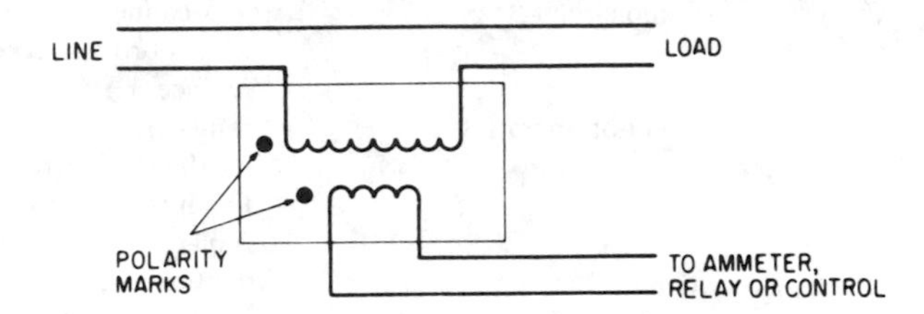

Figure 14-19 Current transformer connection.

For causes and remedies when troubleshooting protective relays, see Table 14-1.

TABLE 14-1 Troubleshooting Protective Relays

Condition	Possible Cause	Suggested Remedy
Relay not operating	Contacts not making	Check and adjust plunger, armature, disc, or cylinder. Check contact arm and adjust. Dress contacts or replace if necessary.
	Poor contacts	Realign contacts. Clean and dress them or replace if necessary. Tighten connections.
	Coil not operating	If wrong coil, replace. If short-circuited turns, replace.
	Open circuit	If broken wire, repair. If poor connection, tighten.
Relay operating too fast or too slow	Settings incorrect	Check and adjust to proper settings.

TABLE 14-1 (cont.) Troubleshooting Protective Relays

Condition	Possible Cause	Suggested Remedy
	Poor contacts	Realign contacts. Clean and dress them or replace if necessary. Tighten connections.
	Improper contact motion	Check restraining spring tensions; adjust or replace. Check dashpot or bellows relief vent. Clean and adjust. Replace oil in dashpot if necessary.
Auxiliary relays not functioning	Poor contacts	Check and adjust contacts; clean and dress. Replace if necessary.
	Open circuit	Check connections; tighten. If wire is broken, repair or replace.
	Coil not operating	Wrong coil; voltage too low; insufficient current supply. Open coil or shorted turns; replace coil.
	Failure to release	Armature out of adjustment; low voltage or current; wrong coil; replace.
Relay coils overheated	Overload	Wrong coil; replace.
	Shorted turns in coil	Check cause; repair or replace if necessary
Contacts overheated	Corroded, pitted, or burned contacts	Clean and dress; replace if necessary.
	Insufficient pressure	Check spring tension; check contact alignment; adjust.
Noisy relay	Loose connection	Tighten connection.
	Armature or plunger mechanism binding or dragging	Realign or replace if necessary.

REVIEW

1. Protective relays must coordinate with other protective devices such as fuses, reclosers, lightning or surge arresters. As a practical matter, coordination generally involves a series of approximation and compromises.
2. Electric systems can be adequately protected by dividing them into protective zones (Fig. 14-5), and subdividing them further as desired. The zones may comprise: sources of generation and their outlet circuits; subsources, substations; local or distribution circuits—coordinated in reverse order.
3. Distribution circuits are most often *radial* circuits, supplied from only

one source (Fig. 14-1). The circuit may be sectionalized with fuses and reclosers that provide points for clearing faults without interrupting the entire circuit. Both fuses and reclosers may be coordinated with each other and with the protective relays operating the circuit breakers at the points of supply.

4. Transmission supply protective relays must coordinate with protective devices on the circuits connected to them. This is more difficult as fault currents may flow from two or more sources. One solution is to assume only one source supplies the fault current; then assume each of the others does so one at a time. Then the probable magnitudes and direction of the fault currents may be found by combining all of the individual determinations and compromise settings can be made after the maximum and minimum fault currents at all locations are calculated (Fig. 14-2 A and B).
5. The protective zones for generators, buses and circuit breakers may be accomplished by differential relaying (Fig. 14-6). Coordination with transmission lines may be accomplished by directional-overcurrent relaying. Radial feeders from these sources require only overcurrent relaying.
6. *Pilot protection* employs separate protective circuits to operate breakers at both ends of a line simultaneously; the protective circuits are called pilot wires. The same general protection may also be accomplished by the use of carrier high-frequency currents using the transmission system in place of the pilot wires, or by the use of radio signals (Fig. 14-10).
7. In three-phase wye connected lines, protection may be accomplished by measuring the current in the neutral or ground wire. An overcurrent relay, in this case called a ground relay (Fig. 14-12), accomplishes this purpose.
8. Relays may be of the electromagnetic (Fig. 14-13), or induction type (Fig. 14-14), or electronic type. In all cases, the current magnitude and time settings to close or open the contacts must be checked periodically to retain their performance. The importance of clean (Fig. 14-16), smooth, properly aligned contacts and satisfactory operation of springs and moving parts cannot be overemphasized.

STUDY QUESTIONS

1. Name several types of faults that may occur in a three-phase ac system.
2. What is meant by the zones of protection in a relaying system? Describe them in a typical electrical power system.
3. Where zones of protection overlap, how is the protection of the overlapped circuit or equipment accomplished?

4. Why must protective relays be coordinated with other protective devices? Name several of these protective devices.
5. What is meant by pilot protection? Name three means of accomplishing it. What are its advantages and disadvantages?
6. What is a ground relay and how does it function?
7. What two characteristics of protective relays determine their performance?
8. Name five important items that should be checked on inspections.
9. What kinds of relay contacts are there? What features are to be considered and what precautions should be taken in their maintenance?
10. How are protective relays tested?

15

STORAGE BATTERIES

CONTROL SYSTEMS

The need for continuity in the supply of energy for the control of circuit breakers and other auxiliary equipment has led to reliance on storage batteries. Although relatively little energy is consumed by the control system, its reliability must be of the highest order since a failure may result not only in serious damage to equipment but to the entire circuit as well. The control system provides for the electrical operation of such relays, circuit breakers, pilot lights, alarms, recording meters, and other such instruments that are all connected to a dc "control bus." This bus is supplied with energy from a storage battery.

Demand on the control system may be divided into two classes. The first is a small, practically constant load, comprising the position-indicating and pilot lights, relay and meter coils, and other similar devices requiring a constant supply of current. The second class of load, which constitutes the supply to solenoids and motors of electrically operated equipment, including circuit breakers, is of a momentary nature but of heavy demand. If the control battery is also required to supply emergency lighting to the area in which the electric equipment and controls are housed, a third class of load is added. This third class may be of comparatively greater magnitude and longer duration.

Control systems may be operated at nominal voltages of 6, 12, 24, 48, or 120 V, although 24-, 48-, and 120-V systems are preferred. They are usually designed for ungrounded operation, with some type of ground detection provided.

FUNCTION

A storage battery is a device that may be used repeatedly for storing electrical energy at one time for use at another. It does not directly store electrical energy as such, although energy is put into it in the form of direct-current electricity. Rather, this energy is stored as "chemical" energy by means of a chemical process. By a reversal of this process, the chemical energy is converted back to direct-current electrical energy and delivered for utilization in this form. The process of putting energy into the battery is termed *charging* and of delivering energy, *discharging*.

PRINCIPLE OF OPERATION

A storage battery consists of one or more cells connected in series, each cell having two plates (one positive, the other negative) immersed in a solution called the *electrolyte* (Fig. 15-1). When the battery discharges, the chemical composition of these three elements change; when it charges, they are restored to their original condition. The chemical reactions that take place are different for different types of batteries.

The most frequently used type of battery is the lead-sulphuric-acid (or simply lead) storage battery, principally because of its economic advantages. Another common type of battery is the nickel-iron-alkaline battery, or so-

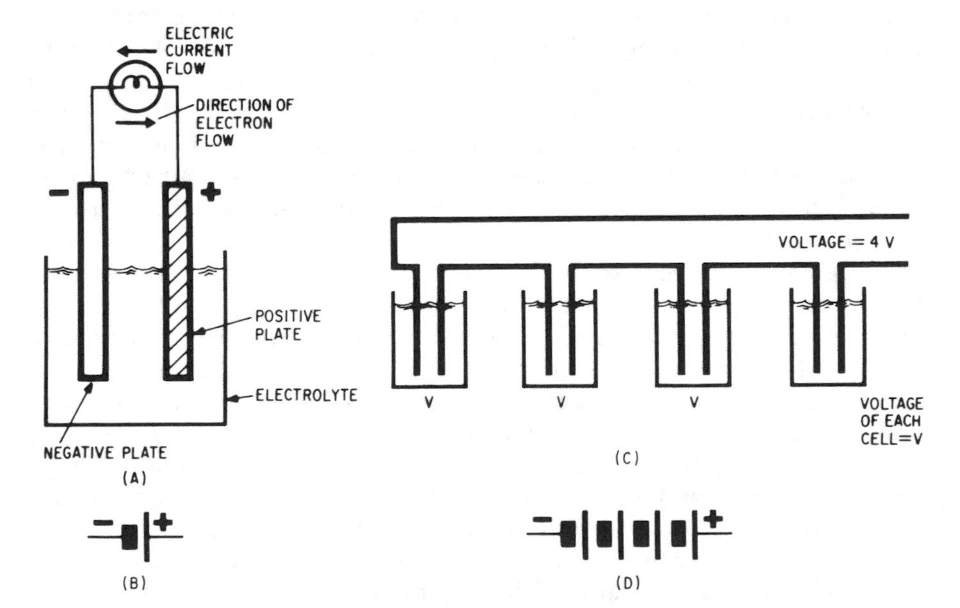

Figure 15-1 Basic cell and battery: (A) Simple cells, (B) symbol of one cell, (C) battery of four cells in series, and (D) symbol of multicell battery.

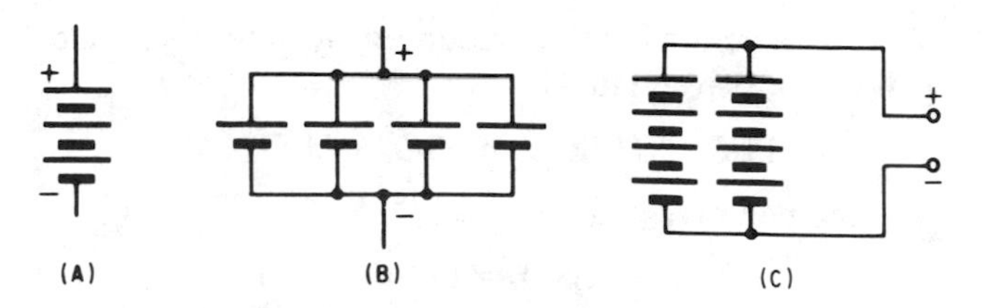

Figure 15-2 Methods of connecting cells: (A) Series, (B) multiple or parallel, and (C) series parallel.

called Edison battery or alkaline battery. It is more expensive than the lead battery but has certain advantages.

The open-circuit voltage of any storage cell depends almost entirely on its chemical constituents and to some extent on the strength of the electrolyte and its temperature. The terminal voltage of the cell rises when the cell is charged and falls when it is discharged. The voltage of the battery is the voltage of one cell multiplied by the number of cells connected in series. The cells may be connected in series, in multiple or parallel, or in series parallel (Fig. 15-2).

The capacity of a cell depends approximately on its plate area. Its discharge rate is given in terms of ampere-hours at a standard temperature of 25°C or 77°F. The capacities of all storage cells and batteries decrease as the rate of discharge increases.

LEAD STORAGE BATTERY

The active material of the lead storage battery is lead peroxide (PbO_2) on the positive plate and finely divided or sponge lead (Pb) on the negative plate. The electrolyte is a solution of sulphuric acid (H_2SO_4) and water (H_2O).

Discharge

When a cell is being discharged (Fig. 15-3A), electric current is produced as the acid in the electrolyte gradually combines with the active material of the plates. The acid in the pores of both plates chemically combines with the active materials and changes them into lead sulphate. Water is also formed at the same time and further dilutes the electrolyte. As the discharge proceeds, additional acid is withdrawn from the electrolyte and the formation of lead sulphate and water continues under the influence of the discharge current. At the point of complete discharge (Fig. 15-3B), the acid in the electrolyte will be reduced to a minimum. It will thus be seen that the formation of lead sulphate is the normal function of discharge.

The chemical reactions are represented by simple chemical equations. The reaction at the positive plate is

$$PbO_2 + H_2SO_4 = PbSO_4 + H_2O + O$$

The reaction at the negative plate is

$$Pb + H_2SO_4 = PbSO_4 + 2H$$

Note the formation of gaseous oxygen (O) and hydrogen (2H). Although some of these gases react to form water (H_2O), some escape into the atmosphere and represent a combustible and possible explosive mixture of the two, for which possibility precautions should be taken. The combined reactions at the two plates are as follows:

$$PbO_2 + Pb + 2H_2SO_4 = 2PbSO_4 + 2H_2O$$

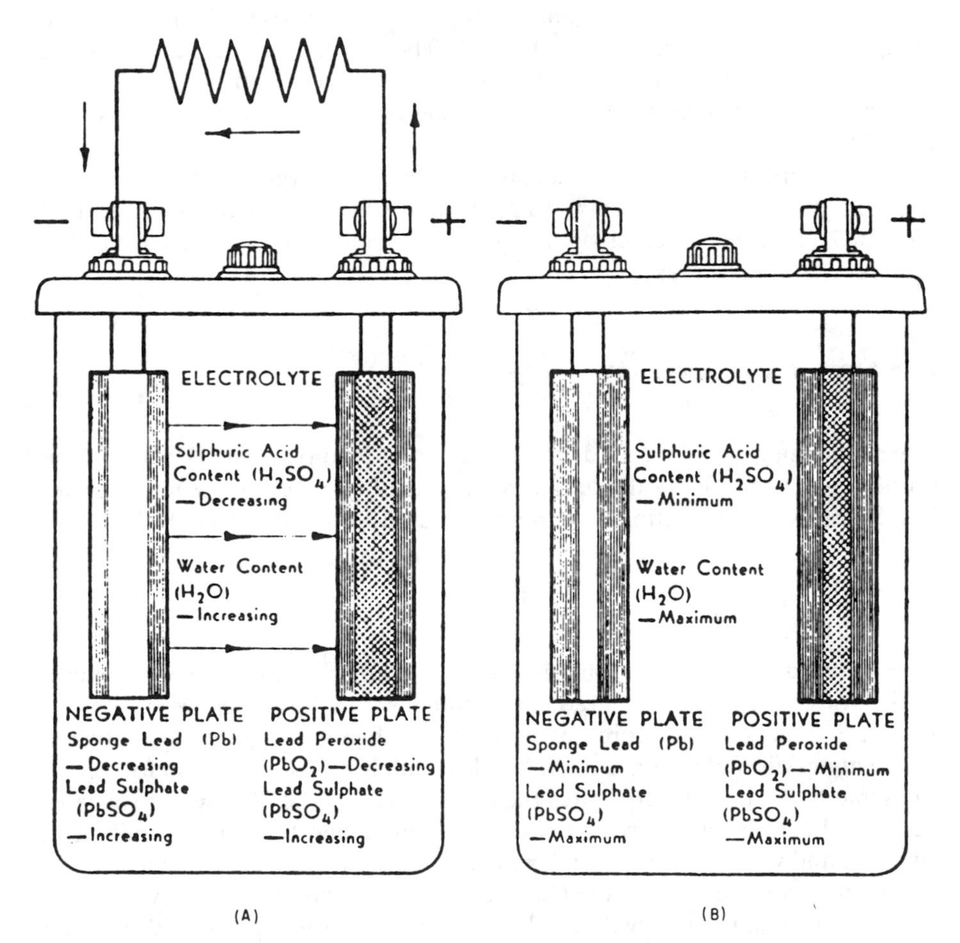

Figure 15-3 Chemical actions of lead-acid cell: (A) Discharging, (B) discharged,

Charge

During the charge (Fig. 15-3C), the direction of the current flow through each cell of the battery is opposite that during discharge, and hence a reverse electrochemical action takes place. The plates are gradually returned to their former state: lead peroxide in the positive plate and spongy lead in the negative plate. All acid previously absorbed in the development of lead sulphate is again set free and returned to the electrolyte. Both the process of desulphation and the increase of acid content in the electrolyte progress slowly throughout the charging period. When the cell is completely charged (Fig. 15-3D), all acid will have been driven out of the plates, and the acid in the electrolyte will be at a maximum.

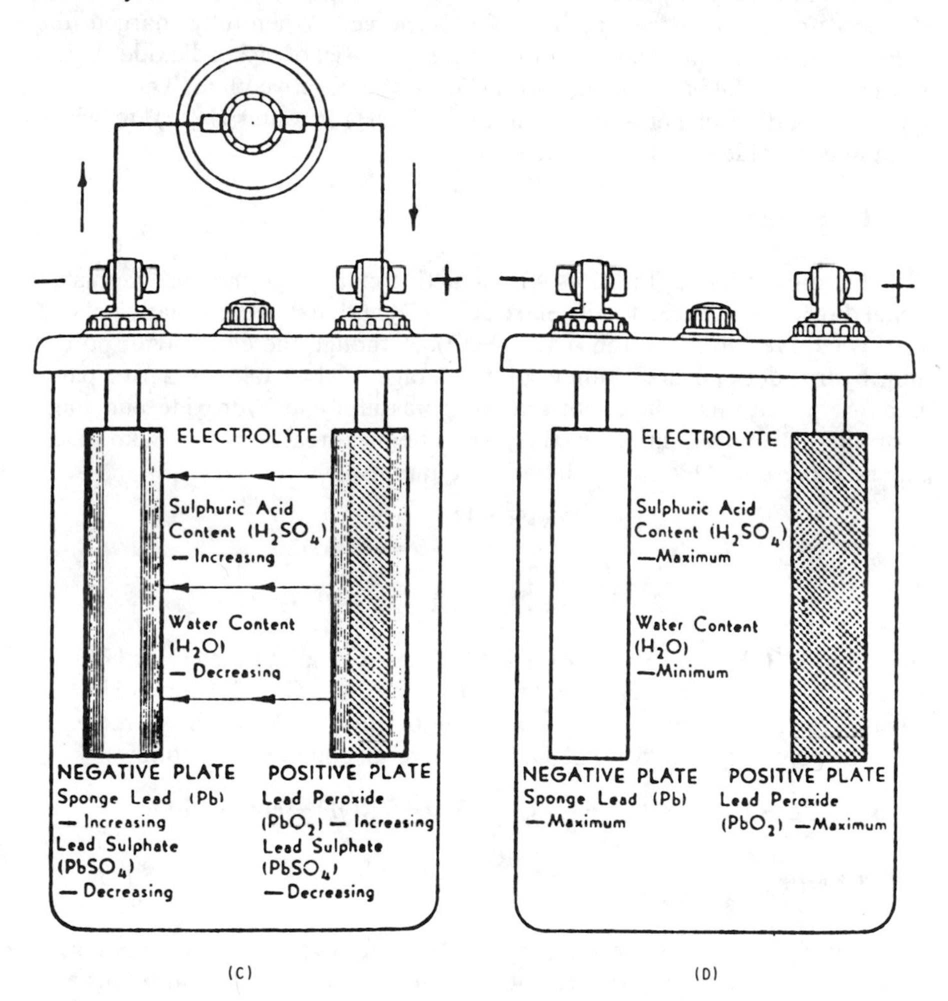

Figure 15-3 (cont.) (C) charging, and (D) charged.

The chemical reactions are represented by the same equations given above for discharge, except that they are now read from right to left to reverse their direction. It was noted that some of the hydrogen and oxygen occasionally escapes before being reformed into water. Water may thus have to be added from time to time to make up for this deficiency. It should also be noted that the equations represent the normal chemical reactions that take place. If impurities are present, other reactions may occur.

ALKALINE STORAGE BATTERY

The chemical reactions that take place in the nickel-iron-alkaline cell are different from those occurring in the lead-acid cell. When fully charged, the active materials of the alkaline storage battery consist of nickel dioxide (NiO_2) in the positive plate and metallic iron (Fe) in the negative plate. The electrolyte is a solution of potassium hydroxide (KOH) in water (H_2O), to which lithium hydroxide is added as a catalyst.

Discharge

During discharge (Fig. 15-4A), the active material in the positive plate, nickel dioxide, is reduced to nickel oxide (NiO) and that in the negative plate, iron, is oxidized to form iron oxide (FeO). Although the electrolyte, potassium hydroxide, appears to undergo no change, it does take an active part, breaking up into its component ions of potassium and hydroxide and then reforming into potassium hydroxide. The chemical reactions that take place may be represented by simple chemical equations:

$$Fe + 2KOH + H_2O + NiO_2$$

ionizes into

$$Fe + 2K + 2O + 2H + 2H + O + Ni + 2O$$

One O of 2KOH joins Fe to form FeO (iron oxide), and one O of NiO_2 is released to form NiO (nickel oxide). When the 2K + O + 2H forms H_2O, it releases the 2K, which then joins with the original H_2O and the O released from the nickel oxide to form 2KOH. The overall equation is as follows:

$$Fe + 2KOH + H_2O + NiO_2 = FeO + 2KOH + H_2O + NiO$$

Charge

During the charging cycle (Fig. 15-4B), the reverse action takes place, with the same breakdown of the potassium peroxide and water into ions and

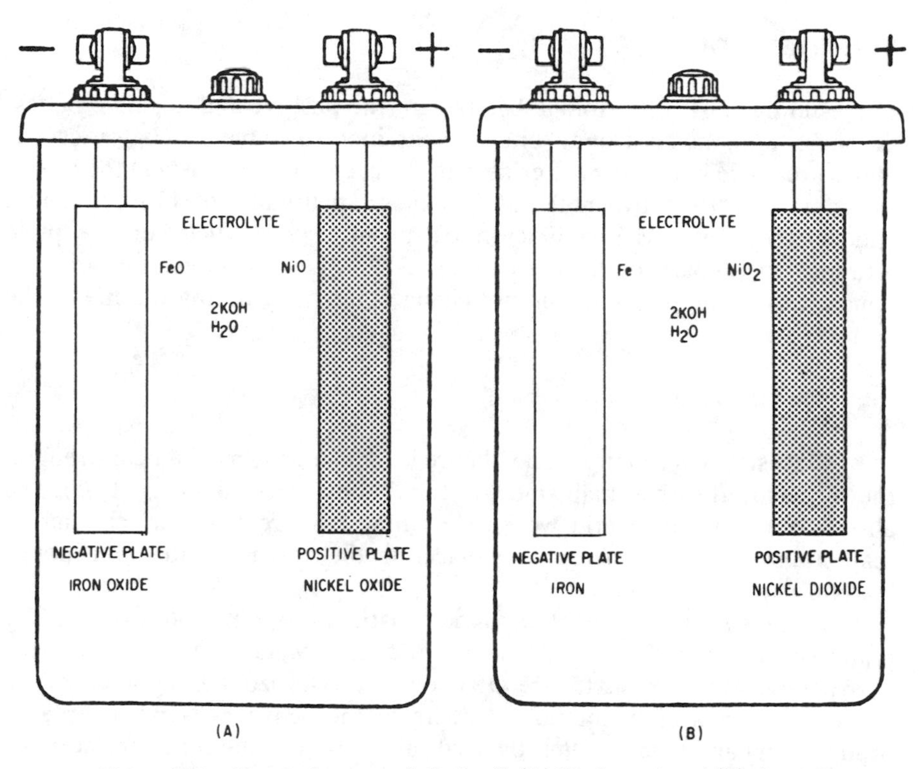

Figure 15-4 Chemical actions of alkali cell: (A) Discharged, and (B) charged.

the same ultimate recombination. The chemical reactions are represented by the same equations as those for discharge except that they are read in reverse order. The overall chemical equation would be read from right to left.

The effect of the reactions can be described as a transfer of oxygen from the negative plate to the positive plate during the charging operation and the reverse transfer during the discharging operation.

The lithium hydroxide catalyst aids but does not take part in the reaction itself. For one thing, it considerably increases battery capacity. Furthermore, it acts as a preservative of ferrous metals and makes possible the use of steel for the structural parts of the cell, both of which services materially increase the life of the battery.

In a modification of the Edison alkaline cell, the iron is replaced with cadmium and the potassium hydroxide electrolyte takes the form of a paste rather than a liquid. The nickel cadmium or Nicad battery overcomes some of the disadvantages of the Edison battery because the cells are hermetically sealed and require no maintenance or tests. It will also deliver the same load current as a lead-acid battery of the same specification but for twice the length of time.

CONSTRUCTION

The essential parts of a storage battery are the positive and negative plates, the separators, the electrolyte, and the container. In all types of batteries, the plates are made in the form of grids, which have two functions: (1) they serve as a support for the active materials since these are usually not self-supporting; and (2) they serve as a conductor to transmit current between the plate terminal and all parts of the active material. For both purposes, the form and dimensions of the plates should not change materially during the life of the battery.

Lead Battery

Plates. Since plates made of pure lead have little mechanical strength, they are alloyed with a small amount of antimony or calcium (Fig. 15-5). The alloy is formed into a grid whose slots may have rectangular, circular or diamond shapes. Lead peroxide is placed in the slots for plates and sponge lead for negative plates.

In the so-called "iron-clad" plate construction (Fig. 15-6), the positive plates consist of vertical parallel spines or shafts. Made of an antimony-lead alloy, these spines or shafts are attached to the horizontal top and bottom sides of the lead alloy frame. Cylinders of the active material are placed around them and held together by hard rubber tubes. The tubes are horizontally slotted to give the electrolyte access to the active material. The negative plates are essentially the same as other imbedded or pasted plates.

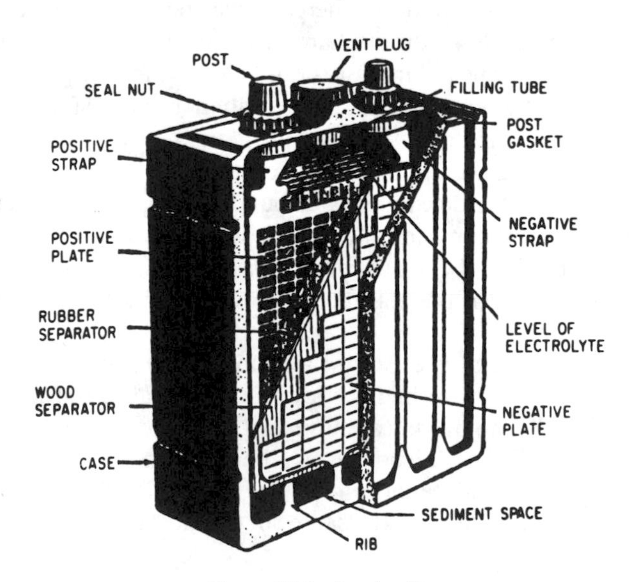

Figure 15-5 Lead cell.

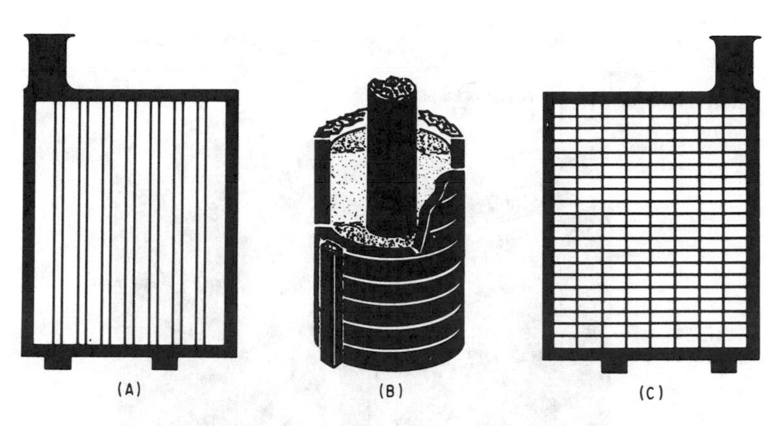

Figure 15-6 Iron-clad plate construction: (A) Positive plate, (B) detail of positive plate, and (C) negative plate.

Separators. The separators inserted between adjacent positive and negative plates have three functions:

1. They serve as mechanical spacers to prevent actual contact.
2. They serve as electrolytic channels or diaphragms that permit current to pass but prevent small particles of lead sediment from lodging on the negative plate and eventually bridging across the plates and causing a short circuit.
3. They act to retain the active material in the slots, preventing it from being dislodged.

Separators are usually made of treated wood, such as cypress, cedar, or redwood, but they are also made of mats of felted glass wool, diaphragms of microporous rubber, or sheets of perforated hard rubber. In the "iron clad" type of plates, the slotted rubber cylinders act as the separators.

Electrolyte. For lead batteries, the electrolyte consists of diluted sulphuric acid. The ratio of water to sulphuric acid for a specific gravity of about 1.2 is approximately 2.4 to 1 by weight or 4.5 to 1 by volume. The ratios will change with a change in specific gravity.

When a storage battery is discharging, sulphuric acid is replaced by water. The electrolyte not only becomes weaker, it also becomes lighter because sulphuric acid is heavier than water. On the other hand, when the battery is charging, sulphuric acid replaces water and the electrolyte becomes correspondingly heavier.

Impurities in an electrolyte may attack the grids of the positive plates, attack the separators, or cause small local actions in the plates that will decrease the output of the batteries.

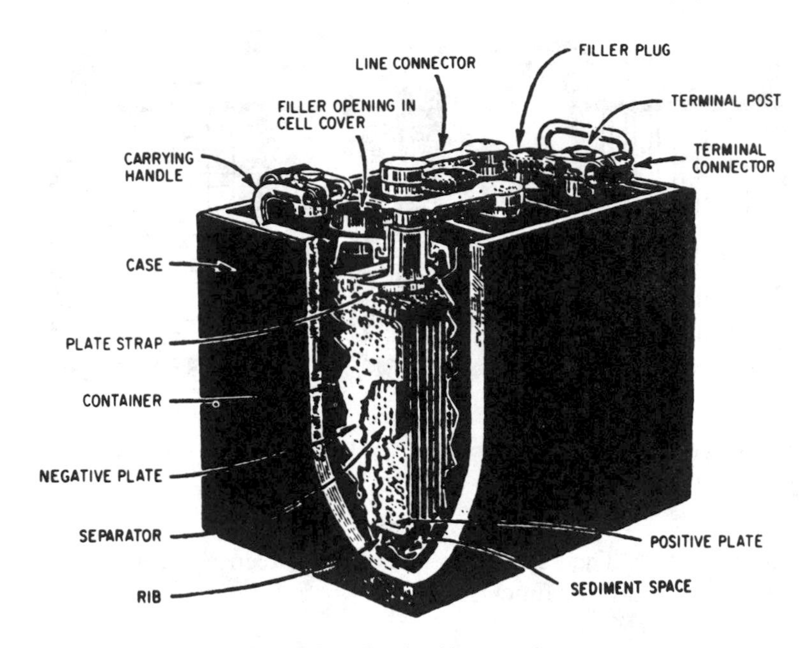

Figure 15-7 Lead-acid storage battery.

Containers. Containers are usually made of glass, hard rubber, plastic, or asphaltic composition. For stationary cells, glass is almost universally used because of its durability, transparency, and low cost. Hard rubber or plastic is used where there is danger of breakage; they are not transparent and also cost more. For small and less expensive portable batteries, asphaltic compositions are used as well as hard rubber. A cross section of a lead battery and its container is shown in Fig. 15-7.

Although sealed, containers provide openings for venting of accumulated gases and to permit water to be added to the cell. A removable vent plug designed to allow for the escape of gas may also serve as a removable stopper to allow filling.

Alkaline Battery

Plates. The positive plate consists of a nickel-plated steel frame or grid fitted with pockets made of perforated nickel-plated steel ribbon. The pockets are filled with alternating layers of nickel hydroxide and flakes of pure metallic nickel. In the so-called *nickel-cadmium battery,* purified graphite is used in place of the nickel flakes. The negative plate is similarly constructed, but its pockets are filled with finely divided iron oxide mixed with a small amount of an oxide of mercury as a catalyst.

Separators. Separators are made of molded hard rubber and support each edge of each individual plate after the two groups of plates have been intermeshed. They thus provide not only the necessary insulation of the plates from one another and the container, but also the physical support for the weight of the plates on the container bottom. Additional insulation between the flat faces of each positive and negative plate is provided by molded hard-rubber pins or spacers.

Electrolyte. The electrolyte is a 21-percent solution of potassium and lithium hydroxides having a normal specific gravity of approximately 1.2 at 16°C or 60°F. Sodium hydroxide is sometimes used instead of potassium hydroxide. The concentration and chemical composition of this electrolyte do not change materially during the periods of charge and discharge.

Container. The container is made of nickel-plated sheet steel with all sides welded together. The cover is fitted with gas- and liquid-tight insulated openings for the terminals. Gas valves in the cell-vent plugs protect the electrolyte from contamination by carbon dioxide from the air. (Carbon dioxide forms carbonates in the electrolyte that increase its resistivity.) The valves allow the escape of generated gases while preventing the entrance of air. Often, when alkaline batteries are being charged, a popping sound is heard as gas displaces the spring-loaded valve.

A cross section of the alkaline cell and its container is shown in Fig. 15-8. Cells are connected together by means of nickel-plated copper straps and must be insulated from one another because the steel containers are electrical conductors.

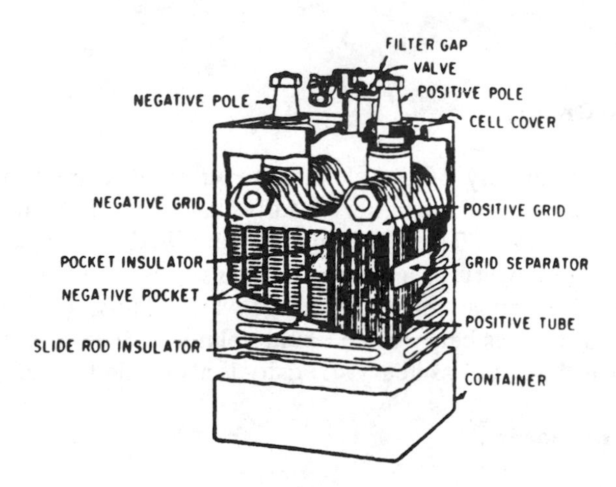

Figure 15-8 Alkaline cell construction (*Courtesy,* Exide Battery Co.)

RATINGS OF BATTERIES

Storage batteries are usually rated in ampere-hours, the rating usually being based on the time period in which the battery will be discharged. Thus, the ampere-hour rating of a battery is equal to the product of the amperes output and the normal time such an output can be supplied. The ampere-hours obtainable from a battery are greater for a long low-rate or intermittent discharge than for a short high-rate discharge because the voltage drops at the higher rates. Moreover, the I^2R losses in the battery are greater for higher currents than for smaller currents.

Battery capacity is usually based on an 8-hour rate. Greater capacity can be obtained if the discharge rate is made lower, and vice versa. A discharge rate above normal will lessen the capacity. For example, a battery rated 40 ampere-hours at the 8-hour rate would have the following outputs:

Amperes	× *Hours*	= *Ampere-Hours*
140	× 1/60	= 2.33
20	× 1	= 20.0
5	× 8	= 40.0
0.8	× 72	= 57.2

To restore a battery to its fully charged condition, it is necessary to return to it as many ampere-hours as have been delivered by it. Since battery efficiencies run from 85 to 90 percent—say 90 percent—then if the battery delivered 20 amperes for 30 minutes, 10 ampere-hours would have been removed from it. The battery would be fully charged again if approximately 11 ampere-hours (10 divided by 90 percent) were returned to it.

OPERATION

Specific Gravity

The specific gravity of the electrolyte provides a convenient way of determining whether a storage battery is fully charged or, if it is not, the extent to which it is discharged. The test is made with a *hydrometer,* an instrument for measuring the density of a liquid. By density is meant the weight of a substance compared to an equal volume of water. Thus, a substance with a density of 2.0 is twice as heavy as an equal volume of water, and a substance whose density is 0.5 is half as heavy. Density is also called *specific gravity.*

The Hydrometer

When a liquid becomes heavier—that is, denser, like the electrolyte of a storage battery when charging—it becomes more buoyant. A floating object

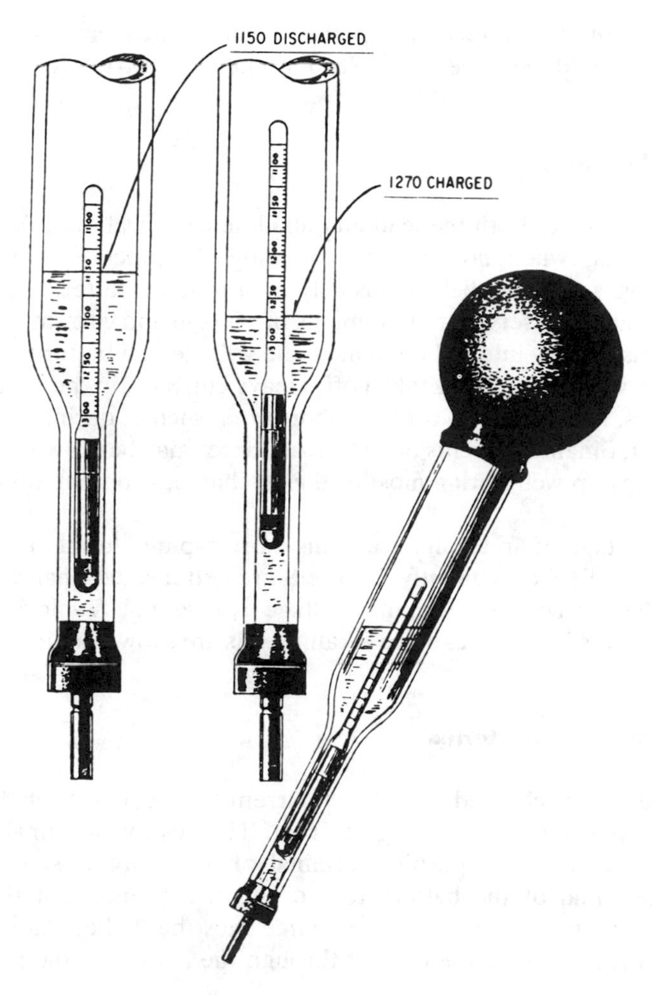

Figure 15-9 Hydrometer battery tester.

will float higher out of the liquid. This is the principle of the hydrometer (Fig. 15-9).

Now since the acid in a lead battery becomes more dilute on discharge, it becomes possible to determine the approximate condition of the cell at any time by a simple hydrometer reading of the specific gravity of the electrolyte. When a lead battery is fully charged, the density of the acid is about 1.220, which means that it is 22 percent heavier or denser than water. When the battery is fully discharged, the density is about 1.150.

In an alkaline battery, the electrolyte remains essentially unchanged throughout its operating cycle, and the measurement of its specific gravity is thus of less importance than for the lead battery. Nevertheless, occasional measurements are made because the weakening of the electrolyte decreases the capacity of the battery. When the specific gravity reaches 1.160, the electrolyte should be renewed. (Normal specific gravity for the alkaline battery is from 1.190 to 1.230.)

Cell Voltages

The voltage of both the lead and alkaline cells is about 2 V in an open circuit, that is, when no current is flowing. In practice, however, cells connected in series-parallel are usually arranged in batteries in sufficient number to meet the service requirements of voltage and current.

Because of low internal resistance, the voltage of a battery will not drop appreciably when a current is taken off. Heavy currents, in the range of 100 to 200 amperes, may be drawn off for a short time, such as during the operation of a breaker. Smaller currents of 10 to 20 amperes may be drawn steadily. The voltage keeps up well during most of the discharge, dropping off only at the end.

The voltage of an alkaline cell falls more rapidly during discharge than that of a lead cell. Consequently, lead cells are better suited than alkaline cells for use where a constant terminal voltage is essential. Typical charge and discharge curves for both lead and alkaline cells are shown in Figs. 15-10A and 15-10B.

Charging of Batteries

Batteries are charged with direct current that is sent through them in a direction opposite to the discharge current. The positive terminal of the battery is connected to the positive terminal of the charging source, and the negative terminal of the battery to the negative terminal of the charging source. The voltage of the charging source must be higher than the battery voltage in order to force the current through the battery in the proper direction.

Lead batteries may be charged at any rate that does not cause gasing. When they are nearly discharged, the permissible charging rate can be considerably greater than when the charge is built up. For minimum charging time, this situation calls for a continual reduction of the charging rate as the charge progresses. When a battery is charged from an essentially constant voltage source, a resistance is placed in the charging circuit. Since the voltage drop in the resistance is greater when larger current flows and lesser when smaller currents flow, the voltage applied at the lead cells tends to follow the optimum charging rate.

Alkaline batteries can be charged at any rate that does not cause overheating. The temperature of the cells should not exceed 46°C or 115°F. Within this limitation, the battery can be charged at a constant rate.

The batteries "float" on their dc supply line, that is, they are per-

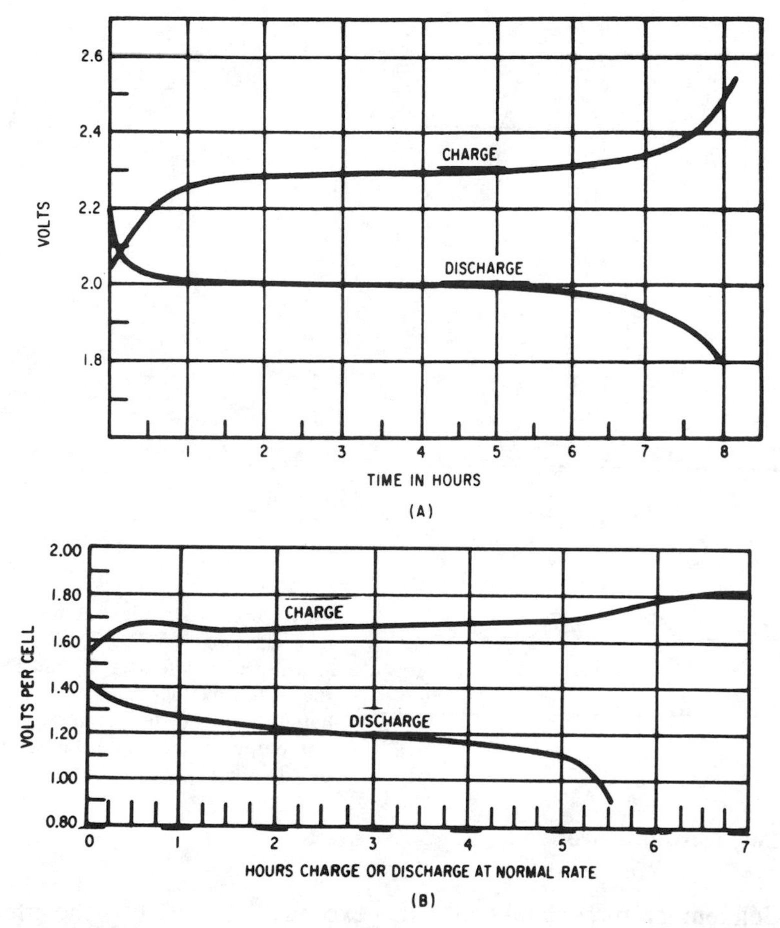

Figure 15-10 Typical charge and discharge curves: (A) For lead storage cell, and (B) for alkali storage cell.

manently connected through the rectifier to the ac source. Practically, they are always fully charged to ensure reliable operation of the equipment and devices that they supply.

Rectifiers

Where alternating-current sources exist, direct current can be provided through rectifiers (Figs. 15-11A and 15-11B). Among a variety of types, the copper-oxide rectifier appears to be the most used for battery recharging (Fig. 15-11C). A property of copper oxide prevents current from flowing in one direction but not in the other. Connections for an arrangement of rectifiers that gives an essentially constant dc voltage are shown in Fig 15-11D. The voltage applied is approximately 2.15 V per cell.

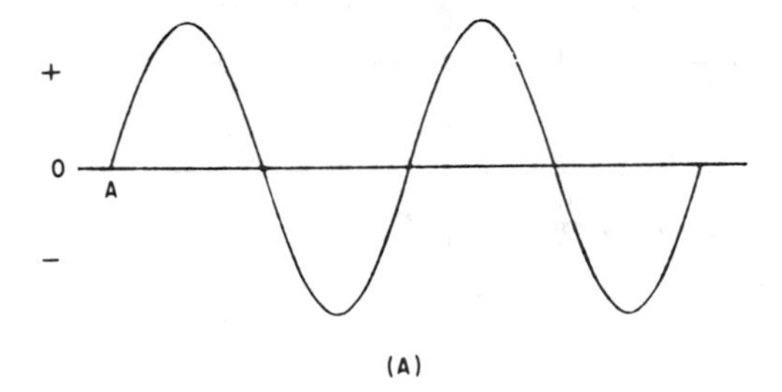

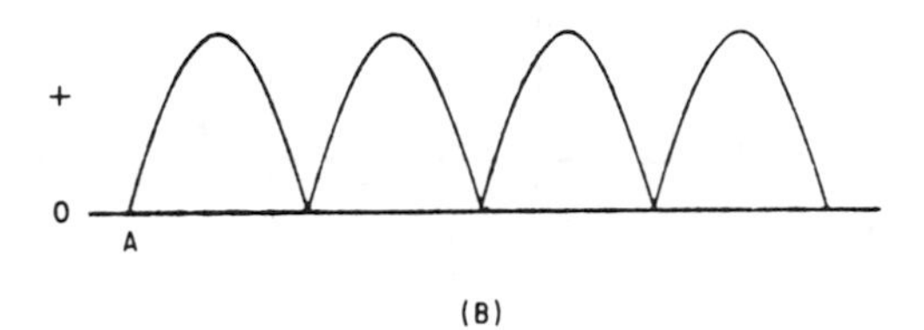

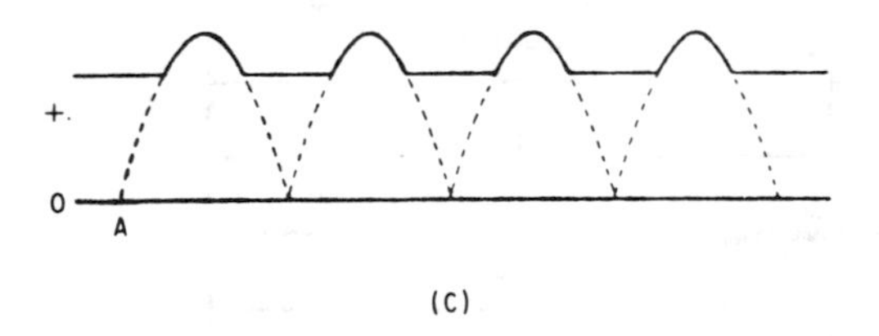

Figure 15-11 Rectifier method of battery charging: (A) A–C voltage wave shape, (B) rectified A–C voltage wave shape, (C) rectifier voltage impressed on battery voltage (solid line shows combination output voltage wave shape with ripples greatly exaggerated).

Cell Temperatures

Cell temperatures should not much exceed 45°C or 110°F. The effect of high temperatures is to shorten the life of the wood separators installed between the positive and negative plates of some cells. There is always a tendency for wood, especially when in contact with sulphuric acid, to become carbonized, and this tendency is greatly increased at high temperatures. Internal losses are also increased.

At low temperatures, the voltage and capacity of a battery are temporarily reduced because of the greater internal resistance at such temperatures. At 20°F, the internal resistance is about double what it is at 80°F.

Gases

The hydrogen and oxygen given off during the charging of a battery form a gaseous combination that will explode violently if ignited. For this reason, battery housings should be well ventilated, and the presence of any exposed flame, spark, lighted matches, or cigarettes must never be permitted.

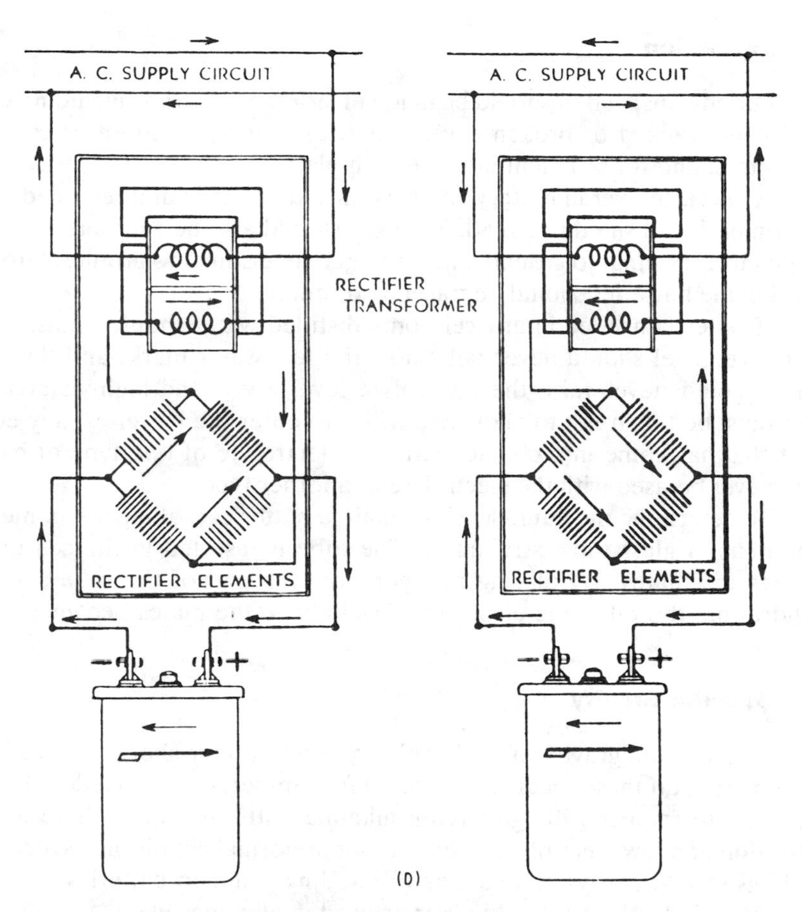

Figure 15-11 (cont.) Rectifier method of battery charging: (D) rectifier connections to give an essentially constant d–c voltage.

MAINTENANCE

Since the electrolyte in both lead and alkaline batteries is not consumed, it never has to be renewed unless some is spilled. Water evaporates or gases off, however, and a little water must thus be added from time to time. This water must be distilled, for impurities or other chemicals in it might cause unwanted chemical reactions and reduce the effectiveness and life of the battery. The level of the electrolyte should always be above the plates by at least one-half inch.

A lead battery should not be left standing in the discharged condition because the lead sulphate coating on both the positive and negative plates will form into hard crystals that will be difficult to break up on recharging. Alkaline batteries, on the other hand, can be left discharged for long periods of time without injury.

Inspection

Periodic inspection should be made of storage batteries, including visual checks for cracked or broken covers, jars, or plates; for loose or corroded connectors; and for sediment in touch with plates.

A check of overall battery voltages should be made and recorded. Voltages on each cell should be read. Voltages should also be checked from both the positive terminal to ground and the negative terminal to ground. Grounds found in the batteries should be traced and removed.

If water is needed in any cell, only distilled water should be used. The electrolyte level should never fall below the low-water mark, and the water added should never raise the electrolyte level above the high-water mark. Care must be taken not to allow impurities to enter the batteries; any equipment that has come into contact with the electrolyte of one type of battery must never be used with the electrolyte of another type.

To determine the liquid level in alkaline batteries making use of metallic containers, a glass tube is required. The tube is inserted in the cell until it touches the plates. By placing a finger over the end of the tube and then withdrawing the tube, the electrolyte level above the plates becomes apparent.

Specific Gravity

The specific gravity of each cell of a lead battery should be read by a hydrometer and these readings corrected for temperature as needed. It is not necessary to read specific gravity on alkaline batteries unless there is some indication of a low electrolyte level or other abnormal condition. Hydrometer readings should always be taken before adding water to batteries. A hydrometer used in lead cells should *never* be used in alkaline cells.

As a guide to acceptable values for battery characteristics, the following table may be used:

	Lead Types	Alkaline Types
Specific gravity	1.210–1.224	1.190–1.230
Water level above plates	0.5–0.75 inch	0.75–1.5 inch
Cell voltage	2.15–2.20	1.43–1.47

For overall acceptable battery voltage limits, multiply the cell voltage listed above by the number of cells in the battery.

Battery Charging

Charger. The output current charging rate should not exceed the maximum dc rating specified on the nameplate. From time to time, the charging rate may decrease as the rectifier stack ages. It can be maintained by advancing the switch shown in Fig. 15-12 by one position.

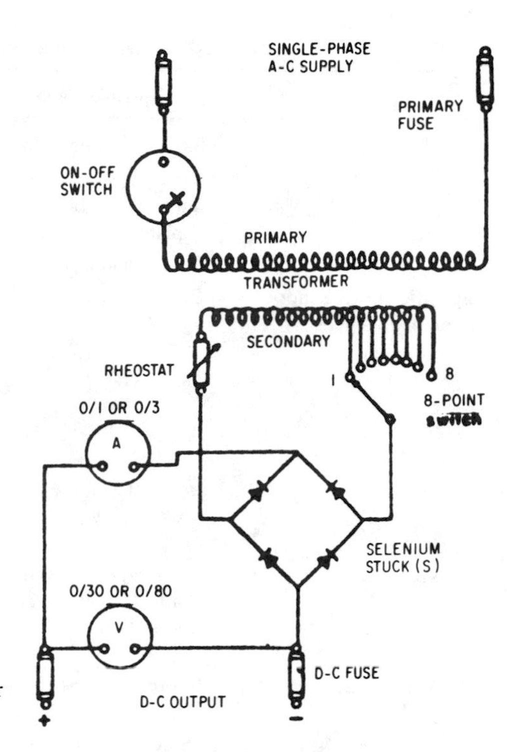

Figure 15-12 Selenium battery charger with manual control.

Equalizing charge. In checking the condition of lead batteries, especially where they are "floating" on a line, it may be found that some cells have a different voltage from others. In such cases, an "equalizing charge" may be applied to the entire battery. This consists of raising the charging rate temporarily by increasing the voltage across the battery to obtain a greater flow of current into it. The equalizing charge is continued until all cells are fully charged, as shown by voltage and specific gravity readings (and also be a certain amount of gasing).

The equalizing charge ensures that all acid (or sulphate) is driven from the plates. If sulphate is allowed to remain for any appreciable time, it will

TABLE 15-1 Troubleshooting Storage Batteries

Condition	Possible Cause	Suggested Remedy
Low electrolyte level in lead battery	Excessive gasing from overcharge or overdrain	Replace water. Check charging and discharging rates.
Excessive gasing	Same as above	Same as above.
Battery or cell voltage above normal	Charging voltage set too high	Adjust charging rate.
Battery or cell voltage below normal	Charging voltage set too low	Adjust charging rate.

TABLE 15-1 (cont.) Troubleshooting Storage Batteries

Condition	Possible Cause	Suggested Remedy
	Sulphate embedded in plates	Apply equalizing charge.
	Short circuit between plates in some cells	Replace separators or entire cell.
	Standing idle	Recharge. Check electrolyte.
Specific gravity cannot be maintained	Electrolyte loss	Add sulphuric acid or potassium chloride until specific gravity is attained. If unsuccessful, drain off all solution, clean case, and replace with fresh electrolyte. Check cover seals and vent and filling plugs.
	Impurities in the electrolyte	Drain electrolyte, clean case, and replace with fresh electrolyte.
Loss of active material in plates	Short circuit between plates; electrolytic attack on the grid; vibration	Replace grids and plates. Check electrolyte for impurities and replace if necessary. Relocate or mount cells on shock absorbing materials.
Buckled plates	Improper support	Check supports and replace both plates and supports if necessary.
	Uneven chemical action on opposite sides of plates caused by clogged plates and separators	Replace plates and separators; replace entire cell.
Excessive cell heating	Low electrolyte from overcharging or overdrawing	Replace electrolyte or water. Check charging and discharging rates.
	Short circuit in cells	Check separators; replace if necessary. Check for impurities in electrolyte; replace if necessary.
Corroded terminals and connecting straps	Chemical action from excessive gas carrying acid fumes; loose connections	Check electrolyte and rates of charging and discharging. Check vent and filling plugs. Clean terminals and connecting straps and apply anticorrosion jelly compound. Tighten connections.
Batteries not charging	Rectifier not operating; fuse blown	Check fuses; replace if necessary. Check connections.
	Defective rectifier	Check rectifier units; replace if necessary. Check connections.
	Loose connections	Check and tighten or repair connections.

crystallize and clog the pores of the plates, causing the battery to lose capacity and deteriorate and making it increasingly difficult to restore it to its normal condition. This procedure is not required for alkaline batteries.

REVIEW

1. Control systems provide for the operation of relays, circuit breakers, pilot lights, alarms, recording meters, and other equipment and require a reliability of electric supply of the highest order; this is supplied by storage batteries. Control systems may operate at nominal voltages of 6, 12, 24, 48, or 120 volts, although 24-, 48-, and 120-volt systems are preferred.
2. A storage battery is a device that may be used repeatedly for storing electrical energy at one time to use at another time. It does not directly store electrical energy as such, although energy is put into it in the form of direct-current electricity. Rather, the energy is stored as chemical energy by means of a chemical process. By a reversal of this process, the chemical energy is converted back to direct-current energy and delivered for utilization in this form. The process of putting energy into the battery is termed *charging* and of delivering energy, *discharging.*
3. A storage battery consists of one or more cells connected in series, each cell having two plates (one positive and the other negative) immersed in a solution called the *electrolyte* (Fig. 15-1). When the battery discharges, the chemical composition of these three elements change; when it charges, they are restored to their original condition. The chemical reaction that takes place is different for each different type of battery.
4. In the lead (or lead-acid) (Fig. 15-7), storage battery, the active material is lead on one plate and lead peroxide on the other, with sulphuric acid and water as the electrolyte. In the discharge, the materials change to lead sulphate and oxygen and hydrogen are developed; in charging, they return to their original state (Fig. 15-3), but some of the oxygen and hydrogen may escape as gases, constituting a possible explosive mixture.
5. In the alkaline storage battery, the active materials are iron (or steel) on one plate and nickel dioxide on the other, with potassium hydroxide and water as the electrolyte. In the discharge, the materials change to iron oxide, nickel oxide and potassium; in charging, they return to their original state (Fig. 15-4). Since no gases are formed, the battery may be hermetically sealed.
6. Batteries are rated in ampere-hours, usually based on the discharge rate, at a standard temperature of 25°C or 77°F. The capacities of all storage batteries decrease as the rate of discharge increases.

7. The condition of the cells or batteries may be checked by a hydrometer (Fig. 15-9), which measures the specific gravity of the electrolyte that indicates how much the electrolyte may be diluted.
8. Batteries are charged by sending a direct-current through them in a direction opposite to the discharge current (Fig. 15-11). The dc is usually supplied from rectifiers connected to ac sources. Rates of charging must be such as to prevent gases being given off in lead batteries and overheating in alkaline batteries.

STUDY QUESTIONS

1. What is a storage cell? What is a storage battery? How does it store energy?
2. What are the principal component parts of a storage cell?
3. Name two types of storage batteries in general use. What are the differences between them? What are the advantages and disadvantages of each?
4. Describe what takes place in the cells while charging.
5. Describe what takes place in the cells while discharging.
6. What is meant by specific gravity? What is a hydrometer? What is the specific gravity of a fully charged cell?
7. How is the capacity of a storage battery expressed? What does the expression mean? What determines the number of cells in a battery, and how are they connected?
8. How are batteries charged? What indicates the completion of charging? What is meant by an equalizing charge?
9. What are the effects of high and low temperatures on a cell?
10. What factors should be taken into consideration and what precautions observed in the maintenance of storage batteries?

16

REACTORS, CAPACITORS, RECTIFIERS

To describe the functioning of reactors, capacitors, and rectifiers—all important types of equipment associated with ac circuits—it is first necessary to understand some of the basic characteristics of ac circuits. Reactors are principally used to limit short-circuit currents, capacitors to reduce current flow and maintain voltage by improving the power factor, and rectifiers to change ac to dc.

For *inductance, capacitance,* and *impedance,* refer to Chap. 1: Theory of Operation.

REACTORS

Function

A reactor is a device used for introducing inductance or inductive reactance into a circuit. Its primary purpose is to limit the current that is allowed to flow through the circuit, particularly under the short-circuit conditions, in order to protect equipment from excessive heat and destructive mechanical forces. It is, therefore, connected in series with the equipment or feeder it is to protect.

Principle of Operation

A reactor usually consists of a coil wire. The current flowing through the coil produces a magnetic field that cuts the turns of the coil and sets up a counter voltage. With normal current flowing, the counter voltage is relatively small. With large short-circuit current flowing, however, this counter voltage becomes relatively large and bucks the applied voltage. The resultant voltage, which will push current into the fault, is small, and the fault current (by Ohm's Law) will also be small. The isolation of the fault from the system is obtained by means of circuit breakers. Since reactors limit short-circuit currents, lower capacity circuit breakers may be installed.

When reactance is placed in series with the equipment under fault, the voltage applied to the fault is reduced, but the voltage on the remainder of the circuit or system is maintained while the fault is being cleared.

Construction

Reactors are of two types, the air-core type and the iron-core type. The first is simply a carefully constructed, mechanically well supported circular coil of suitable proportions for cooling. The second consists of a coil encircling an iron core that makes up a magnetic circuit in which air gaps may exist. Both types of reactors may be air cooled or immersed in oil or similar fluid (in much the same way as transformers).

Types of Reactors

Air-core reactors. An air-core reactor produces a large magnetic field external to the coil. In constructing supporting structures, therefore, it is necessary, to avoid employing metal loops or circuits in which stray currents can form and circulate. Such circuits produce undesirable heat and unnecessary losses. Reinforcing rods in concrete supports, for example, should be isolated, never in contact with each other. Care should be exercised in spacing adjacent reactors and in the fastening of support insulators and bracing units.

Mechanical stresses should be provided for by ample supports. The windings are usually a series of "pancake" coils, mounted one above another, that exhibit as nearly equal electrical characteristics as possible to avoid circulating currents. The complete reactor is normally mounted on a concrete disc supported on porcelain insulating feet (Fig. 16-1).

Iron-core reactors. Coils for these reactors consist of small wires, straps, or cables wound about an iron core. The air gap may consist of several small gaps to keep eddy currents to a minimum.

Saturable iron-core reactors. By saturating the core of an iron-core reactor by means of a controlled dc winding, the reactance of the unit can be made variable. If the core is magnetically saturated, an ac winding will produce only small changes in its magnetic field (or flux), and the inductive reactance of the coil will be small. To keep the ac winding from inducing large voltages of changing magnitudes in the dc winding, the winding is arranged

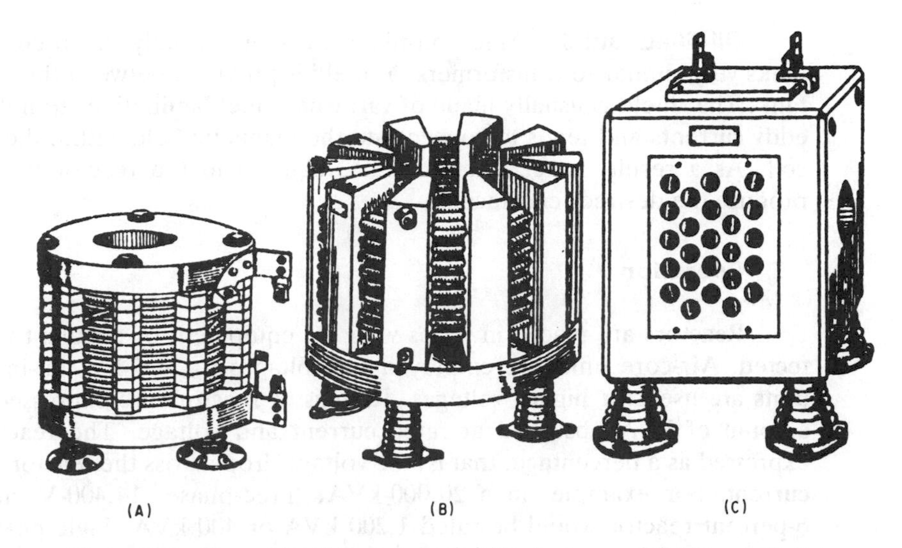

Figure 16-1 Current-limiting reactors: (A) Westinghouse, (B) General Electric, and (C) Metropolitan Device Corp.

as shown in Fig. 16-2. The two ac coils produce magnetic fields in opposite directions through the core of the dc winding, thereby neutralizing one another.

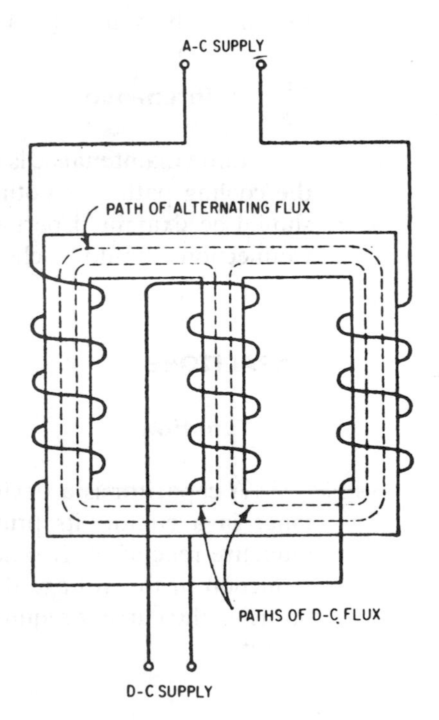

Figure 16-2 Saturable iron-core reactor.

Oil-filled units. Since oil-filled units are usually enclosed in steel tanks very similar to transformers, a shield is provided between the coils and tank. The shield is usually made of varnished steel laminations to hold down eddy currents and tends to concentrate the magnetic field within the reactor coil. As a result, fewer turns may be required in the reactor winding to produce the desired reactance.

Operation

Reactors are placed in series with the equipment or circuits to be protected. Air-core units are employed for voltages to 34.5 kV; oil-immersed units are used for higher voltages. Ratings of reactors are expressed in the amount of kVA absorbed at rated current and voltage. The reactance is expressed as a percentage, that is, the voltage drop across the reactor at rated current. For example, in a 20,000-kVA, three-phase, 14,400-V circuit, a 6-percent reactor would be rated 1,200 kVA or 400 kVA single-phase. The drop would be 6 percent of 14,000 × 0.866 (for single-phase), or 500 V at 800 amperes (400 kVA divided by 800 amperes). The reactance would be 500 volts divided by 800 amperes, or 0.625 ohm.

Reactors are sometimes installed in series with transformers when transformers of dissimilar characteristics are paralleled in order to obtain an equitable distribution of load between them. The saturable iron-core reactor is used to control the value of inductive reactance (in theater lighting, for example).

Maintenance

Little maintenance is required for reactors. Care should be taken to keep the cooling paths for both air and oil units clear. Conductors and supports should be examined periodically for cracks or other signs of excess stress. Connections should be checked for tightness.

CAPACITORS

Function

A capacitor is a device for introducing capacitance or capacitive reactance in a circuit, its primary purpose being to counteract the effects of inductive reactance. It reduces the impedance of a circuit with a consequent reduction in the voltage drop and an improvement in the power factor. By reducing the current required to supply a given load, it decreases losses in the circuit.

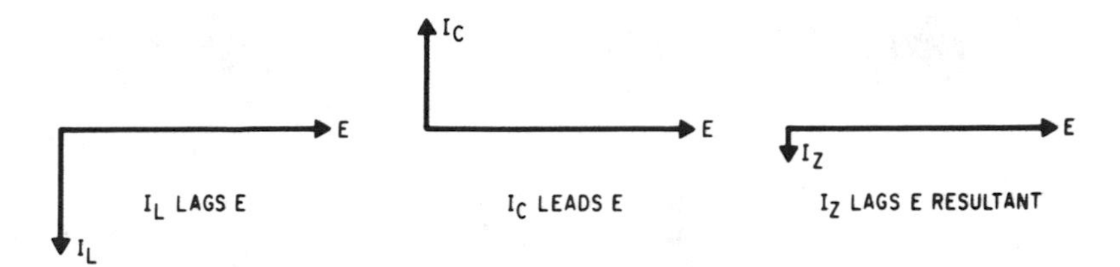

Figure 16-3 Effect of capacitor in circuit.

Principle of Operation

A capacitor consists of two parallel plates that are conductors, separated from one another by an insulating or dielectric material. The interacting magnetic fields that are found in motors, transformers, and overhead circuit wires that parallel each other are all causes of inductance. In an ac system, the current flowing in the circuit supplying the equipment will therefore "lag" the voltage. The alternating movement of electrons to and from each of the plates of the capacitor when an ac voltage is applied across them constitutes a flow of current, but one that will appear to "lead" the voltage. If a capacitor is connected as a load on the circuit, the current it will cause to flow will lead the voltage. If the current in the circuit lags the voltage because of the inductance previously mentioned, the effect of the leading current will be to neutralize the lagging current, thereby improving the power factor of the circuit and reducing the current flow necessary to supply the load (Fig. 16-3).

Another way of looking at this phenomenon is to envision the capacitor as a source of supply for volt-amperes, or *reactive power,* that will tend to neutralize the effects of inductive reactance. Capacitors are therefore normally rated in the kVA of reactive power they supply. Their capability, or capacity, depends on the size of the plates, the distance between them, and the kind of insulating or dielectric material used.

Construction

Capacitor plates may be made of copper, brass, or aluminum, and the insulating material may be air, vacuum or gas, mica, porcelain, or other ceramic, glass, oil-treated paper, or a special plastic. The most practical plates in general use however, appear to be made of aluminum foil separated by oil-impregnated paper or plastic. These are first wrapped into bundles, and a number of such bundles are then connected in series-parallel to attain the required capacitance and voltage rating. The whole unit is usually enclosed in a metal case or tank and the leads brought out through bushings of porcelain (Fig. 16-4).

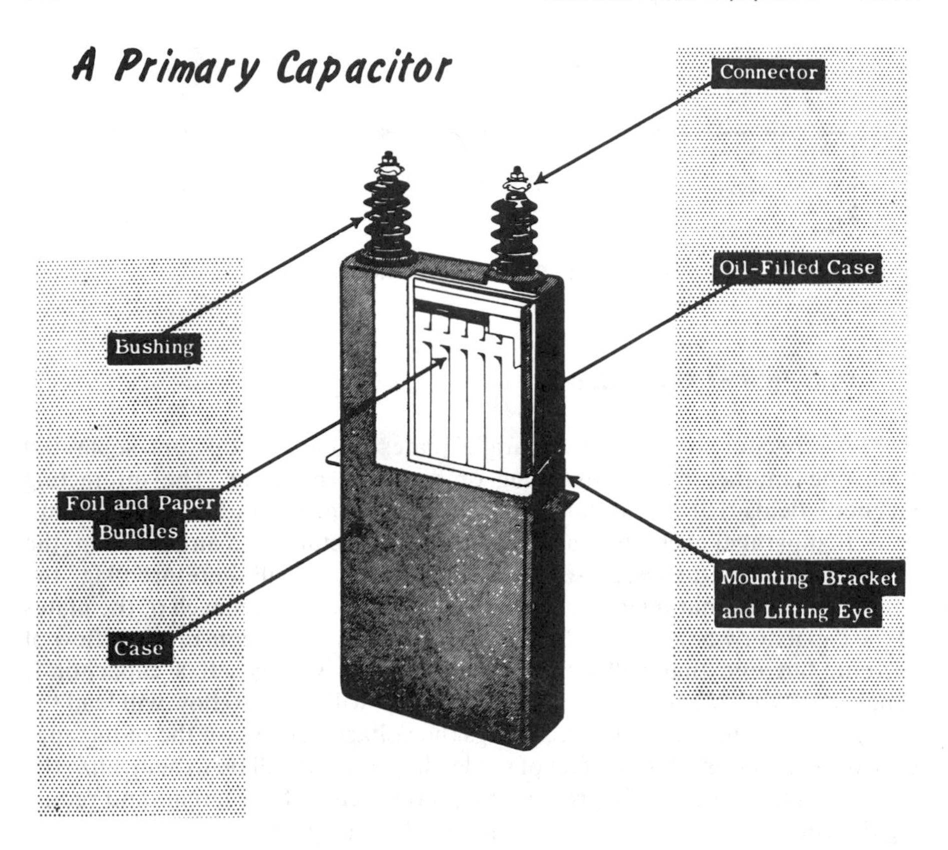

Figure 16-4 Capacitor construction.

Operation

Banks of such units are assembled to provide the necessary capacitance or reactive power supply. They are usually installed as close as possible to the equipment to be regulated, which may be an inductive load. In this way, the current supplied over the circuit will be as close to being in phase with the voltage as possible; that is, the current will neither lag nor lead the voltage. In practice, this perfection is almost never attained. Capacitors may also be installed in the circuit itself, at intervals along the line, to offset the effects of the inductance of the circuit itself.

Capacitors may be installed outdoors on poles, in racks, or enclosures (Fig. 16-5). They may also be installed indoors in vaults or other suitable enclosures (Fig. 16-6). They may be protected by circuit breakers, switches, fuses, lightning arresters, or a combination of any of these. Each capacitor unit is usually individually fused, with the case of each unit connected to ground.

Since loads on a circuit may be constantly varying (as motors start and

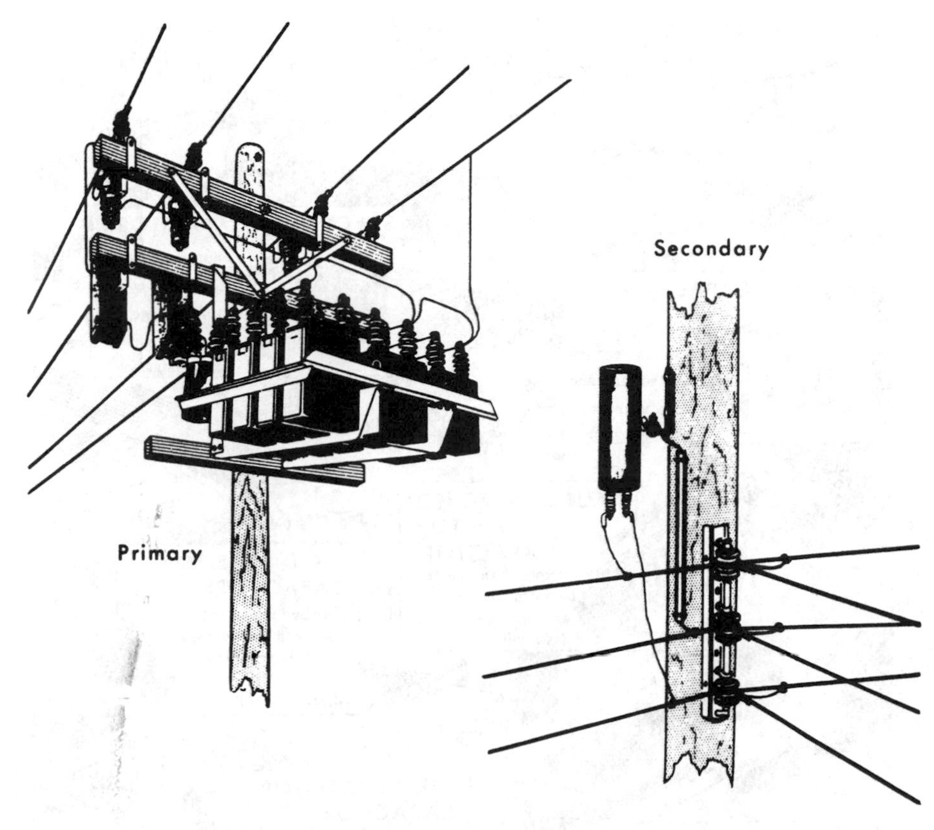

Figure 16-5 Capacitor installation on poles.

stop, for example), capacitors may be connected or disconnected from the system in order to supply just enough capacitance (approximately) to cancel out the inductance of the system. Banks of capacitors may be switched on and off by means of time-clocks, by over- and under-voltage relays, or by other devices.

One relatively large block of capacitance that is almost continually connected to electrical systems is that of underground cables, especially those with metallic sheaths.

Capacitors connected in parallel with loads correct the component of current caused by inductive reactance. Capacitors connected in series with the line compensate for the reactive voltage drop in the circuit. Although shunt capacitors are used widely for power-factor correction, series capacitors are used in low-voltage, heavy-current applications, such as furnaces and welders, to compensate for the voltage drop in the conductors. Series capacitors are also used in high-voltage transmission lines to compensate for the voltage drop caused by inductive reactance and to increase the amount of power transmitted.

Figure 16-6 Where and how capacitors are used.

Dielectric absorption. When capacitors are disconnected from the supply, they remain generally in a charged state. Considerable energy is stored in a capacitor in this condition, and there is a voltage present between its terminals. This phenomenon is called *dielectric absorption.* If the capacitor were left in this charged state, anyone in contact with equipment might receive a dangerous shock, or the equipment itself might be damaged by an accidental short circuit. Capacitors, therefore, are provided with means of draining the stored energy. The discharge equipment may consist of resistors or inductive coils (reactors) permanently connected to the terminals of the capacitor bank. If the capacitors are connected directly to other equipment without the interposition of switches or over-current devices, however, no discharge equipment is required, since the charges will drain off rapidly through the windings of the equipment.

Safety interlocks. To prevent exposure to energized or charged capacitor banks, interlocks are sometimes provided for the vaults or enclosures in which the banks are installed. These interlocks may be mechanical, requiring the associated breaker or switch to be opened and some time delay introduced before access to the capacitors can be obtained. Other systems provide electrical interlocks to accomplish the same purpose.

Maintenance

Very little maintenance is required for capacitors. Regular inspection of capacitor installations should include a check of bushings, ventilation, fuses, voltages, and ambient temperature as well as a cleaning of surfaces for removal of dust or other foreign material. It is particulary important that bushings and other insulating surfaces be kept clean. Capacitors that are leaking insulating fluid should be taken out of service immediately. If minor, the leak may be soldered over; major leaks, broken bushings, or other internal or external damage can best be repaired in the shop or factory.

RECTIFIERS

Function

A rectifier is a device that converts alternating current into a unidirectional current by virtue of a characteristic that permits an appreciable flow of current in only one direction. More popularly, it is a device that converts ac to dc.

Principle of Operation

The general principle of operation of all rectifiers is that the action of the device permits current to flow freely in one direction but presents a great resistance to its flow in the opposite direction. There are many types of rectifiers, but two are used extensively: the mercury-arc rectifier for large loads and the contact or disc rectifier for smaller loads.

Mercury-arc rectifier. The mercury-arc rectifier consists of a metal or glass chamber with a pool of mercury at the bottom and one or more anodes made of graphite or iron supported above the surface of the pool at a proper distance (Fig. 16-7). A vacuum pump maintains a low pressure within the chamber. A movable auxiliary starting anode is used to strike an arc, which produces mercury vapor from the pool. The ionized mercury fills the chamber, and gaseous conduction is established between the mercury pool and the anodes. This conduction tends to sustain itself as long as the current does not drop to zero and a positive potential (with respect to the mercury pool) is maintained.

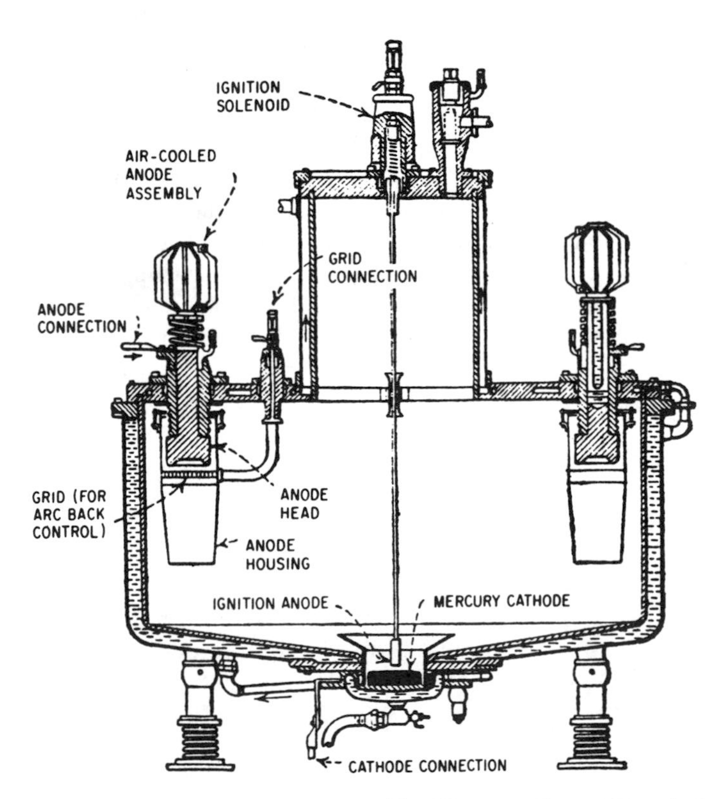

Figure 16-7 Tank-type mercury-arc rectifier (*Courtesy,* General Electric Co.)

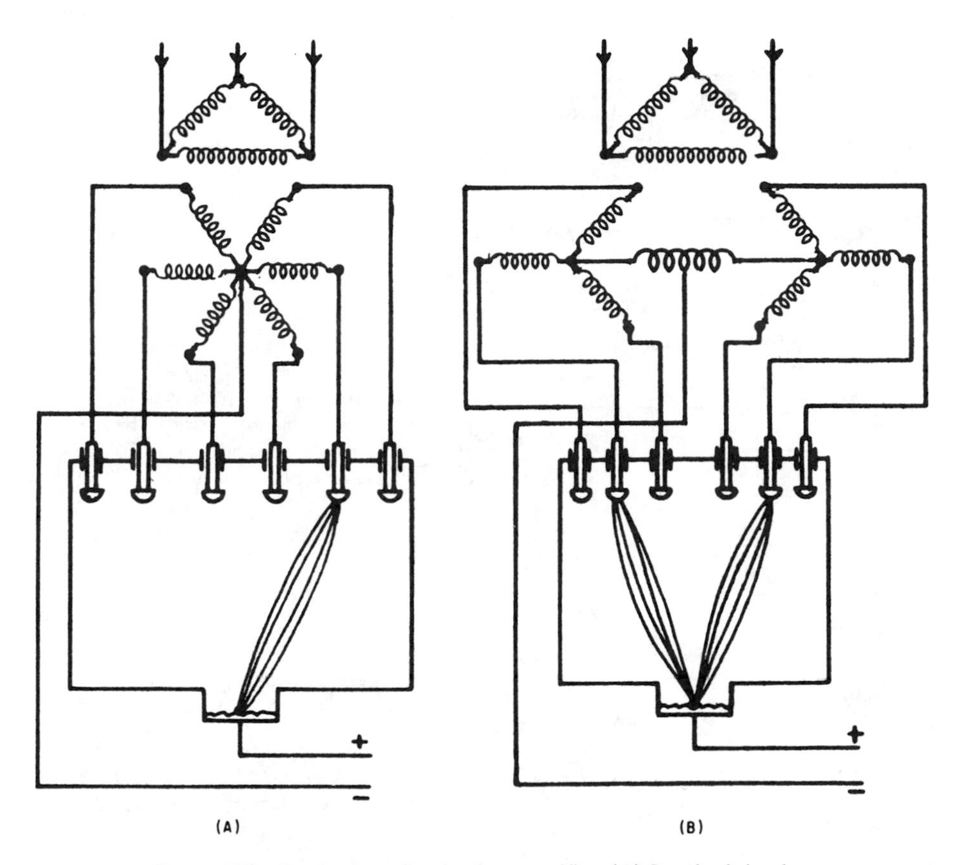

Figure 16-8 Connections for six-phase rectifier: (A) Standard six-phase connection, and (B) double three-phase connection.

In three-phase or six-phase rectifiers (Fig. 16-8) under normal operating conditions, conduction will shift from one anode to another so that current flow will be maintained.

In single-phase rectifiers (Fig. 16-9), the current drops to zero twice in a cycle, and an auxiliary set of anodes is provided. These anodes are connected to a supply voltage in series with an inductance. The inductance makes the current lag the voltage so that conduction will be maintained by the auxiliary set until the regular anode can continue on from the zero point of the cycle. This arrangement is sometimes known as the *keep alive circuit*.

Since the ionized mercury permits current to flow in only one direction, only half the ac cycle flows through the rectifier. A second rectifier, with polarity reversed, can be used to permit both halves of the cycle to be converted to dc. The dc voltage thus produced will be serrated. Its sharp peaks can be smoothed somewhat by placing a reactor in the circuit. Further smoothing can be obtained by using a three-phase supply that will produce dc from six

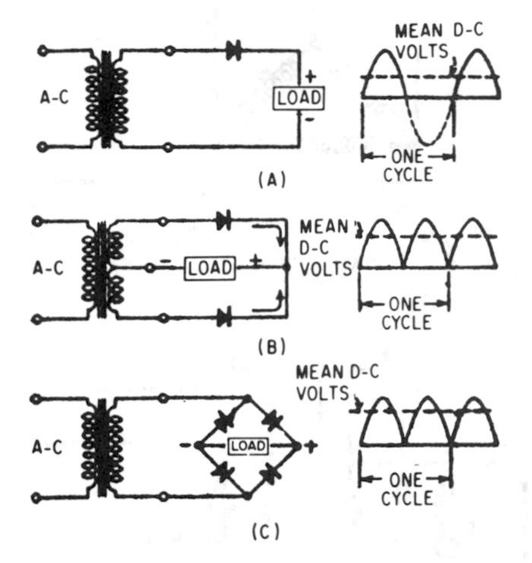

Figure 16-9 Single-phase rectification: (A) Single-phase, half-wave circuit; (B) single-phase, full-wave circuit; and (C) bridge circuit (single-phase, full-wave).

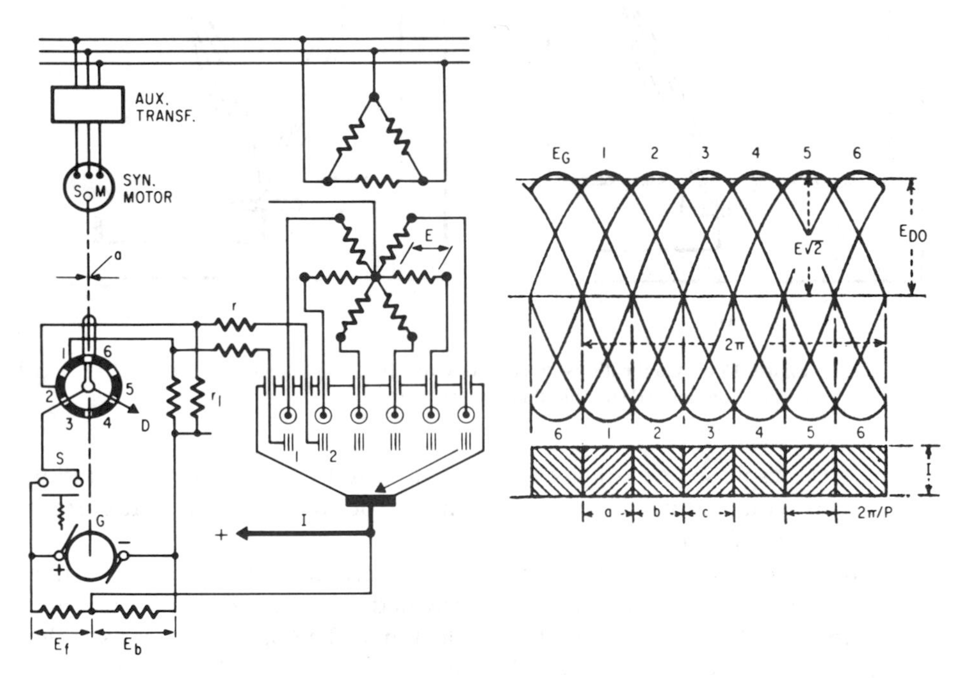

Figure 16-10 Six-phase rectification with twelve half cycles.

half cycles. The three-phase supply can be transformed into a six-phase supply that will produce dc from twelve half cycles (Fig. 16-10).

Contact or disc rectifier. These devices (Fig. 16-11) have the property of allowing current to pass through the contact surface of two materials much more easily in one direction than in the other. The contact surface

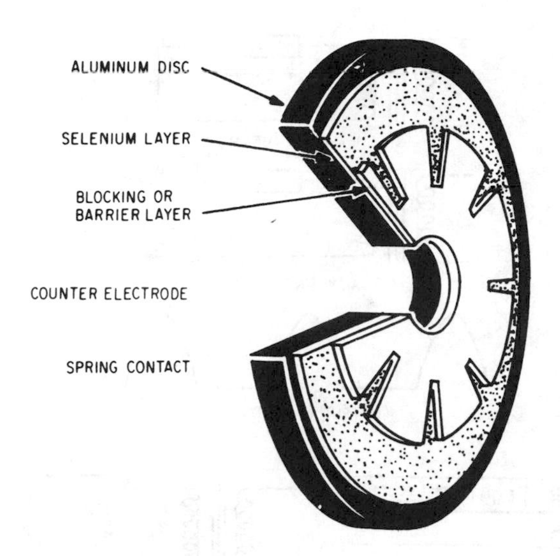

Figure 16-11 Cross section of selenium contact rectifier cell (*Courtesy,* General Electric Co.)

between the two materials is called the "barrier" or "blocking" layer. Typical dc voltage characteristics are shown in Fig. 16-12.

Two common types of these rectifiers are the copper oxide and the selenium rectifiers. Both types are made up of series or series-parallel combinations of blocking layer units. In the copper-oxide unit, the blocking layer is the juncture between a thin film of copper oxide sandwiched between two copper discs. In the selenium unit, the blocking layer is the juncture between a thin layer of selenium sandwiched between the iron disc and a disc of some conducting material.

Since such rectifiers transform only a half cycle of ac to dc, combinations can be arranged for rectification of both halves of the ac cycle (Fig. 16-13). Since these units are usually used for only small quantities of dc power, no three-phase and six-phase supply circuits are employed.

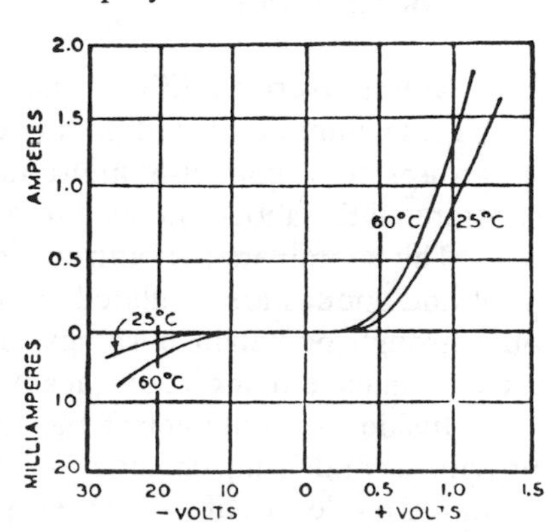

Figure 16-12 D–C voltage characteristics of selenium rectifier cell (*Courtesy,* General Electric Co.)

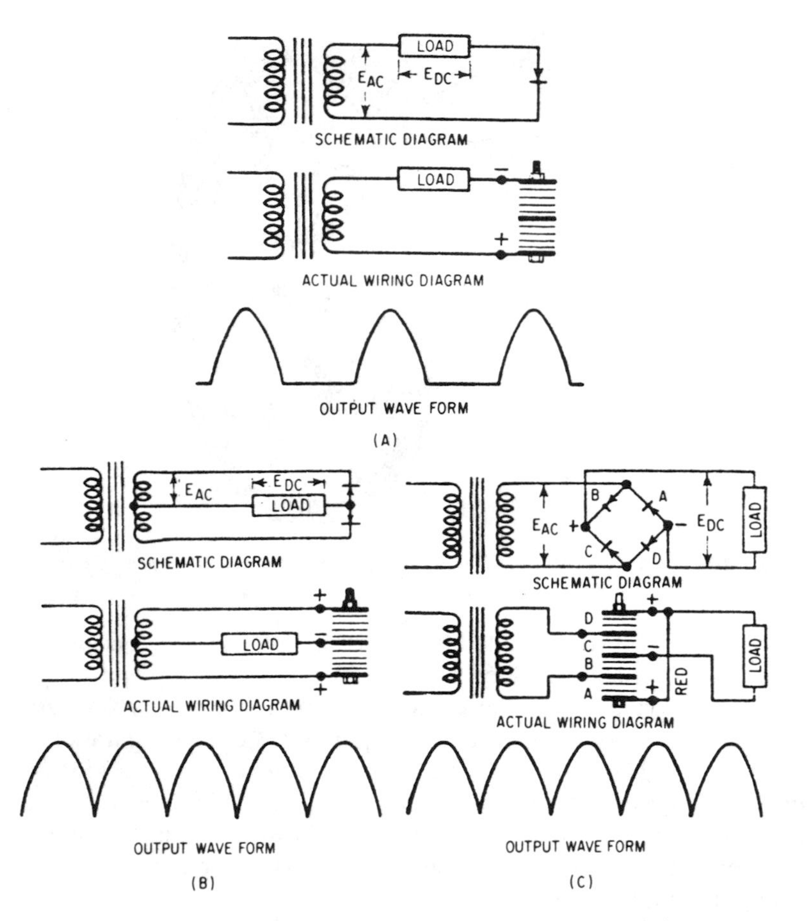

Figure 16-13 Methods of connecting contact or disc rectifiers: (A) Single-phase, half-wave rectifier, (B) single-phase, full-wave rectifier, and (C) full-wave bridge rectifier (*Courtesy,* General Electric Co.)

Construction

Mercury-arc rectifier. Large-capacity mercury-arc rectifiers may be of the single-anode or multiple-anode types. Single-anode types are combined in groups to accomplish multiphase rectification. Multianode types concentrate all the anodes in one large container.

The containers for both types are built of steel, and both the mercury pool and anodes are insulated from the container. Connections are brought out through porcelain bushings. The container is normally constructed to include an enveloping water jacket for cooling.

Inside the container at the top, a space is left within a cooling coil to provide a condensing chamber for the mercury. Anodes made of graphite and mounted on sleeves of steel are suspended from the cover and so arranged that

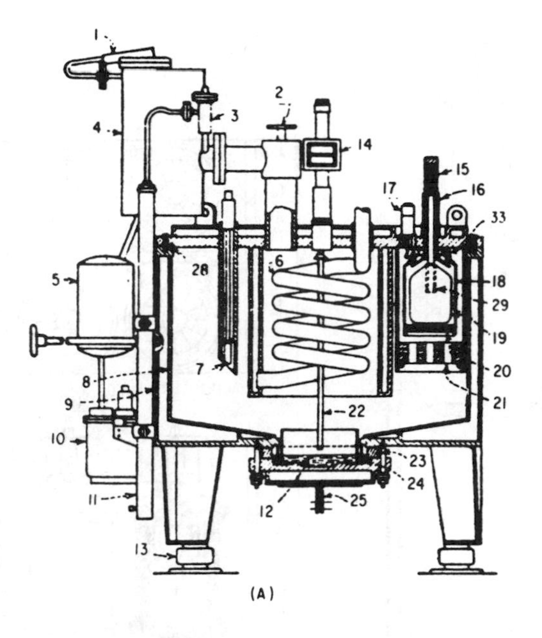

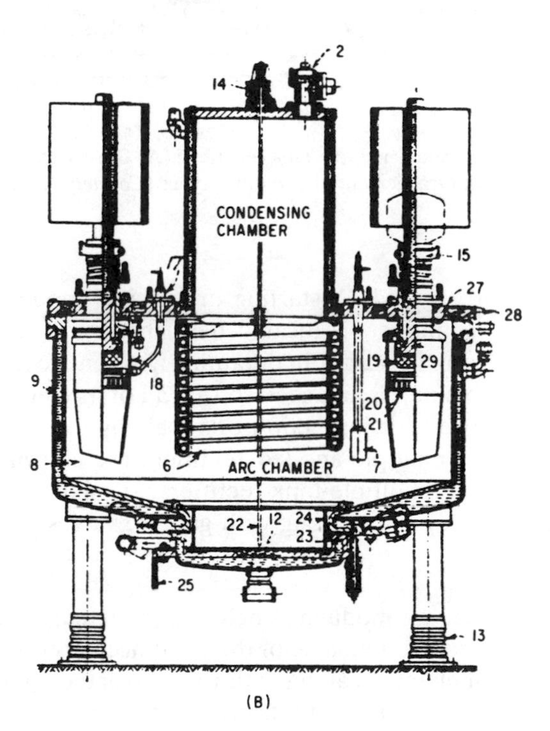

Figure 16-14 Two designs for single-tank, multiple-anode rectifier (*Courtesy*, General Electric Co.)

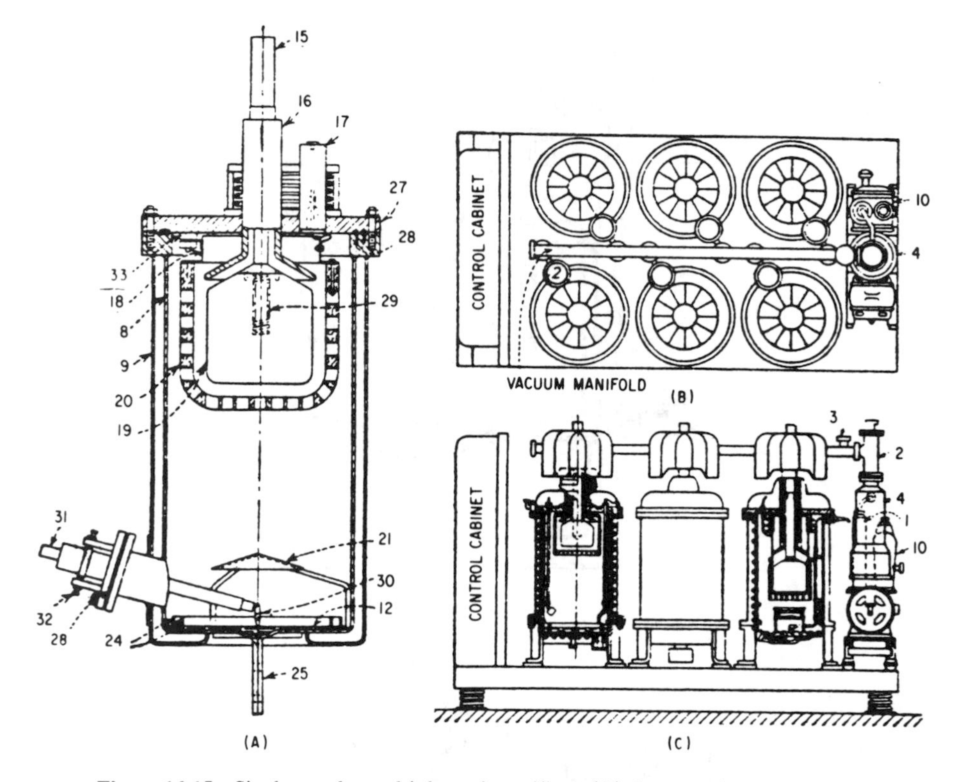

Figure 16-15 Single-anode, multiple-tank rectifiers: (A) Cross section of one type, and (B) and (C) tank arrangement of two other types (*Courtesy,* General Electric Co.)

they can be raised or lowered. A starting or ignition anode, operated by a solenoid outside the container, is located near the center of the container. Other auxiliary anodes and grids for maintaining ignition are fastened near the main anodes. The whole unit is arranged so that all of its elements, except the mercury, can be removed for inspection or repair by lifting the cover. Two designs of single-tank, multiple-anoded rectifiers are shown in Fig. 16-14. Designs for single-anode, multiple-tank rectifiers, are shown in Fig. 16-15.

Water and vacuum pumps, along with gauges, valves, and controls, are mounted outside the unit.

Ignitrons. These are modular, single-phase mercury-arc rectifier units in which the ignition and maintenance of the arc is accomplished by electronic means. Their essential elements along with the associated control circuits are shown in Fig. 16-16. Banks on such units provide for polyphase supply and for the required capacity of rectification.

Contact or disc rectifier. Both copper oxide and selenium rectifier units are arranged in stacks with a specified number of them connected in series, depending on voltage requirements. They are usually assembled on an insulated bolt (Fig. 16-17). The units are designed to have radiating fins and spacers between them for cooling purposes. For most applications, a natural circulation of air is sufficient. For rectifiers of about 500-watt capacity, however, fans provide forced cooling.

Operation

The operation of rectifiers has already been described in general terms. Banks of mercury-arc and contact-disc types, are used to supply capacity requirements. Standard breakers and protective equipment can be used with mercury-arc rectifiers. Breakers on both the ac and dc sides should be trip-free throughout the closing cycle; their tripping times, including arcing and relay time, should be 8 to 12 cycles. For the sake of economy, it is possible to use a multiple breaker scheme to open the associated transformer secondary circuit to each anode. For smaller units, fuses are sometimes specified.

Before mercury-arc rectifiers are placed in service, or after long periods of idleness, they must be "heated out." Heat is applied to the container to drive out any moisture and entrapped gasses that might affect the operation of the rectifier. More specifically, this precaution is taken to prevent "arc back,"

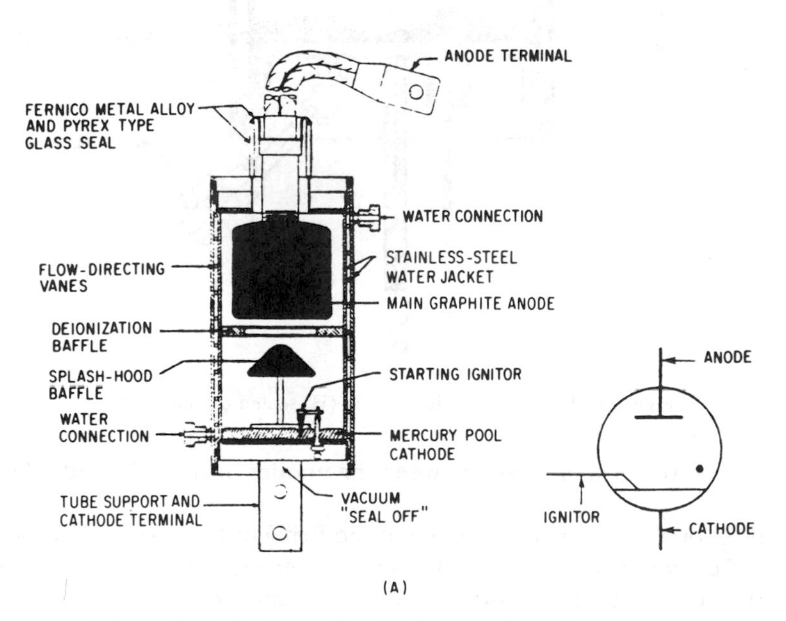

Figure 16-16 Ignitron tube: (A) Cross section and schematic.

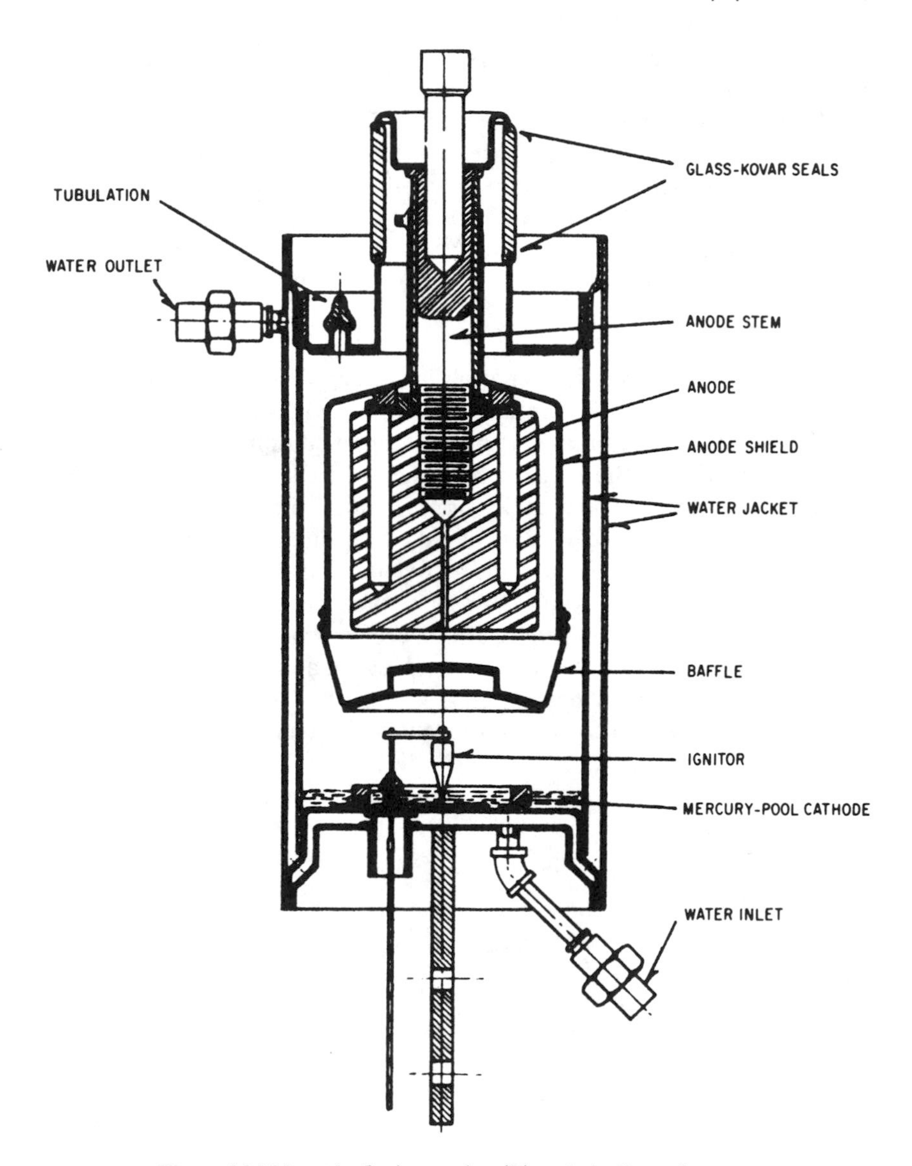

Figure 16-16 (cont.) Ignitron tube: (B) sealed-off metal type.

that is, the formation of an arc between the anodes or from the cathode to the anode.

By proper control of the timing of the firing of the anodes, the mercury-arc rectifier can be made to operate as an "inverter," that is, it can change dc to ac. Furthermore, it can also be used to convert ac from one frequency to another.

Rectifiers are used for welding purposes and for the control of motors in

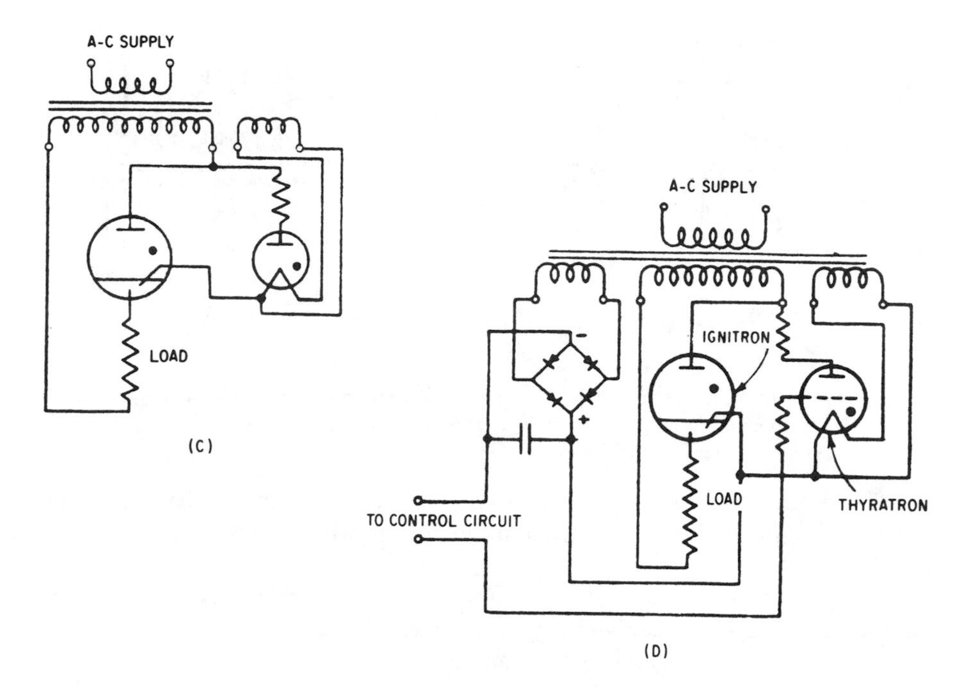

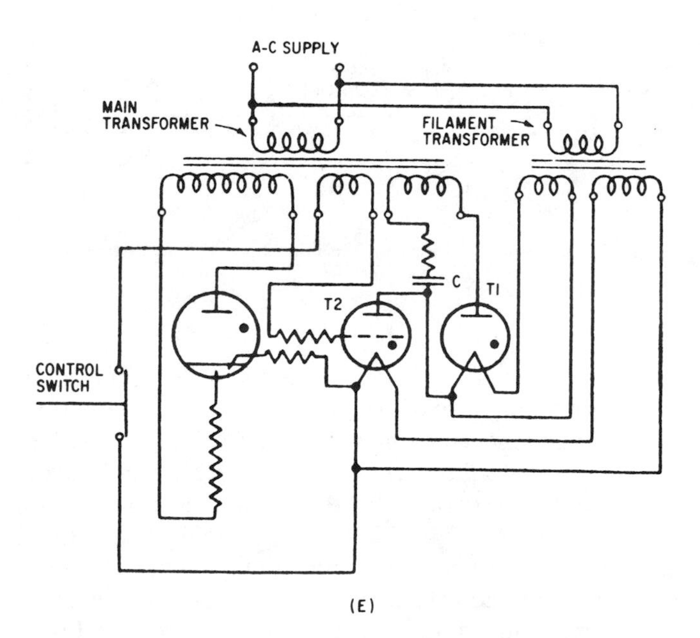

Figure 16-16 (cont.) Ignitron tube: (C) Circuit of tube controlled by a gas-filled diode (load-current ignition). (D) circuit of tube controlled by a thyraton (load-current ignition), and (E) circuit of tube controlled by a capacitor discharge (separately excited ignition) (*Courtesy,* General Electric Co.)

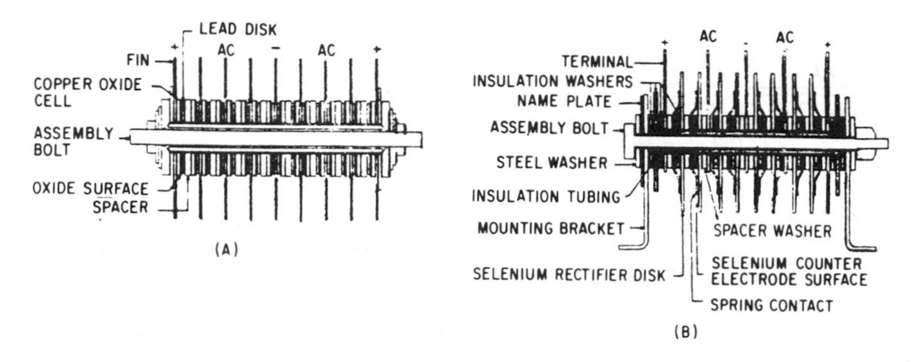

Figure 16-17 Contact or disc rectifiers: (A) Assembly and cross section of copper-oxide rectifier stack, and (B) assembly and cross section of selenium rectifier stack (*Courtesy,* General Electric)

such applications as elevators, electrolytic processes, traction systems, steel and textile mills, mines, and other applications where rapid regulation of speed is desirable.

The ability to convert ac to dc and dc to ac is taken advantage of by high-voltage electric transmission systems (Fig. 16-18). Such systems eliminate the inductance and capacitance effects of ac circuits.

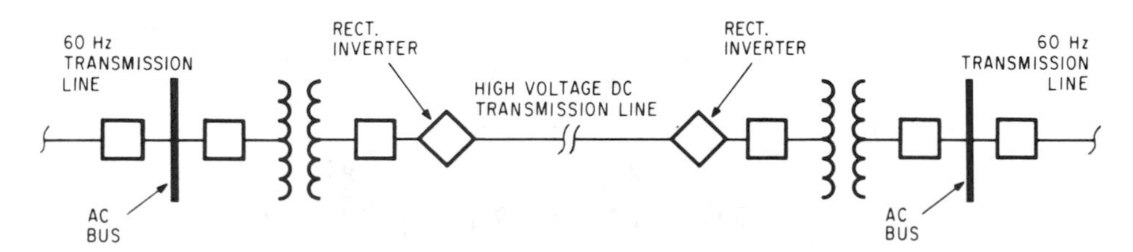

Figure 16-18 Direct current transmission over very long distances.

Maintenance

Maintenance of water pumps and jackets and coils to prevent corrosion in cooling systems and taking care of vacuum pumps and containers constitute a great part of the maintenance of mercury-arc rectifiers. Dressing and replacement of anodes requires the shutdown of the rectifier and the opening of the container. The unit must then be heated out and evacuated before being restored to service. Cleaning of bushings, and scraping and painting of the container, replacement of gaskets, checking and tightening of water and vacuum lines, lubrication of motors, and other normal routine maintenance activities are all necessary in maintaining mercury-arc rectifiers. Practically no maintenance, except for an occasional dusting, is required for contact-disc rectifiers. For causes and remedies when troubleshooting reactors, capacitors, and rectifiers, see Table 16-1.

TABLE 16-1 Troubleshooting Reactors, Capacitors, and Rectifiers

Condition	Possible Cause	Suggested Remedy
	Reactors	
Overheating of coils	Overloads	Reduce load or replace.
	High ambient	Relocate or provide forced ventilation.
	Dirt or dust	Clean.
	Short circuit	Locate and repair.
	Capacitors	
Overheating of unit	Overload	Reduce load or add units.
	High ambient	Relocate.
	Short circuit	Replace unit.
Fuse blown	Short circuit	Replace unit.
	Overcurrent	Check voltage and protective devices.
	Excessive temperature	High ambient; relocate.
		Short circuit; replace and improve ventilation.
Bushing flashover	Lightning	Clean or replace unit if necessary.
	Dirt and atmospheric pollution	Clean or replace unit if necessary.
	Rectifiers	
High operating temperature	Defective water cooling; system plugged	Remove rectifier from service. Check strainer, cooling coil, and jacket. Clean with chemicals as specified. Replace with clean, distilled water.
	Water regulator not functioning	Remove, clean, and repair. Replace if necessary.
	Water pump not functioning	Check fuses, motor, and impeller. Repair or replace.
	Insufficient water	Check water gauges. Check for leaks. Add water in surge tank.
Poor vacuum	Faulty vacuum regulator	Check motor, copper-oxide supply rectifier, circuit, connections, broken wire. Repair or replace.
	Failure of mercury pump (from overheating, too little mercury, or faulty water cooling)	Repair or replace.

TABLE 16-1 (cont.) Troubleshooting Reactors, Capacitors, and Rectifiers

Condition	Possible Cause	Suggested Remedy
	Gassing of rectifier (from excessive load)	Reduce load and check gauges.
	Air leak in rectifier	Remove load from unit and check for leak. Repair. Heat out before putting back in service.
Anode misfire	Igniter misfire	Replace with new igniter.
Arc back	Temperature	Check cooling systems.
	Vacuum	Check vacuum system.
	Tank problem	Check tank for leaks, dirt, or impurities. Clean tank, heat out, and restore to service, or replace.
Vibration	Vacuum pump	Check pump and mounting. Repair or replace if necessary.
	Water circulating pumps	Check pumps and mounting. Repair or replace if necessary.
	Loose bolts, clamps, etc.	Tighten.
Overheating of contact-disc rectifier	Overload	Reduce load or increase capacity.
	Dirt or dust	Clean; ventilate.
	Defective unit or units in stack	Replace stack.

REVIEW

1. Reactors are principally used to limit short-circuit currents (Fig. 16-1); capacitors to reduce current flow and maintain voltage by improving power factor (Fig. 16-3); and rectifiers to change alternating current into direct current.
2. A reactor consists of a coil of wire. Current flowing through it produces a magnetic field which in turn produces a counter-voltage in the coil that acts to limit the flow of current. They are normally connected in series with the line or equipment.
3. Reactors may have an air-core or an iron-core; some may also be immersed in oil (as are transformers). They are built ruggedly to withstand the heavy mechanical strains imposed on them when heavy short-circuit currents flow through them. They require little maintenance.
4. Capacitors counteract the effect of inductance in a circuit, thereby reducing the voltage drop and improving power factor (Fig. 16-3). By reducing the current required to supply a given load, they decrease the losses in a circuit.

5. A capacitor consists of two parallel plates that are conductors, separated by insulating or *dielectric* material. The electrostatic interaction between them produces effects that are in opposition to those produced by magnetic fields and hence, tend to cancel them. They are rated in volt-amperes, or more practically, in kilovolt-amperes, or kVA. Little or no maintenance is required.
6. A rectifier is a device that converts ac to dc. The device generally permits a free flow of electricity in one direction and a great resistance to its flow in the opposite direction.
7. In the mercury arc type (Fig. 16-7), vaporized mercury provides the conducting path which permits current to flow only in one direction. When the ac supply is single-phase (Fig. 16-9), the current drops to zero twice in each cycle and difficulty may be experienced in sustaining the arc. With a three-phase supply connected for six-phase operation, twelve such paths produce a more even flow of direct-current (Fig. 16-10). Pumps and other auxiliary equipment require regular maintenance.
8. For smaller amounts of energy conversion, contact or disc type rectifiers may be used. Here the alternating current is made to pass through the contact surfaces of two materials, much more in one direction than the other (Fig. 16-11). The materials usually employed are copper oxide and selenium. These type units require practically no maintenance.

STUDY QUESTIONS

1. What two effects are produced when a current flows through a wire?
2. What is meant by inductance? What is the effect of inductance on the relation between voltage and current in an ac circuit?
3. What is the function of reactors? How are they constructed and where are they used?
4. What is meant by capacitance? What is the effect of capacitance on the relation between voltage and current in an ac circuit?
5. What is the function of a capacitor? How are they constructed and where are they used?
6. What is meant by impedance? Power factor? Resonance?
7. What is the function of a rectifier? Where is it used?
8. Name two basic types of rectifiers. What are their principles of operation?
9. What are ignitrons? How are they controlled?
10. What precautions should be taken in connecting and disconnecting reactors? Capacitors? Rectifiers?

17

INSTRUMENTS

FUNCTION

Instruments are needed to determine whether electrical circuits and equipment are operating safely and efficiently. By measuring the values of the electrical quantities involved, operators can pinpoint what is occurring within circuits and equipment. Comparison of these values with predetermined design specifications and with previously experienced values, allows abnormal or troublesome conditions to be detected before they occur. The data thus obtained not only improves the effectiveness of circuits and equipment but also may be used to develop adequate designs for future demands and requirements.

PRINCIPLE OF OPERATION

Modern instruments for the detection and measurement of current and voltage are all based on the interaction between two or more magnetic fields. These fields may be created by a conductor or conductors carrying a current or by a permanent magnet (Fig. 17-1). By combining these basic current and voltage elements, instruments are able to determine values of electrical resistance, power, energy, power factor, and other desired quantities.

In their simplest form, many of these instruments are small but accurate motors. In the measurement of most quantities, the tendency for rotation of the small motor is pitted against a spring. In measuring other quantities, the motor is allowed to turn, and the revolutions are counted on a register.

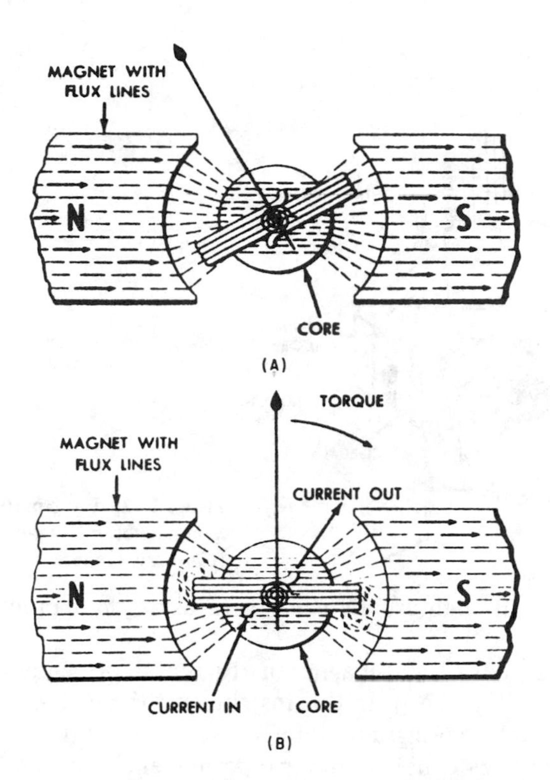

Figure 17-1 Operating principle of permanent-magnet, moving-coil instrument: (A) No current in coil and hence no torque, and (B) direct current in coil with a clockwise torque developed by interaction of coil flux with magnetic flux (broken lines and arrows show direction and distribution of flux) (*Courtesy,* General Electric Co.)

Developments in solid-state technology may make practical corresponding instrumentation based on "electronics" rather than on electromagnetics. Present type electromagnetic-type instruments, however, will continue to be employed in industry for a long time to come.

AMMETERS

Permanent Magnet-Coil Instrument

An ammeter makes use of the interaction between the fields of a permanent magnet and a current-carrying coil (Fig. 17-2). The interaction tends to produce a rotation in the coil while springs on either side resist the rotating force. The amount the pointer of the ammeter moves depends on where the two forces balance; its final position is a measure of the current flowing. This

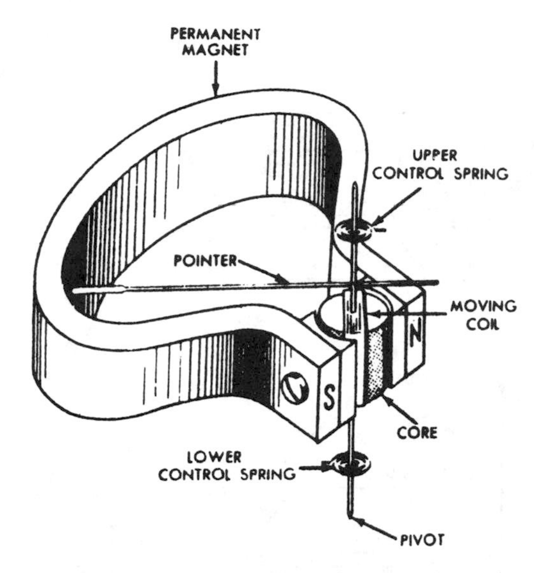

Figure 17-2 Basic parts of permanentmagnet, moving-coil instrument (*Courtesy,* General Electric Co.)

type of ammeter, although usually used to measure dc, can also be used to measure ac.

Related to the permanent-magnet instrument in principle is the moving-magnet instrument (Fig. 17-3). In this instrument, the coil is stationary, and the magnet is movable. Springs are not necessary here as the moving element is kept from turning by the field of the two permanent control magnets. When current flows through the coil, it sets up a magnetic field that interacts with the magnetic field of the control magnets. The magnetic field of the rotating magnet will align itself with the resultant magnetic field, and the instrument pointer moves up the scale.

Electrodynamometer Instrument

In the electrodynamometer, the permanent magnet is replaced with a stationary coil (Fig. 17-4). The strength of its magnetic field, therefore, is not constant but varies with the amount of current in the coil. The tendency for rotation depends not only on the field set up by the current in the rotating coil but also on that set up by the stationary coil. This type of element provides more accuracy in the measurement of low values of current. Curves showing the torque or rotating tendency it develops for current and voltage measurement are shown in Fig. 17-5. Use of an electrodynamometer as an ammeter is shown in Fig. 17-6.

VOLTMETERS

A voltmeter is simply an ammeter to which a fixed (and usually large) resistance is connected in series with the coils. Deriving from Ohm's Law,

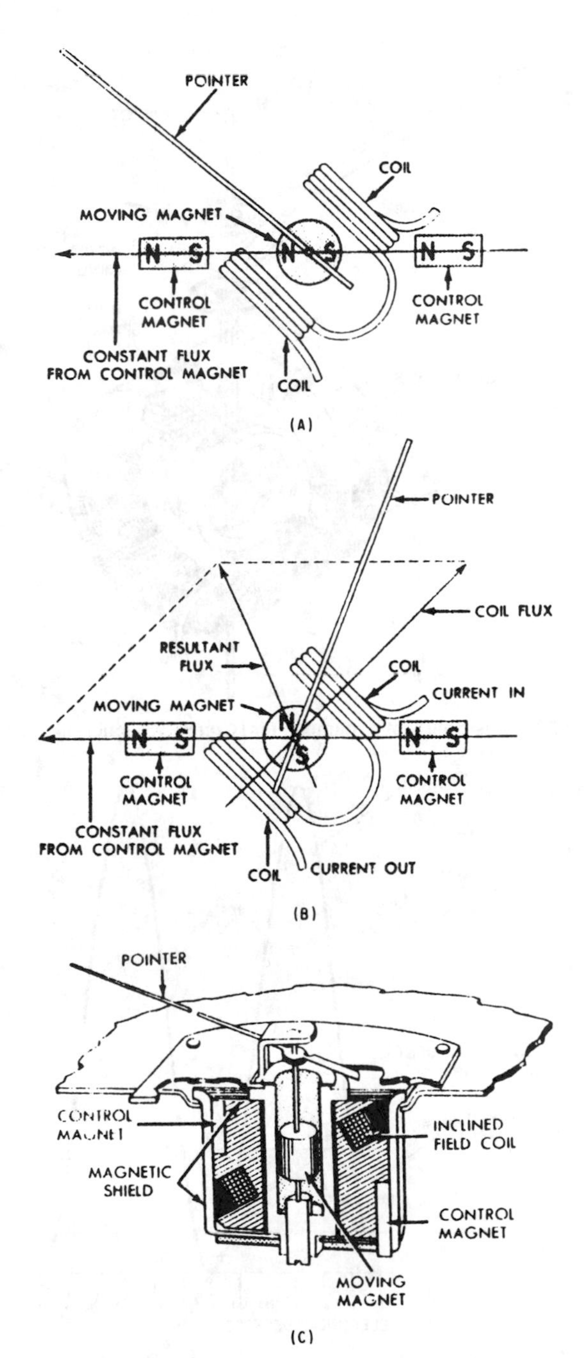

Figure 17-3 Moving-magnet instrument: (A) Cross section, (B) without current in the coil, and (C) vector diagram with current in the coil (*Courtesy,* General Electric Co.)

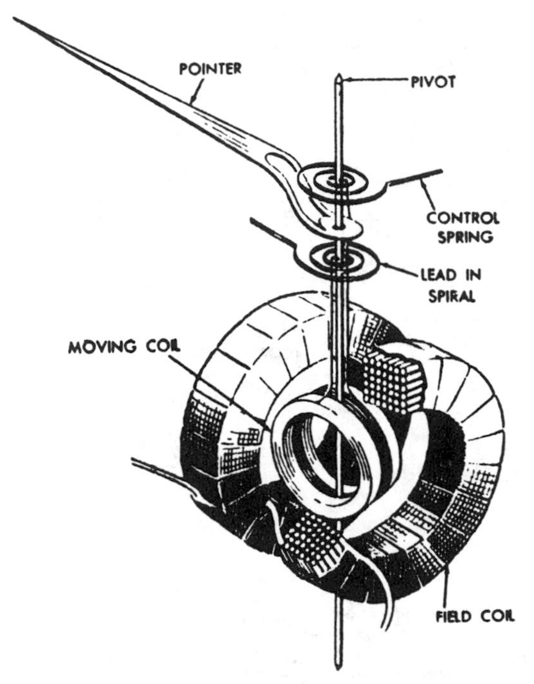

Figure 17-4 Basic parts of dynamometer (*Courtesy,* General Electric Co.)

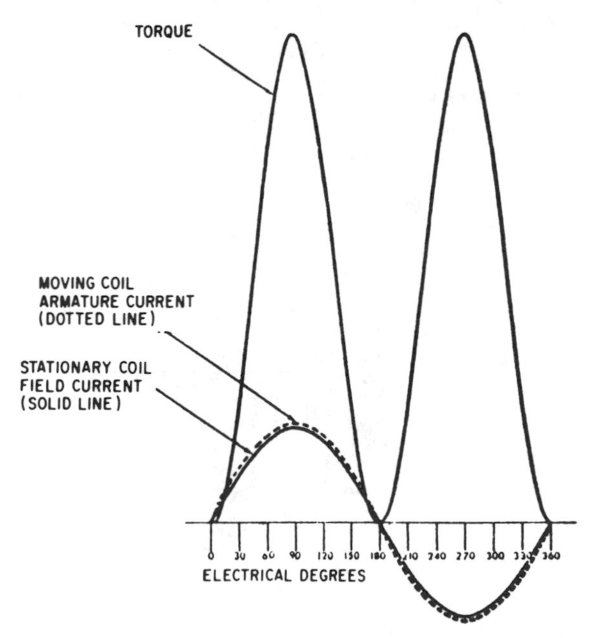

Figure 17-5 Current and torque curves for dynamometer-type ammeter or voltmeter (*Courtesy,* General Electric Co.)

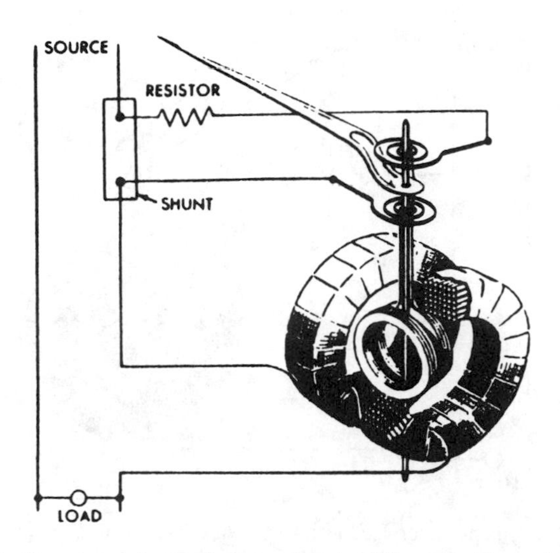

Figure 17-6 Dynamometer used as an ammeter (*Courtesy,* General Electric Co.)

voltage is the product of current and resistance. If the resistance is constant, then the value of the current flowing can be taken as a measure of the applied voltage. The magnetic fields set up by this current, therefore, will reflect voltage values. The scale of the ammeter in this case is calibrated to read volts directly, and the instrument is known as a voltmeter. Figure 17-7 shows diagrammatically the connection of a dynamometer type of voltmeter.

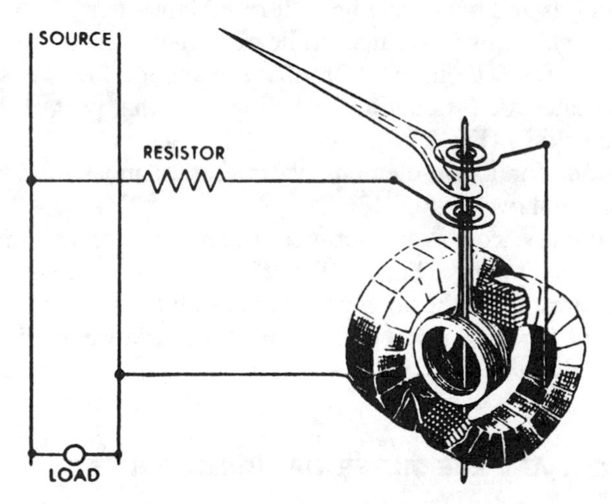

Figure 17-7 Dynamometer used as a voltmeter (*Courtesy,* General Electric Co.)

WATTMETERS

The measurement of power by means of an electrodynamometer element is possible because the tendency for rotation (torque) varies with changes in the

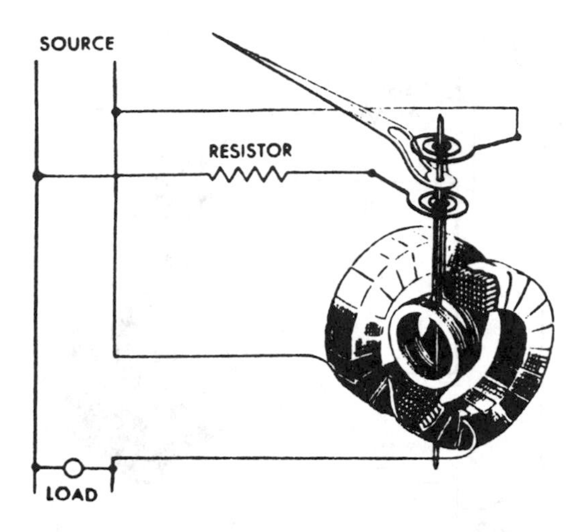

Figure 17-8 Dynamometer used as a wattmeter (*Courtesy,* General Electric Co.)

magnetic fields set up by both coils. These fields, in turn, are dependent on the currents flowing through them. In the simple wattmeter (Fig. 17-8), the magnetic field of the stationary coil is proportional to the line current, that of the moving coil is proportional to the voltage, and the torque is proportional to the product of the line current and voltage. Since this product represents the instantaneous value of power in the circuit, the instrument reads the average of the power pulses and hence can be calibrated in watts.

In an ac circuit, this instrument indicates "real" power, not the product of volts and amperes. The interaction of the magnetic fields is such that the instrument indicates the product of the voltage and that part of the current in phase with the voltage (Fig. 17-9).

Connections for a two-element wattmeter to measure a three-phase, four-wire load are shown in Fig. 17-10.

The stationary coil is sometimes provided with a laminated-core structure of magnetic alloy (Fig. 17-11). This type of element has a greater magnetic concentration in the air gaps, which produces a greater torque. Since it requires less power for a given deflection, it is therefore applicable to small portable instruments.

REACTIVE VOLT-AMPERE METER OR VARMETER

A wattmeter can be made to measure the reactive volt-amperes or *vars* in a circuit. Since the wattmeter indicates the product of the voltage and the component of the current in phase with the voltage, it is necessary to shift the voltage only 90° so that the current in the movable coil will lag that in the stationary coil by 90°. This measurement is accomplished by providing a

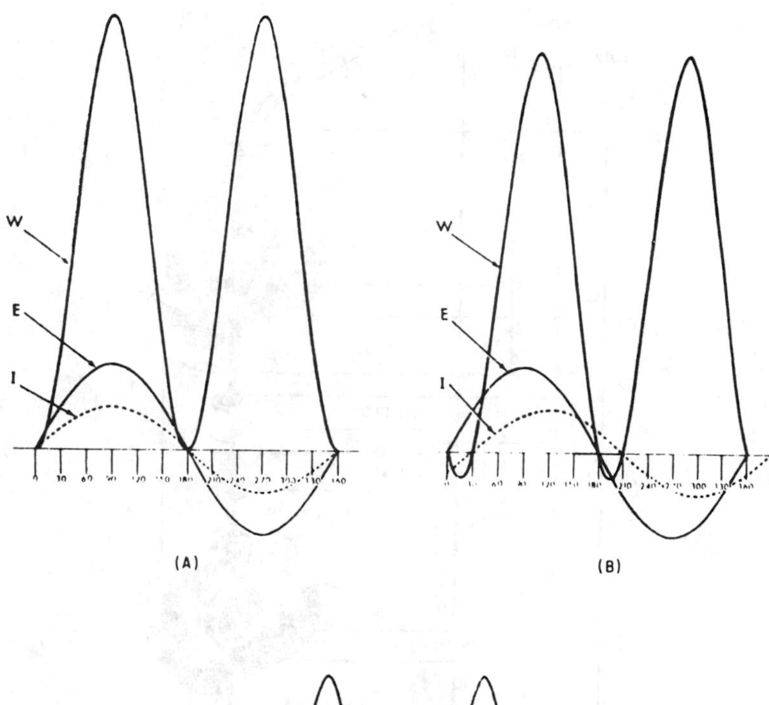

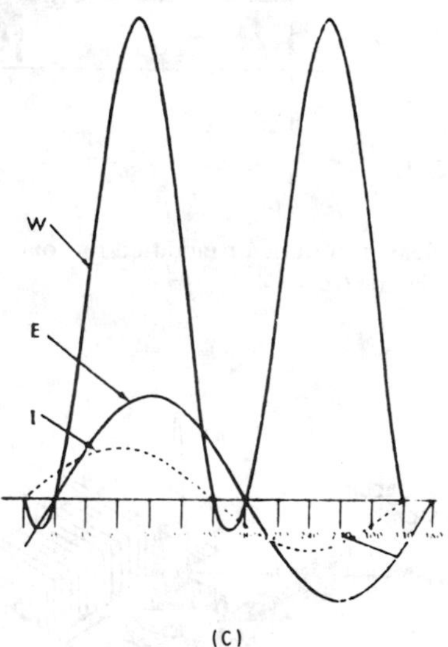

Figure 17-9 Curves of voltage, current, and power for single-phase circuit: (A) Current in phase with voltage, (B) current lagging voltage by 30 deg, and (C) current leading voltage by 30 deg (*Courtesy,* General Electric Co.)

reactance that shifts the phase of the voltage across the potential or movable coil. The magnetic field of this coil will then be in phase with that produced by the reactive component of the current in the field coil.

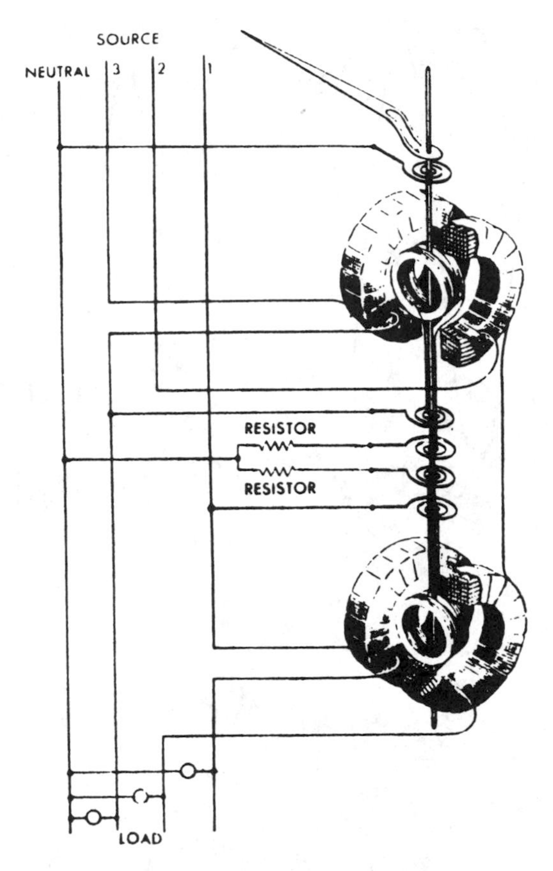

Figure 17-10 Two-element wattmeter modified for four-wire, three-phase circuit (*Courtesy,* General Electric Co.)

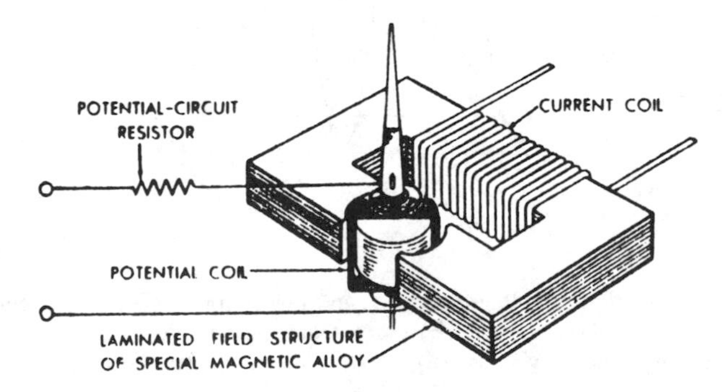

Figure 17-11 Iron-core dynamometer for wattmeter (*Courtesy,* General Electric Co.)

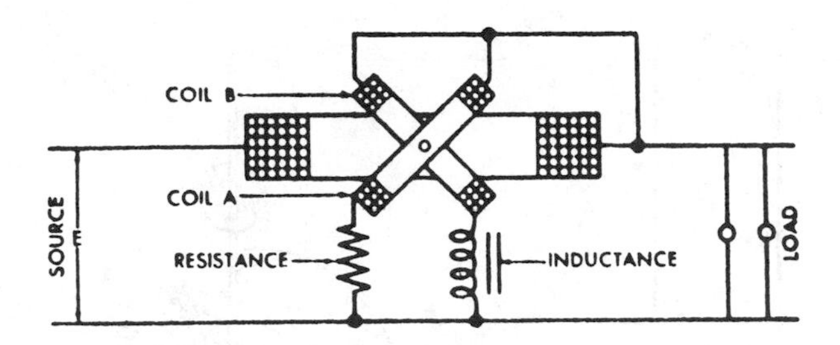

Figure 17-12 Single-phase, crossed-coil power factor meter (*Courtesy,* General Electric Co.)

POWER-FACTOR METER

An adaptation of the electrodynamometer construction for measuring the power factor is shown in Fig. 17-12. In the so-called crossed-coil type, there is one fixed coil and two moving coils mounted on a common shaft at an angle to one another. The instrument is designed for either single-phase or three-phase service. The connections are shown in Fig. 17-13.

FREQUENCY METER

Another adaptation of the electrodynamometer construction for measuring frequency within certain limits is shown in Fig. 17-14.

WATT-HOUR METERS

In the watt-hour meter for measuring electrical energy, a disc takes the place of the movable coil (Fig. 17-15). An electromagnet having two coils is constructed so as to allow the disc to rotate between a pair of poles. One coil is connected to measure current while the other is connected to measure voltage. The magnetic fields created cut the disc, thereby inducing eddy currents within it (Fig. 17-16). The eddy currents, in turn, set up a magnetic field that reacts with the other magnetic fields, causing it to act as a motor.

The disc will therefore rotate since it does not have springs to prevent it from doing so. The speed at which it rotates will depend on the power flowing. A train of gears and a set of dials are arranged in a register to count the number of revolutions made by the disc, this number being a measure of the electrical energy.

Two types of single-phase watt-hour meters are shown in Fig. 17-17 A and B. Schematic diagrams for single-phase two-wire and three-wire meters are shown in Fig. 17-18. Registers for these meters are made in two types, the conventional and the cyclometer type, both of which are shown in Fig. 17-19.

Solid-state technology has made possible adding components to the

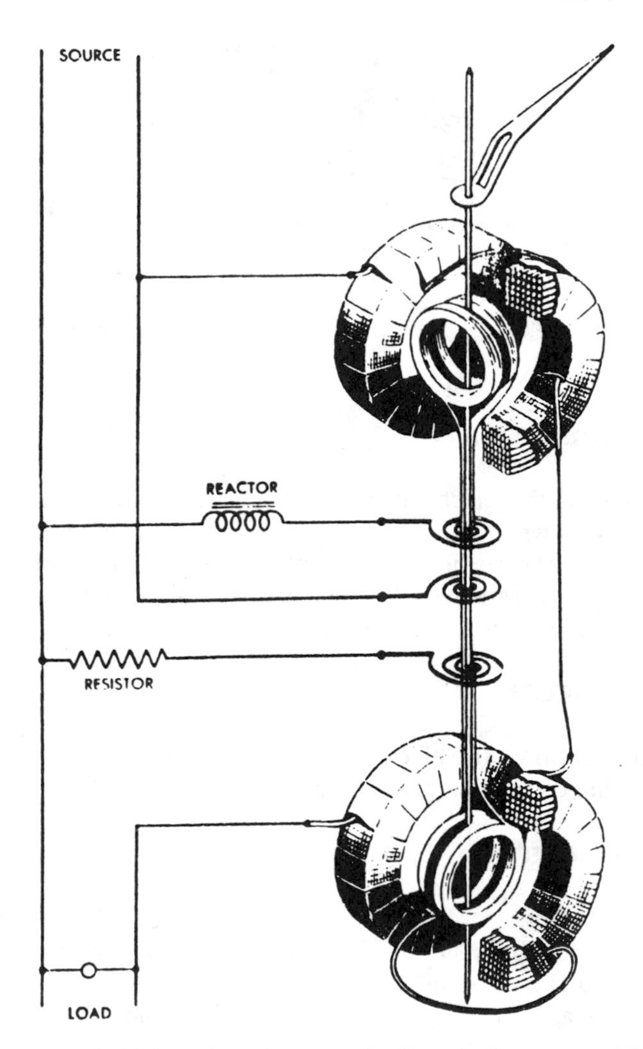

Figure 17-13 Mechanism of another crossed-coil power factor meter (*Courtesy,* General Electric Co.)

registers enabling the display of several functions, including not only watt-hour consumption, but instantaneous demands, time of utilization rates, both daily and seasonal. Encoders permit consumer cooperation in advising when demands are exceeded, with possibilities of control of operation of equipment and appliances. They also bring closer the remote reading of meters via telephone or utility lines (Fig. 17-19C).

Watt-hour meters for measuring energy in two-phase and three-phase circuits operate on the same principle as the single-phase meter but may have

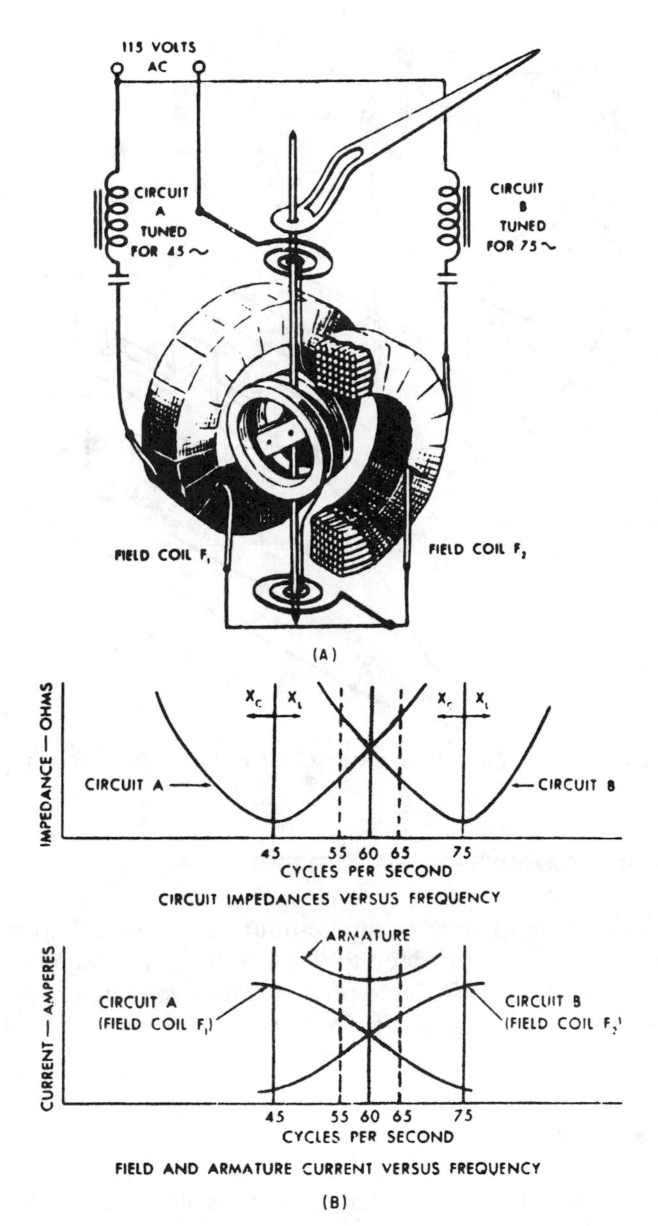

Figure 17-14 Resonant-circuit frequency meter: (A) Schematic, and (B) electric characteristics (*Courtesy,* General Electric Co.)

two or three current coils and a similar number of potential coils. Schematic diagrams comparing them with the single-phase meter connection are shown in Fig. 17-20.

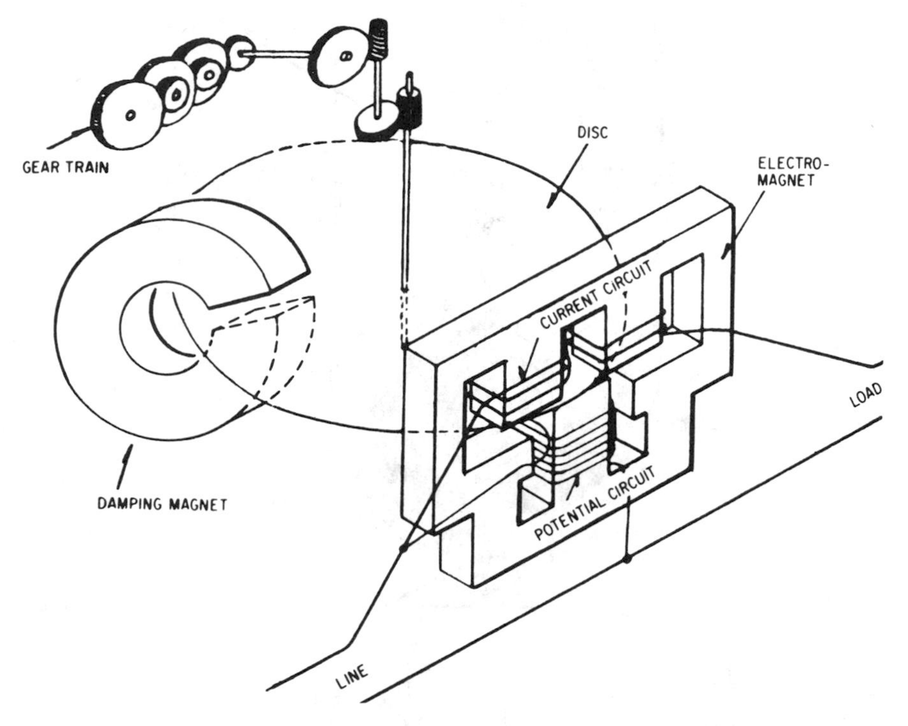

Figure 17-15 Typical watthour meter (*Courtesy,* Westinghouse Electric Co.)

REACTIVE VOLT-AMPERE-HOUR METERS

Meters to measure reactive volt-ampere-hours are essentially ordinary watthour meters with reactances added to displace the voltage applied by 90°. A schematic diagram for a basic single-phase reactive volt-ampere-hour meter is shown in Fig. 17-21. Similar principles apply to two-phase and three-phase meters.

DEMAND METERS

The maximum demand of an installation may be obtained by adding another dial and set of gears to the watthour meter. There are two hands on this dial, one of which is connected to the actuating mechanism and the other of which just "floats," being pushed along by the first hand. The first hand measures the number of revolutions made in a standard period of time, usually 15 minutes (in special instances, a half hour or hour). At the end of this period, another mechanism operates to return the first hand to zero, and the process starts all over. In the meantime, the second hand having been pushed by the

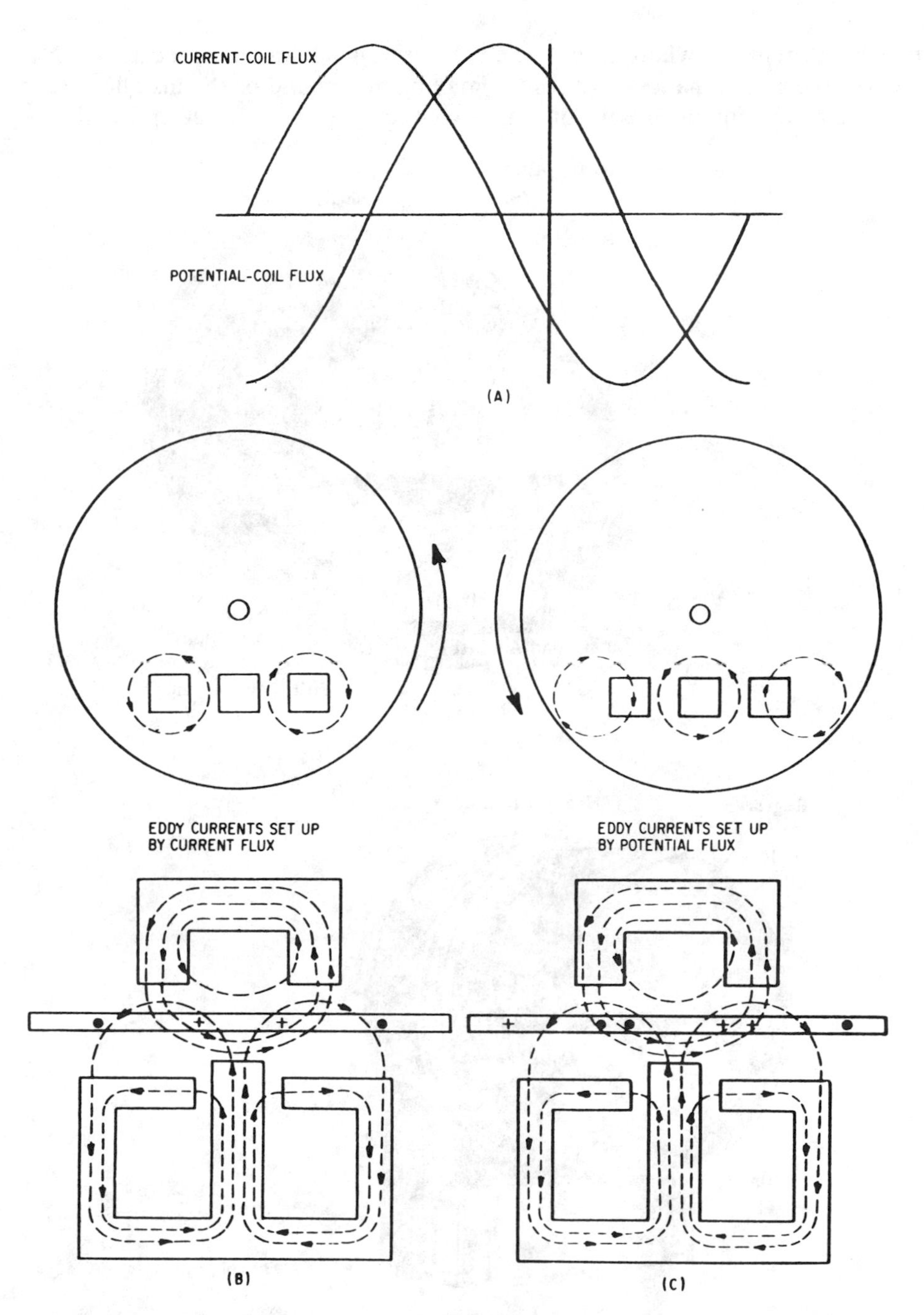

Figure 17-16 Relations of magnetic fields in a watthour meter: (A) Between the fluxes that traverse the gap and cut the disc, (B) potential coil flux on the eddy currents set up by the current-coil flux, and (C) current-coil flux on the eddy currents set up by the potential coil flux (*Courtesy,* Westinghouse Electric Co.)

first hand, remains where it was when the first hand was returned to zero. Hence, the second hand measures the maximum demand of the installation.

A meter equipped with only this second type of gear set and dial is

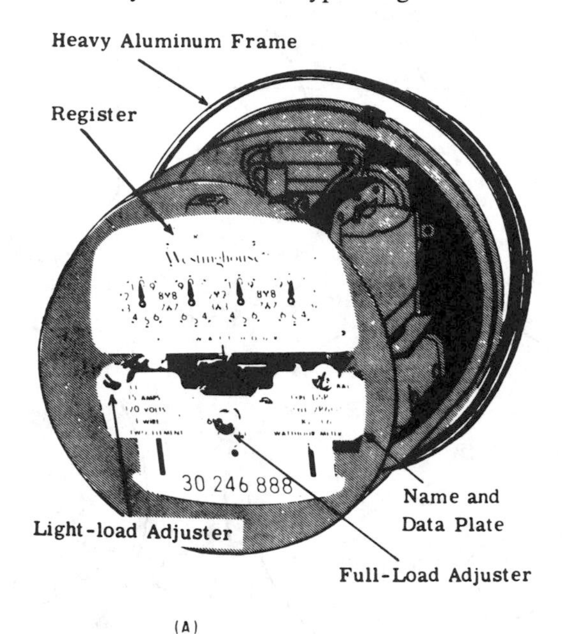

(A)

Register
Heavy Aluminum Frame
Dials
Regulating
Wheel
LINE
LOAD

(B)

Figure 17-17 Two types of watthour meters (above): (A) Socket S-type, and (B) A-base (*Courtesy,* Westinghouse Electric Co.)

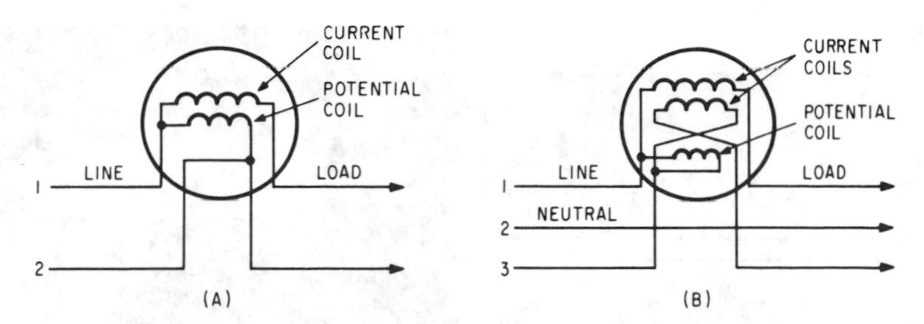

Figure 17-18 Single-phase meters: (A) Two-wire, and (B) three-wire.

known as a *demand meter.* If the meter is equipped with both sets of gears and dials (registers), it is known as a *watt-hour-demand meter.*

In this latter type of meter, the regular four dials are read for the

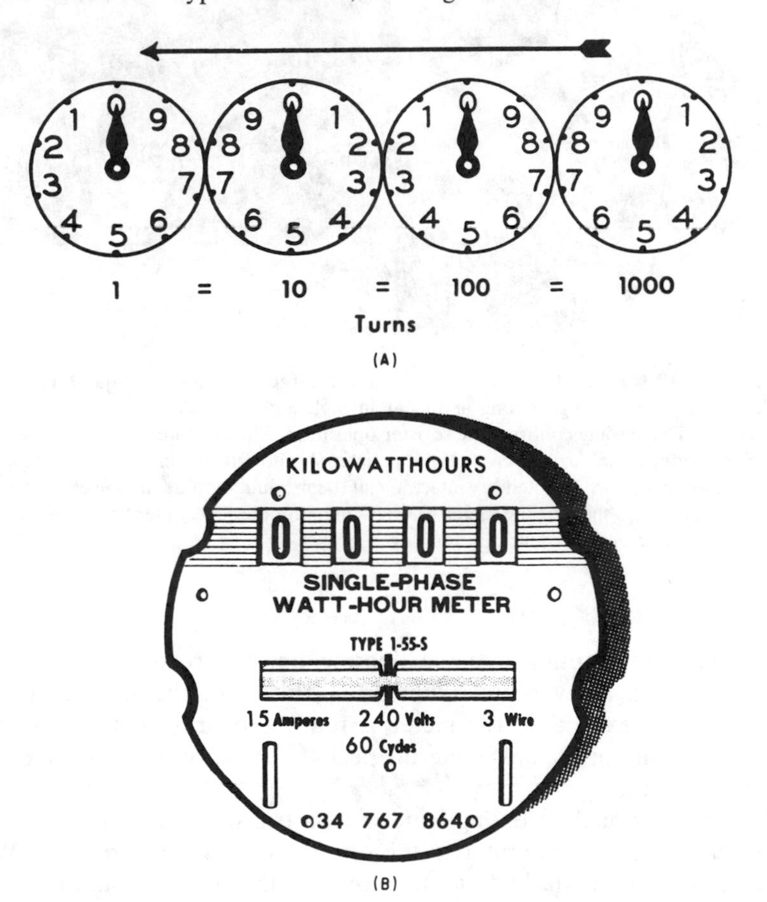

Figure 17-19 Watthour meter registers: (A) Dial register, which works from right to left on a ten-to-one ratio set of gears, (B) cyclometer register, which simplifies meter reading (C) register with solid-state components.

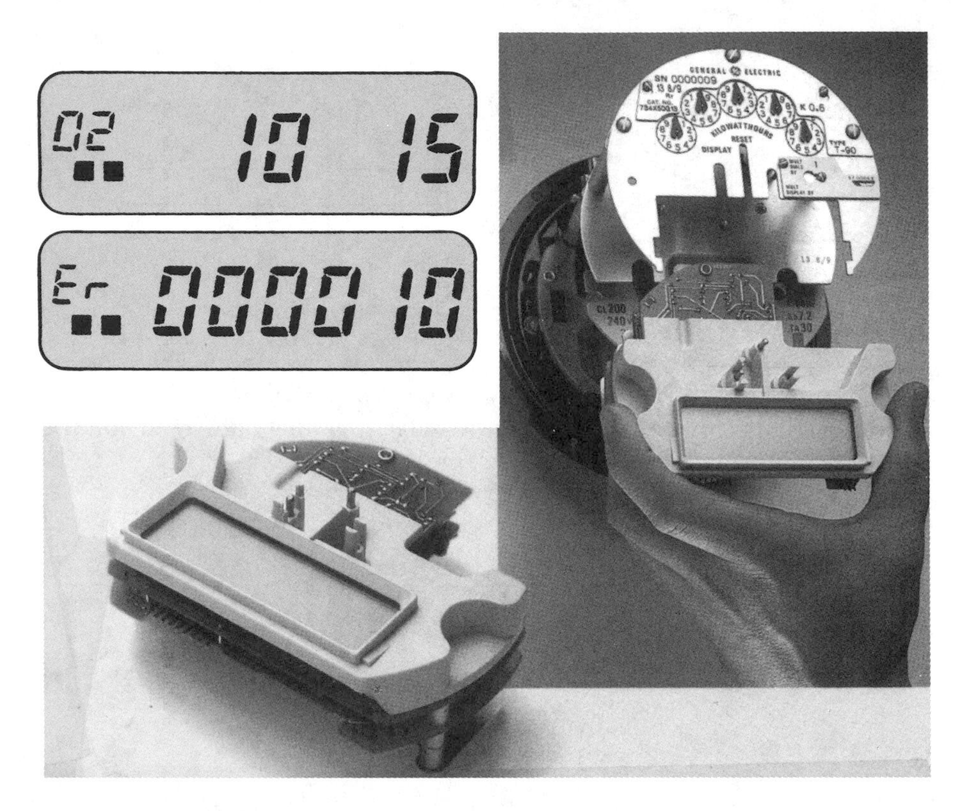

Figure 17-19 (cont.) The T-90 displays data in large, easy-to-read liquid crystal digits that shine though strong and clear in high ambient light, including direct sunlight. The display confirms the register operation. Up to 17 items are available for selection in the display scroll. If the register should fail, the electronic module can be removed with the memory intact. Or, if the module should fail, you can put a new module into the existing meter without changing the meter identification. All electronic modules are identical.

kilowatt-hour consumption and the floating hand is read for the kilowatt demand. Through a lever protruding from the case, the floating hand is returned to zero each time the meter is read, preparing it to register the maximum demand for the next metering period. This lever is usually locked to prevent tampering.

The reading of the floating hand is also transmitted to another set of dials on another register similar to the one measuring consumption. When the lever or plunger is pushed and thus returns the floating hand to zero, it also causes the new maximum demand to be added to the original reading. The new reading provides a means to check and record demand readings.

Figure 17-19 (cont.) Register equipped with solid-state components enabling utilization of register for several other purposes. (*Courtesy,* General Electric Co.)

Schematic diagrams of both types of demand meters and their registers are shown in Fig. 17-22.

Demand meters for two- and three-phase circuits are similar to the single-phase-demand meter, the dials being used to read the two- and three-phase demands.

Similar demand meters for measuring reactive volt-amperes and reactive volt-ampere-hours operate in a similar fashion, being basically watt-hour meters with the voltage displaced 90° by means of added reactances.

OTHER TYPES OF METERS

Recording Meters

Almost all the meters described above are fitted with recording mechanisms. A recording device, either spring- or motor-driven or motor-

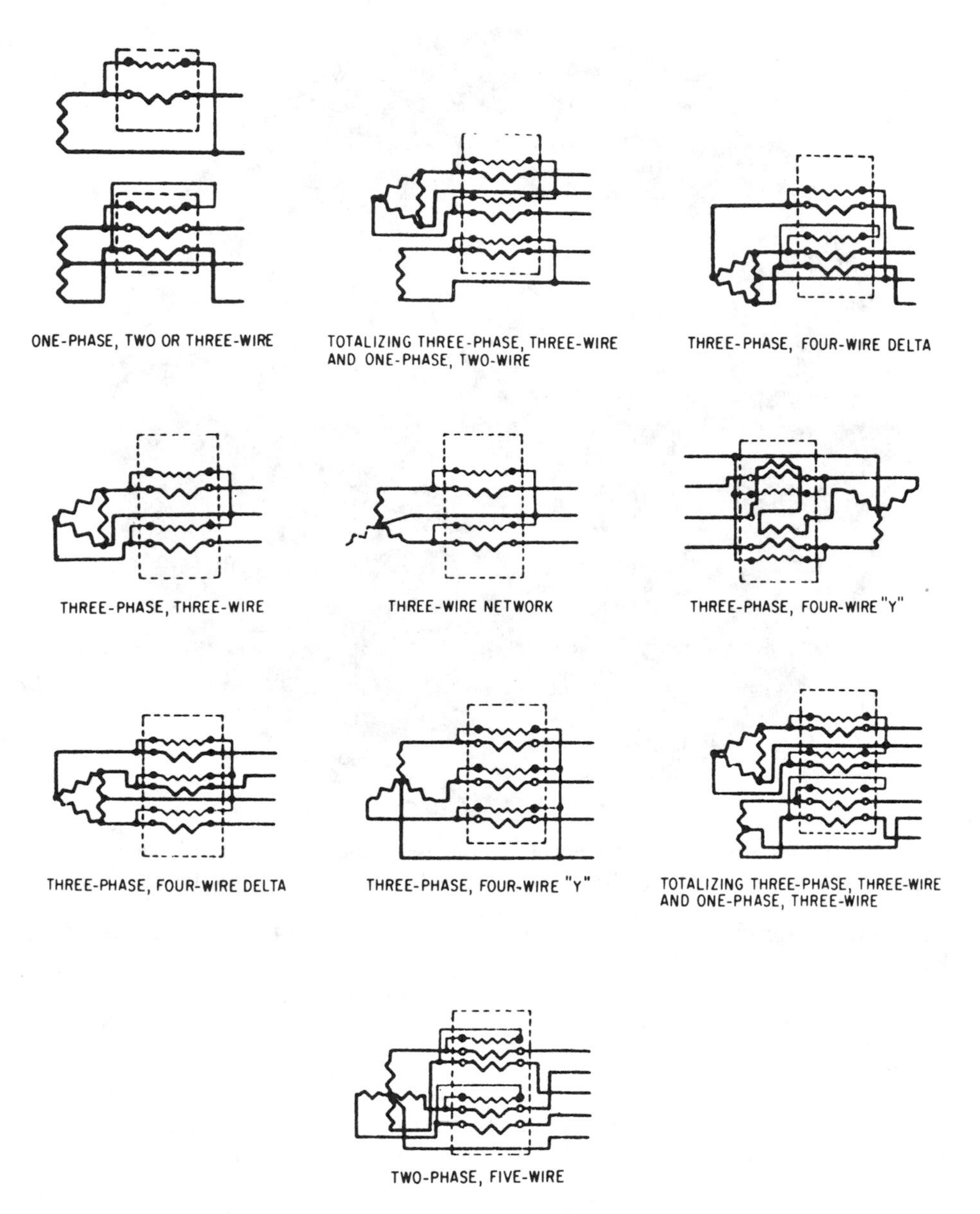

Figure 17-20 Two- and three-phase compared with single-phase meter applications.

wound spring-driven, causes a chart to travel. The indicating element of the meter has an inking device attached to it. As the chart travels, the inking device leaves upon it an ink track, thereby creating a continuous record of instantaneous values. The charts may be of the circular or strip type. Typical recording meters that use a circular chart are shown in Fig. 17-23.

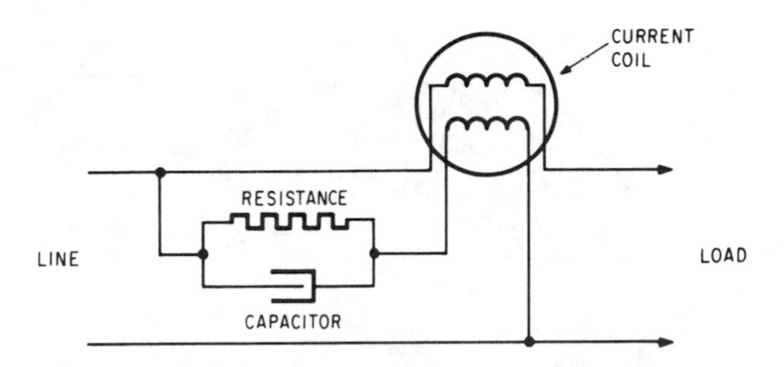

Figure 17-21 Single-phase, reactive volt-ampere-hour meter.

Maximeters

Ammeters that register the maximum current flowing in a circuit are also available. In these *maximeters,* a floating indicator is pushed forward by the ammeter needle. When the reading recedes, the floating indicator stays where it was pushed. After it has served its purpose, the floating indicator is returned to its starting position, ready for the next occasion when a maximum reading is desired. Similar meters are designed to read maximum values of voltage, but they are less frequently used than the ammeter type.

Clamp-On Meters

The clamp-on meter consists essentially of a current transformer with a split core and a rectifier type of instrument connected to the secondary (Fig. 17-24). The rectifier converts the ac to dc so that the measuring instrument is the single moving-coil type that is not only more rugged but has an accuracy equal to or better than that of other types. The primary of the current transformer is the conductor through which the current to be measured flows. The split core permits the instrument to be "hooked on" the conductor without disconnecting it.

This instrument is usually constructed so that voltages may also be measured. Either a switch on the meter or another separate set of terminals is provided to read voltages. Care must be taken that the leads are connected properly to the terminals across which the voltage is to be measured.

The instrument provides for a series of current and voltage ratings. Both current and voltage circuits may be energized at the same time, the pointer being deflected by the service indicated on a selector switch. Extreme care must be exercised when using this instrument on high-voltage circuits. So-called hot-line equipment, such as hot sticks and rubber gloves, should be used. When current or voltage values are unknown, the highest value on the range selectors should be used initially. In general, the use of this instrument should be limited to currents not exceding 1000 amperes on services operating

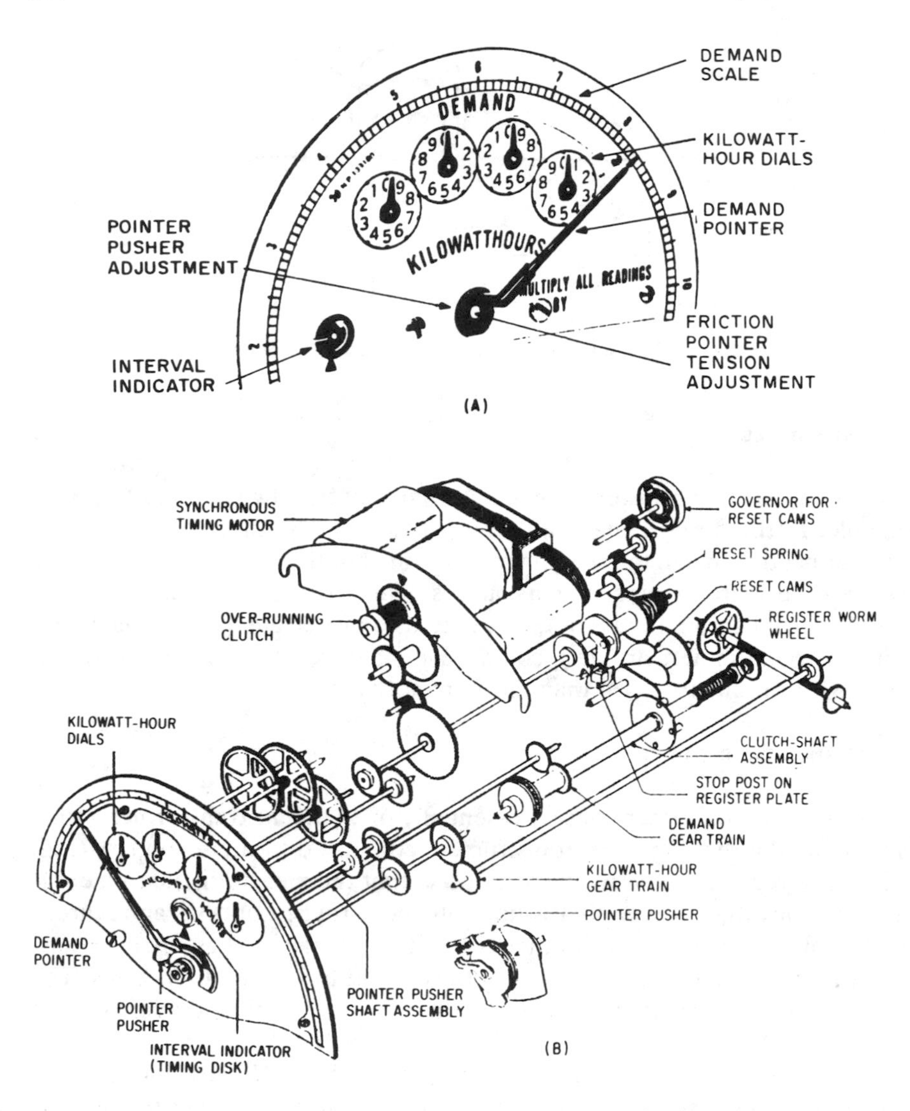

Figure 17-22 (A) External details, and (B) internal details of watt-hour demand register.

at 5000 V or less. The voltage coil may be connected directly to voltage sources up to about 700 V.

METHOD OF CONNECTING INSTRUMENTS

Since an ammeter measures current, it must be connected in series with a circuit so that the current flowing in the circuit will also flow through it. A

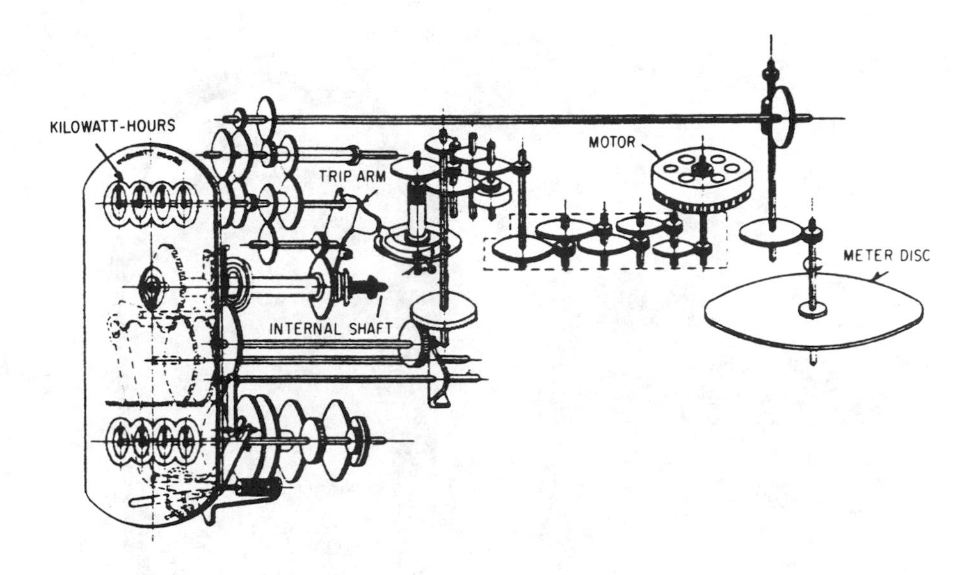

(C)

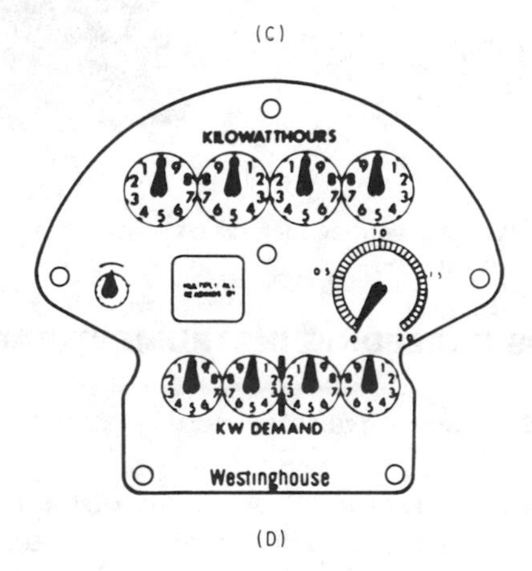

(D)

Figure 17-22 (cont.) (C) External details, and (D) internal details of cumulative demand register (*Courtesy,* Westinghouse Electric Co.)

voltmeter, on the other hand, must be connected across or in parallel with the circuit if it is to measure the voltage of the circuit. Similarly, the current coil in wattmeters, watt-hour meters, reactive volt-ampere-meters, and power factor and other meters must be connected in series with the coil. The voltage coil in these instruments must be connected across or in parallel with the circuit. A schematic diagram showing how some of these meters are connected in a circuit is shown in Fig. 17-25.

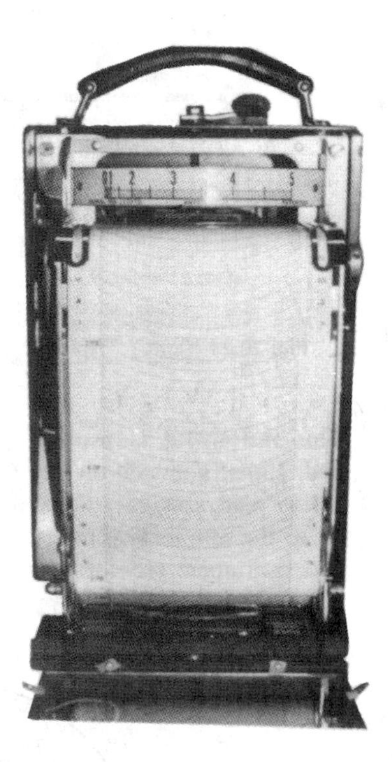

Figure 17-23 Typical recording meters: (A) single-pen, single-range voltmeter, and (B) strip-chart ammeter.

MEASUREMENTS EXCEEDING INSTRUMENT RANGES

Shunts and Current Transformers

If the values of current involved in any of the measuring devices might possibly exceed the range of the instruments, two schemes are used to measure only a smaller but definite proportion of the total current. One makes use of a shunt whose resistance is a known proportion of the internal resistance of the ammeter (Fig. 17-26A). For example, if the internal resistance of the ammeter is 9 ohms, and the resistance of the shunt is 1 ohm, then 9/10 or 90 percent of the current will flow through the shunt and 1/10 or 10 percent through the ammeter. The ammeter reading is then multiplied by 10 to obtain the current in the circuit.

If the current is associated with a high-voltage circuit, a current transformer (Fig. 17-26B and 17-26C) is usually used instead of a shunt. The standard current transformer (CT) secondary is usually rated at 5 amperes,

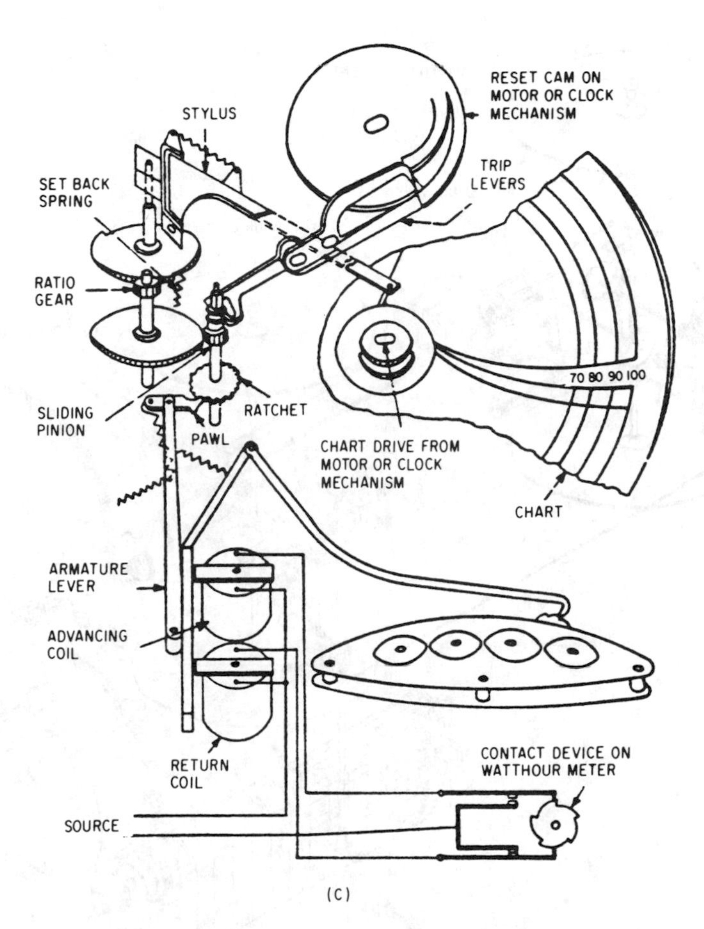

Figure 17-23 (cont.) Typical recording meters: (C) graphic demand meter.

regardless of the primary rating. At lower ratings, the primary of the CT is connected directly in series with the conductor. At higher ratings, the line itself is used as a single-turn primary and the CT may have its secondary coil shaped as a doughnut through which the conductor passes.

Since a current transformer has a small number of primary turns (as low as one) and many secondary turns, the result is to step down the current and step up the voltage. Normally, a high voltage is short-circuited through the low impedance of the instruments and does not build up. If the instruments are disconnected and the secondary of the CT left open, high voltages will occur. The secondary of a CT, therefore, should never be opened while the circuit is energized.

Shunts are usually confined to dc circuits but may be used on ac circuits. CTs are used only with ac circuits. One-to-one ratio CTs are sometimes used if

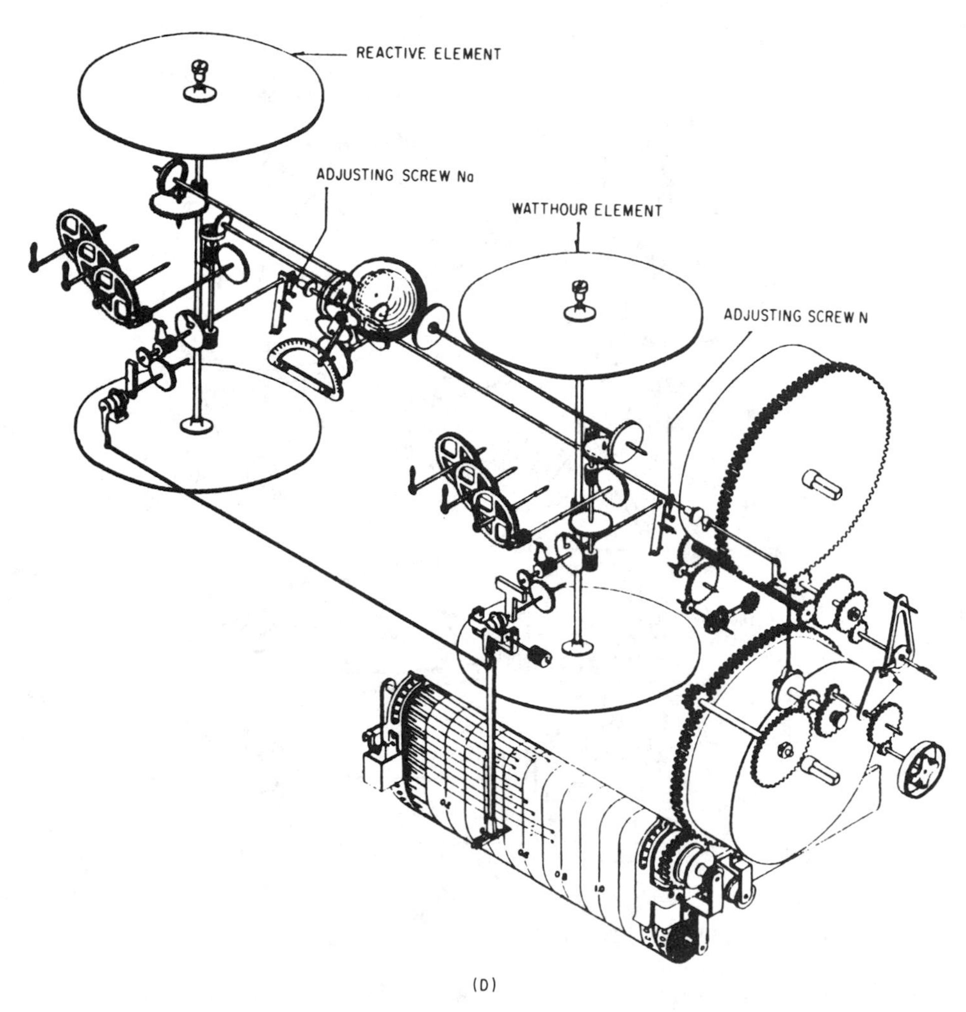

Figure 17-23 (cont.) Typical recording meters: (D) kilovolt-ampere demand meter.

it is desirable to isolate one circuit electrically from another. For additional information regarding CTs, refer to Chap. 6.

Potentiometers and Potential Transformers

If the values of voltage involved might exceed the range of the instruments, two schemes are used to measure only a smaller but critical proportion of the total voltage. Either *potentiometer* or a *potential transformer* may be used for this purpose.

A potentiometer (sometimes called a *multiplier*) consists of a coil of very high resistance connected across a line. A voltmeter is connected across a

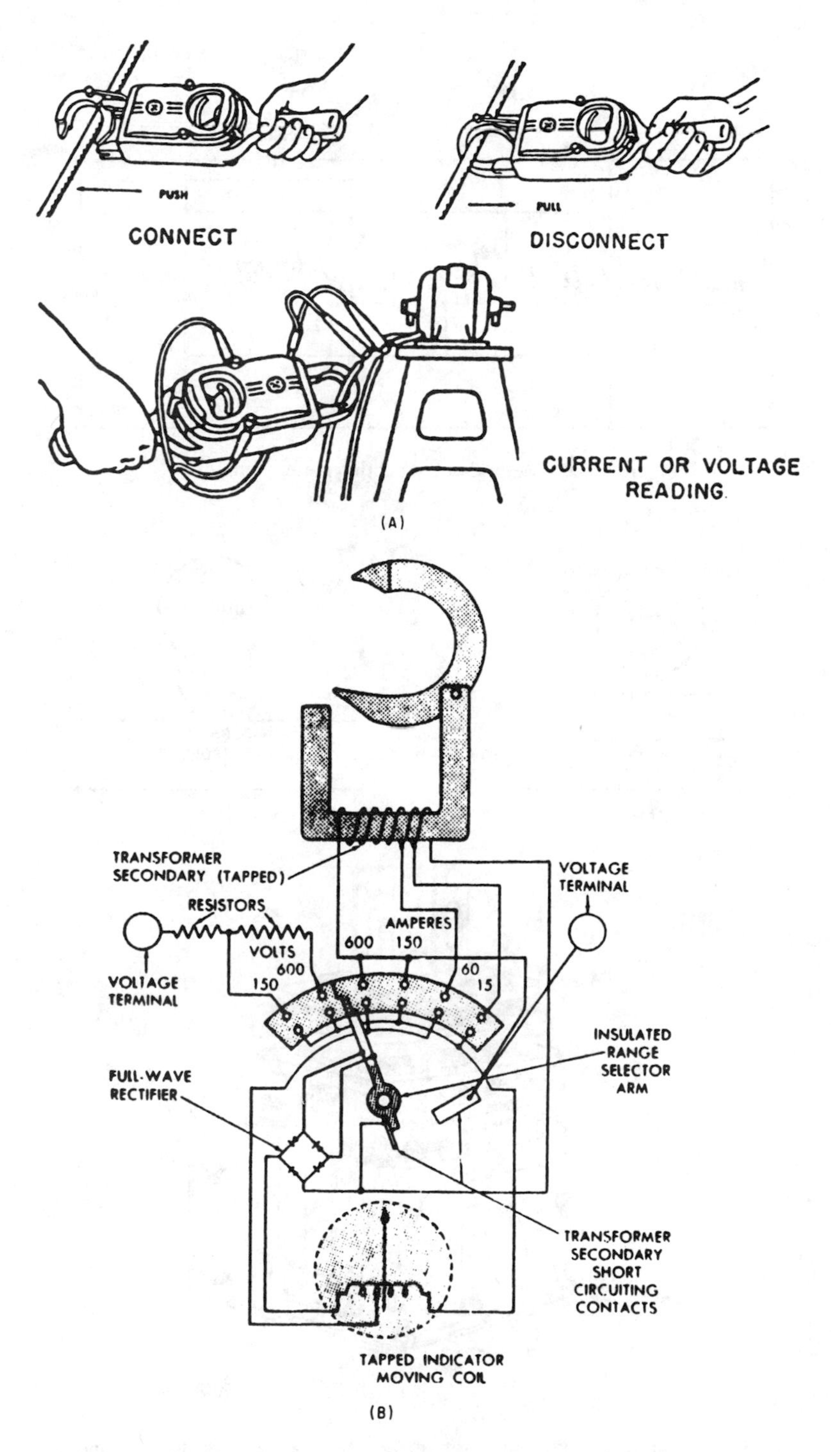

Figure 17-24 Clamp-on volt-ammeter: (A) Connecting, disconnecting, and reading; and (B) internal connections.

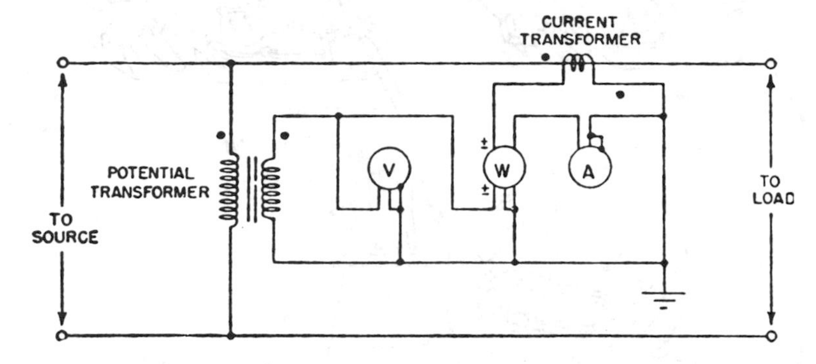

Figure 17-25 Method of connecting meter elements.

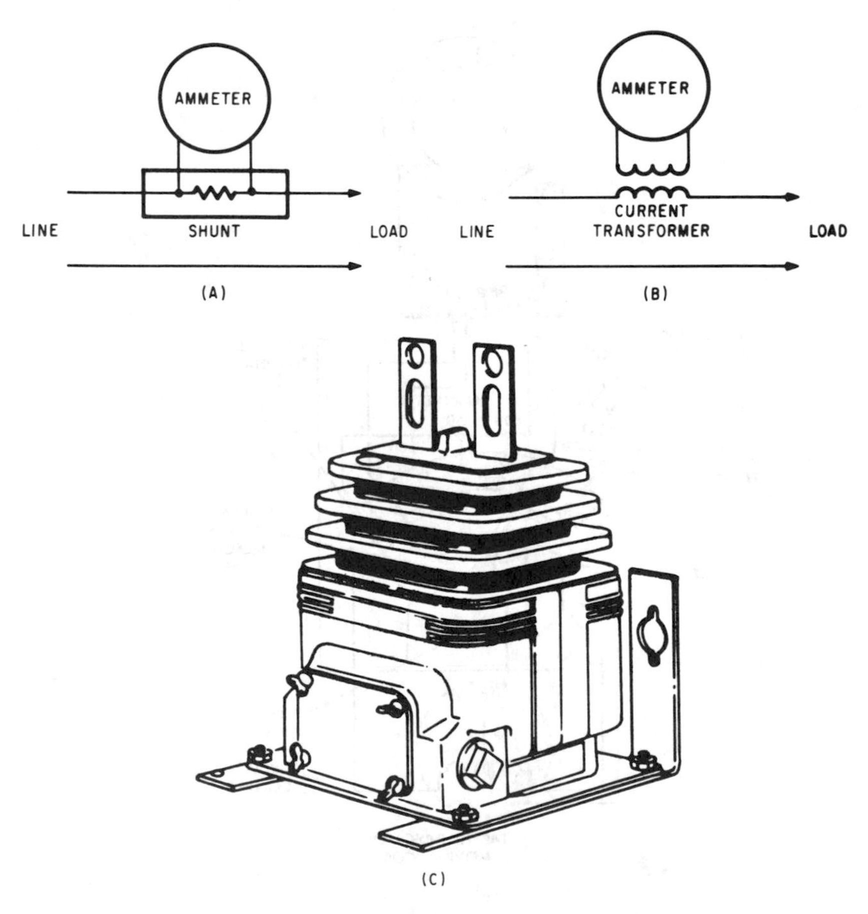

Figure 17-26 Ammeter connections for currents beyond range of the instrument: (A) shunt, for a–c and d–c; (B) current transformer, for a–c only; and (C) typical current transformer.

specific percentage of the total coil (Fig. 17-27A), and its reading multiplied by the ratio of the portion of the coil to the total. For example, if a 100,000-ohm resistance is connected across a 500-V circuit and a voltmeter is connected across 10,000 ohms of that coil, the instrument will read 50 V. Multiplying this value by 10 (the ratio between the total coil and the voltmeter portion) results in a value of 500 V. Although this scheme is usually limited to dc circuits, the connections of a potentiometer are the same as those of an autotransformer, and it is therefore applicable to ac circuits as well.

Potential transformers (Fig. 17-27B and 17-27C) are used with many of the instruments mentioned earlier; they are also used for pilot and signal lights. Standard potential transformers (PTs) are usually rated from 50 to 300 volt-amperes, with a secondary voltage of 150 or 300 V. One potential transformer can supply a number of instruments at the same time provided that the value of secondary volt-amperes, called the *burden,* does not exceed the PT capacity rating. They may be either air-filled (dry type) or oil-filled, depending on the magnitude of the voltage involved.

In order to protect against sustained short circuits, fuses (sometimes equipped with current-limiting resistors) are connected in series with the primary winding. The resistors limit the short circuit of the PT to values that a fuse can safely interrupt.

For additional information regarding potential transformers, refer to Chap. 6.

RESISTANCE MEASUREMENT

Two instruments are used to measure the resistance of a circuit or circuit element and to check the continuity of a circuit: the ohmmeter and the Megger. The former is limited to measuring a few megohms (millions of ohms), whereas the range of the latter extends to over 1000 megohms.

Ohmmeter

The ohmmeter is essentially a milliammeter in series with a dc voltage source of a known fixed value (Fig. 17-28A and 17-28B). It is placed in series with the circuit to be measured. Since, deriving from Ohm's Law, resistance is equal to the voltage divided by the current, the current reading represents the resistance, and the meter scale is so calibrated.

Megger

The Megger consists of two main elements, both of which are provided with individual magnetic fields from permanent magnets. One element, a hand-driven dc generator, supplies the necessary current for making the

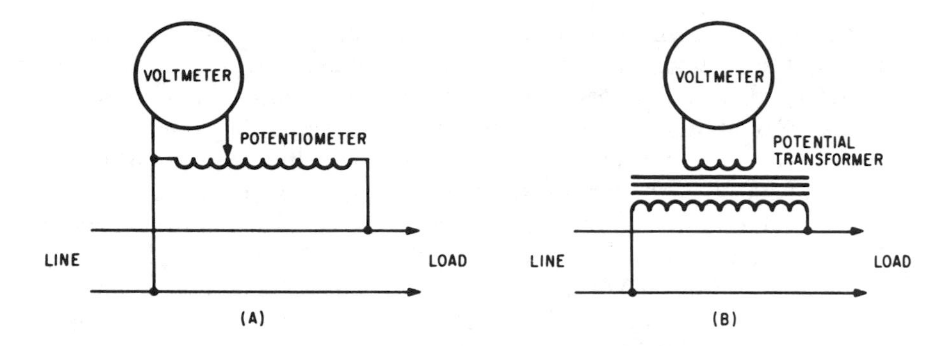

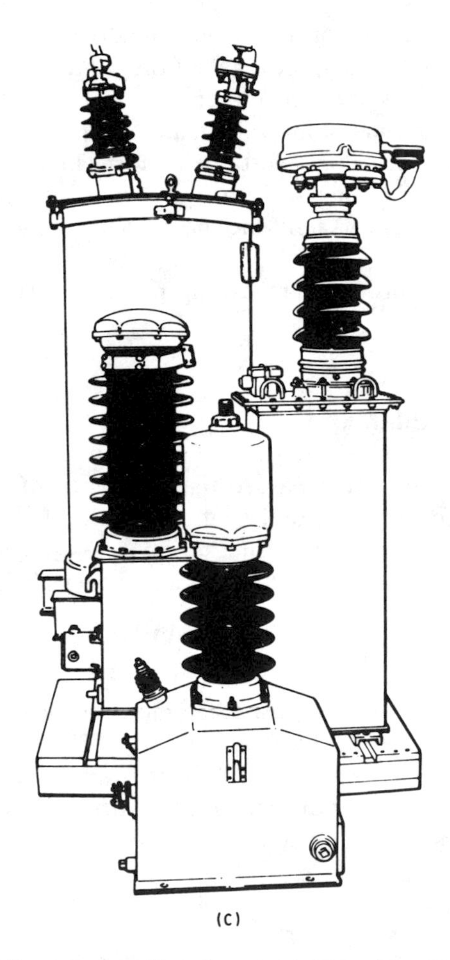

(C)

Figure 17-27 Voltmeter connections for currents beyond range of the instrument: (A) Potentiometer, for a–c and d–c; (B) potential transformer, for a–c only; and (C) typical potential transformers.

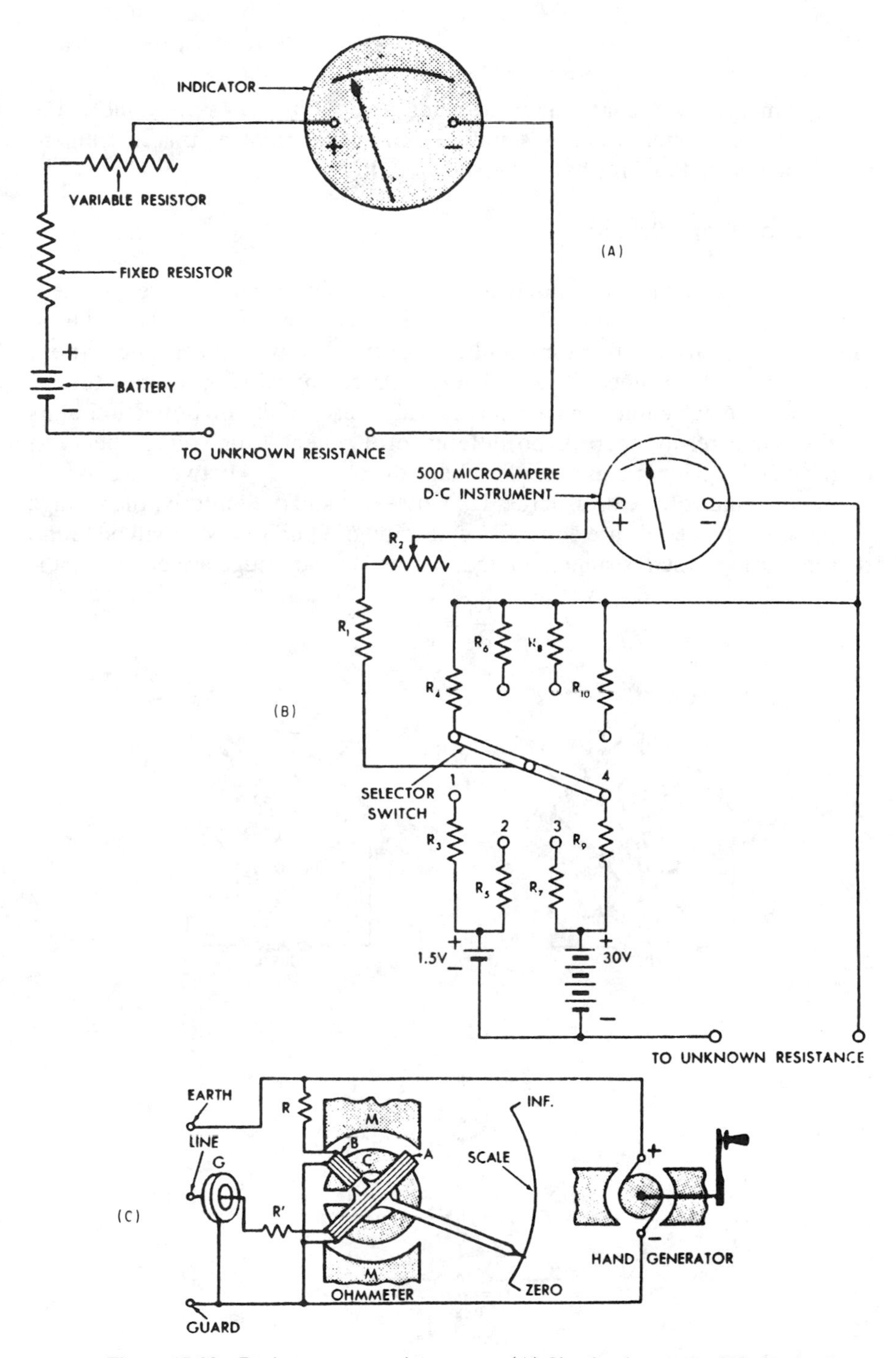

Figure 17-28 Resistance measuring meters: (A) Simple ohmmeter, (B) circuit of four-range ohmmeter, and (C) circuit of Megger insulation tester (*Courtesy,* J. G. Biddle Co.)

measurement. The second, the instrument portion, indicates the value of the resistance being measured. A simplified circuit diagram of this instrument (described more fully in Chap. 7) is shown in Fig. 17-28C.

Wheatstone Bridge

A type of circuit used widely for precision measurements of resistance is the Wheatstone bridge (Fig. 17-29A). Three of its elements—R_1, R_2, and R_3—are precision variable resistors, and the fourth, R_x, is the resistance whose value is to be determined. The bridge is balanced by varying the resistors so that when the galvanometer switch is closed, there will be no deflection of its needle. As a result, there is no difference of potential or voltage between points b and d. This means that the voltage drop across R_1, between a and b, is the same as the voltage drop across R_3, between a and d. Similarly, the voltage drops across R_2 and R_x are equal. By extension of Ohm's Law, it will be found that the ratio of the resistances in both "arms" of the bridge are equal, that is

$$\frac{R_1}{R_2} = \frac{R_3}{R_x}$$

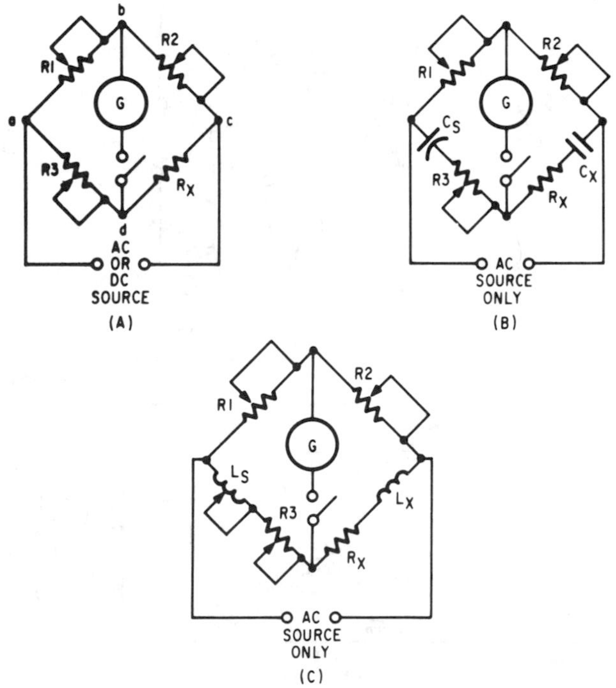

Figure 17-29 Wheatstone bridge circuits: (A) Resistance bridge, (B) capacitance bridge, and (C) inductance bridge.

from which,

$$R_x = \frac{R_2 \times R_3}{R_1}$$

The Wheatstone bridge may also be used with an ac source to measure capacitive reactance (Fig. 17-29B) and inductive reactance (Fig. 17-29C) in a similar fashion. The equations are as follows:

$$C_x = \frac{R_1}{R_2} \times C_s \qquad \text{and} \qquad L_x = \frac{R_2}{R_1} \times L_s$$

PHASE ROTATION INDICATOR

The rotation or sequencing of phases in a three-phase system is often desired, particularly when pieces of equipment are connected in parallel. Two methods may be used. In Fig. 17-30 two lamps and a highly reactive coil (such as the potential coil of a watt-hour meter) are needed. The bright lamp indicates the particular phase sequence (x bright indicates sequence 1, 2, 3; y bright indicates sequence 3, 2, 1). In Fig. 17-30B, a noninductive resistance and a reactive coil of equal impedance are used in conjunction with a lamp. The brightness of the lamp indicates the phase sequence (bright light indicates sequence 1, 2, 3; dim light indicates sequence 3, 2, 1).

CONSTRUCTION FEATURES

Electrical instruments of all types of construction contain five basic elements: (1) the moving element, (2) the stationary element, (3) the controlling devices, (4) bearings and pivots, and (5) case, terminals, dial, and so forth.

Since the moving element is the most susceptible to damage from electrical and mechanical shocks and disturbances, it is built ruggedly. Moreover, since it has to carry not only the pointer, but other damping devices as well, its balance in its pivots is critical.

The stationary element presents few problems. The selection of permanent magnets and the insulation of coils and leads are easily accommodated.

Controlling devices usually consist of damping magnets or air dampers or vanes that keep the moving element from oscillating. In general, their speed of response involves the inertia of the moving element and the spring or restoring force. Springs are made of special alloys as they also serve as connections and are susceptible to any heat that may be produced.

Bearings and pivots are perhaps the most critical element since they must sustain mechanical shocks while maintaining extremely low values of friction. Consequently, pivots of very hard metal have points with only a slight

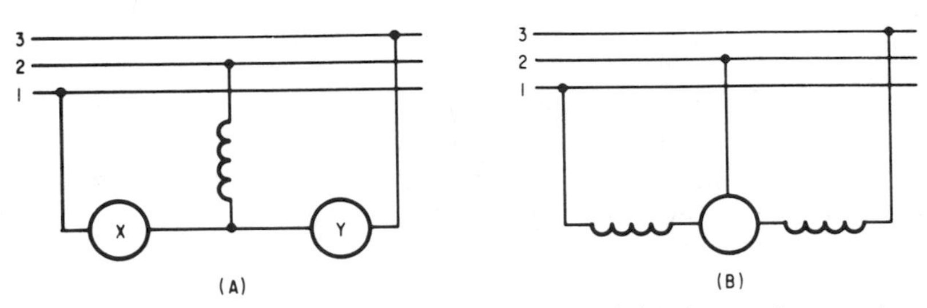

Figure 17-30 Phase rotation or sequence indicators: (A) Using two lamps and a highly reactive coil; and (B) using one lamp, a noninductive resistance, and a reactive coil.

roundness, with a radius of from one ten-thousandth to three thousandths of an inch (0.0001 to 0.003 in.). Such dimensions, in turn, create tremendous pressures in terms of pounds per square inch. Polished jewels of extreme hardness and requiring little or no lubrication must therefore be used.

The remaining elements involve many separate coordinations but present few problems as a rule. Case design, including terminal arrangement, dial, glass, and cover depend on the application in question. Switchboard instruments differ entirely from portable ones. Size is, to a degree, associated with accuracy. More structural weight limitations are imposed on portable instruments than on a switchboard installation. Ventilation is important, especially for instruments carrying heavy currents, but it may also present problems of dust accumulation. In general, these concerns have been standardized satisfactorily.

Electronically activated instruments, more accurate and flexible, are supplanting electromechanical instruments; refer to Electronic Relays in Chapter 14.

INSTRUMENT CARE: OPERATION AND MAINTENANCE

All instruments should be handled carefully to avoid injury to bearings and pivots.

If an instrument has more than one range, connection should be made to the highest range first in order to avoid damage should the load or voltage be underestimated.

Instruments should be used in their normal operating position—portable units horizontally and switchboard units vertically.

Check the instrument zero setting. If the pointer is slightly off, reset it by using the external zero shifter. Never try to compensate for a bent pointer by moving the zero shifter.

In connecting a voltmeter to a circuit, connect the leads to the instrument before connecting the leads to the voltage source. This method avoids loose leads that may be energized. Be sure that the voltmeter leads are properly insulated for the voltage to be measured.

Make sure that the secondary circuit of a current transformer is never opened while current is flowing in the primary winding; otherwise, dangerously high voltage will result. The secondary winding should be short-circuited when instruments are being changed or when they are not in use. Disconnect the secondary winding from the line before making primary connections.

If a clamp-on ammeter is used, make sure that the split transformer core is completely closed or joined together.

To overcome slight stickiness, an instrument may be tapped lightly. If movement of a pointer is erratic and tapping moves it appreciably, the instrument bearings or pivots may need attention. If necessary, they should be repaired by a competent facility or returned to the manufacturer.

Do not use instruments in strong magnetic or electrostatic fields (such as near cables, buses, unshielded capacitors) unless they are designed for such treatment.

Try to take the instrument reading between the one-half and the three-quarters scale value.

Read directly at right angles to the instrument so that the needle and its image in the mirror (if one is provided) coincide.

Use a damp cloth to clean the glass of an instrument. A dry cloth may induce a static charge and affect the instrument reading. Breathing on the glass may discharge the static charge.

Do not slam instrument covers. Make sure that binding post nuts are tight, even when an instrument is not in use.

If an instrument is overloaded or dropped or its accuracy is in doubt, check it against another at several points on the scale before using it again.

Slide, do not drop, an instrument into its carrying case.

Store instruments in a place free from dust, corrosive fumes or other polluted atmosphere, and excessive humidity.

Arrange instruments and leads to avoid knocking the former from a table or tripping over the leads. In measuring high voltages, make sure that no one will accidentally come in contact with any energized parts.

Do not carry more than one instrument in one hand. If means are provided for locking the moving element, it should be locked before the instrument is transported.

REVIEW

1. By measuring the values of electrical quantities involved, operators can determine whether circuits and equipment are operating safely and efficiently. Comparing these values with predetermined design specifications and with previously experienced values, troublesome conditions may be detected before they occur.

2. Most instruments depend on the interaction of two or more magnetic fields. These fields may be created by electromagnets or by permanent magnets (Fig. 17-1). By combining the fields measuring current and voltage, instruments may determine values of resistance, power, energy, power factor, and other quantities in addition to the basic ones of current and voltage.
3. Instruments comprise ammeters (Figs. 17-2 and 17-3), voltmeters (Fig. 17-7), wattmeters (Figs. 17-8, 17-10, and 17-11), volt-ampere meters, power-factor meters (Figs. 17-12 and 17-13), frequency meters (Fig. 17-14), watt-hour meters (Figs. 17-15, 17-17, and 17-18), volt-ampere-hour meters (Fig. 17-21), demand meters (Fig. 17-22), and other special purpose meters.
4. Meters may be of the indicating type, maximum reading type, recording type (Fig. 17-23), or clamp-on type (Fig. 17-24).
5. Ammeters and current elements of other meters are connected in series with the circuit (Fig. 17-25) or equipment under observation. Voltmeters and voltage elements are connected in parallel.
6. The range of instruments for measuring current values may be increased by the use of current transformers for ac and by shunts for dc (Fig. 17-26). For voltage values, potential transformers are used for ac and potentiometers for ac and dc (Fig. 17-27).
7. For measuring resistance, ohmmeters (Fig. 17-28A and B), Meggers (Fig. 17-28C), and Wheatstone bridges (Fig. 17-29) may be used.
8. Phase rotation indicators (Fig. 17-30) show the rotation or sequence of phases in a polyphase circuit, and are used when connecting such circuits or equipment in parallel.
9. Instruments may be of the portable type or of the stationary or mounted type.
10. All instruments should be handled very carefully and generally used or installed away from magnetic fields which may affect their operation.

STUDY QUESTIONS

1. What is the principle of operation of an ammeter? Of a voltmeter? How are they connected in a circuit?
2. Describe two types of instruments for measuring current. What is the difference in their mode of operation?
3. What other instruments employ the electrodynamometer principle?
4. What is a wattmeter? A watt-hour meter? What is the difference between them, and how are they connected in a circuit?
5. What is the principle of operation of a watt-hour meter? What other quantities may be measured by this type of instrument?

6. What is the function of a watt-hour meter register? What two types are commonly used and what is the difference between them?
7. What is a demand meter? A demand watt-hour meter? How do they operate?
8. What means may be employed to measure values beyond the range of an instrument for current? For voltage? What precautions should be taken in their employment?
9. Name three methods of measuring the resistance of a circuit or a circuit element. What are the differences between them?
10. What are the five basic elements contained in the construction of electrical instruments? Explain their functions.

Appendix A

INSULATION: PORCELAIN VS POLYMER

For many years, porcelain for insulation purposes on lines and equipment has exercised a virtual monopoly. It was perhaps inevitable that plastics, successful as insulation for conductors since the early 1950'S, should supplant porcelain as insulation for other applications in the electric power field.

The positive properties of porcelain are chiefly its high insulation value and its great strength under compression. Its negative features are its weight (low strength to weight ratio) and its tendency to fragmentation under stress. Much of the strength of a porcelain insulator is consumed in supporting its own weight. Figure A-1a&b.

In contrast, the so-called polymer not only has equally high insulation value, but acceptable strength under both compression and tension. It also has better water and sleet shedding properties, hence handles contamination more effectively, and is less prone to damage or destruction from vandalism. It is very much lighter in weight than porcelain (better strength to weight ratio), therefore more easily handled. Figure A-2, Table A-1.

Economically, costs of porcelain and polymer materials are very competitive, but the handling factors very much favor the polymer.

Polymer insulation is generally associated with a mechanically stronger insulation, such as high strength fiberglass. The fiberglass insulation serves as an internal structure around which the polymer insulation is attached, usually in the form (and function) of petticoats (sometimes also referred to as bands, water shedders; but for comparison purposes, however, here only the term petticoat will be used). The insulation value of the polymer petticoats is equal to or greater than that of the fiberglass to which it is attached.

The internal fiberglass structure may take the form of a rod (or shaft), a tube, cylinder, or other shape. It has a high comparable compression strength as a solid and its tensile strength, equally high, is further improved by stranding and aligning around a fiber center. The polymer petticoats are installed around the fiberglass insulation and sealed to prevent moisture or

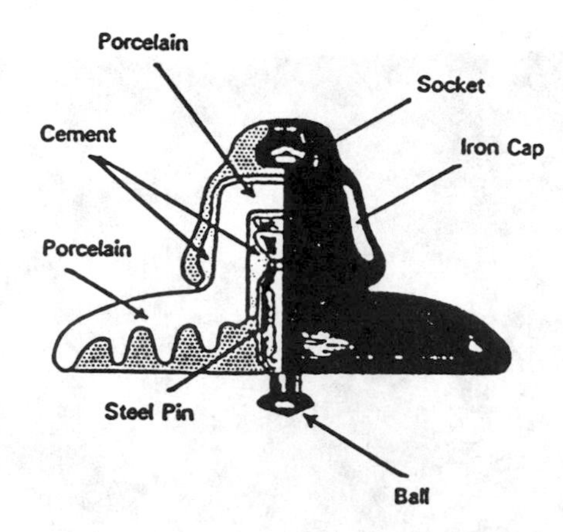

Figure A-1 a Ball and socket type suspension insulator.

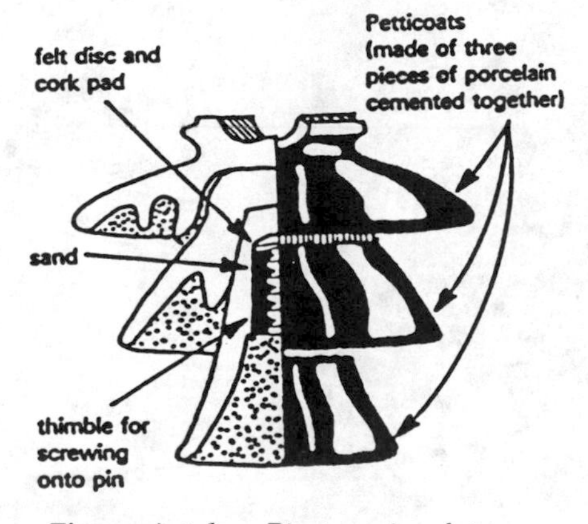

Figure A-1 b Pin-type insulator.

Table A-1 Polymer insulation weight advantage

Product	Type	Voltage (kV)	Porcelain Weight (lbs)	Polymer Weight (lbs)	Percent Weight Reduction
Insulator	Distribution	15	9.5	2.4	74.7
Arrester	Distribution	15	6.0	3.8	36.7
Post Insulator	Transmission	69	82.5	27.2	67.0
Suspension	Transmission	138	119.0	8.0	93.2
Intermediate Arrester	Substation	69	124.0	28.0	77.4
Station Arrester	Substation	138	280.0	98.9	64.7

Figure A-2 A variety of typical polymer insulator shapes. (*Courtesy* Hubbell Power Systems)

contamination from entering between the petticoats and fiberglass; Figure A-3. The metal fittings at either end are crimped directly to the fiberglass, developing a high percentage of the inherent strength of the fiberglass. It should be noted that fiberglass with an elastometric (plastic) covering has been used for insulation purposes since the early 1920's.

The polymer petticoats serve the same function as the petticoats asso-

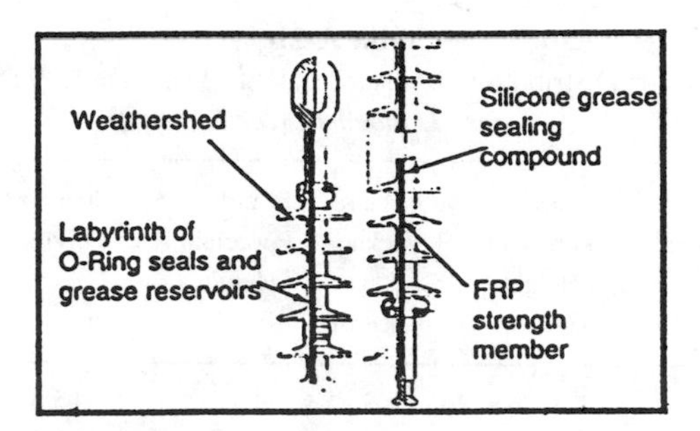

Figure A-3 Polymer insulator showing fiberglass rod insulation and sealing (*Courtesy* Hubbell Power Systems).

ciated with porcelain insulators, that of providing a greater path for electric leakage between the energized conductors (terminals, buses, etc.) and the supporting structures. In inclement weather, this involves the shedding of rain water or sleet as readily as possible to maintain as much as possible the electric resistance between the energized element and the supporting structure, so that the leakage of electrical current between these two points be kept as low as possible to prevent flashover and possible damage or destruction of the insulator. Tables A-2a&b.

When the insulator becomes wet, and especially in a contaminated environment, leakage currents begin to flow on the surface; if the current becomes high enough, an external flashover takes place. The rate at which the

Table A-2a
Polymer improvement over porcelain
(*Courtesy* Hubbell Power Systems)

Watts Loss Reduction*
(watts per insulator string)

Voltage	Relative Humidity				
kV	30%	50%	70%	90%	100%
69	0.6	0.8	0.9	1.0	4.0
138	1.0	2.4	4.5	7.2	8.0
230	1.0	2.5	5.7	14.0	29.0
345	2.5	4.2	8.5	15.0	30.0
500	2.8	7.8	11.5	33.0	56.0

*Power loss measurements under dynamic humidity conditions on I-strings.

Table A-2b
Polymer Distribution Arrester Leakage Distance Advantage
(Courtesy Hubbell Power Systems)

Standard MCOV	Standard Porcelain Leakage Diatance (in)	Special Porcelain Height (inches)	Special Porcelain Leakage Distance (in)	Standard Porcelain Height (inches)	Standard Polymer Leakage Distance (in)	Polymer Height (inches)
8.4	9.0	9.4	18.3	15.9	15.4	5.5
15.3	18.3	15.9	22.0	20.0	26.0	8.5
22.0	22.0	20.0	29.0	28.9	52.0	17.2

insulation dries is critical. The relationship between the outer petticoat diameter and the core is known as the form factor. The leakage current generates heat (I^2R) on the surface of the insulator (eddy currents). In addition to the effects of the leakage current, the rate at which the petticoat insulation will dry depends on a number of factors. Starting with its contamination before becoming wet, the temperature and humidity of the atmosphere and wind velocity following the cessation of the inclement weather. In areas where extreme contamination may occur (such as some industrial areas or proximity to ocean salt spray), the polymer petticoats may be alternated in different sizes, Figure A-4, to obtain greater distance between the outer edges of the petticoats across which flashover might occur. When dry, the leakage current (approximately) ceases and the line voltage is supported across dry petticoats, preventing flashover of the insulator. It is obviously impractical to design and manufacture comparable porcelain insulators as thin as polymers and having the same form factor. Table A-3.

In addition, in porcelain insulators, the active insulating segment is usually small and, when subjected to lightning or surge voltage stresses, may be punctured. In subsequent similar circumstances, it may breakdown completely, not only causing flashover between the energized element and the supporting structure, but may

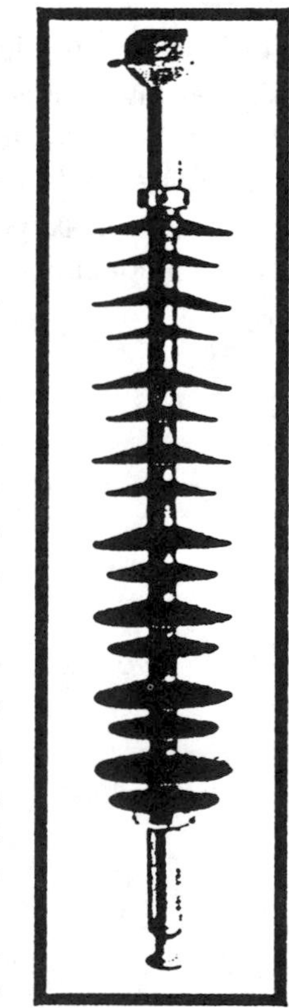

Figure A-4. Polymer insulator arrangement areas of high contamination where flashover between petticoat edges is possible (*Courtesy* Hubbel Power Systems)

Table A-3
Comparison of Contamination Performance of
Polymer versus Porcelain Housed Intermediate Class Arresters (*Courtesy Hubbell Power Systems)*

			Partial Wetting Test		5 Hr. Slurry Test— Maximum Currenr After Slurry Number			
MCOV (kV)	Housing Material	Housing Leakage Distance (in)	Max. Current (mA crest)	Max. Disc. Temps. (°C)	5	10	15	20
57	Polymer	81	<1	<38	35	42	44	44
66	Porcelain	54	68	>163	—	—	—	—
84	Polymer	109	<1	<38	50	52	60	60
98	Porcelain	122.4	18	<82	143	160	175	185

Tested using the 5-hour uniform slurry test procedure. This test consists of applying a uniform coating of standard 400 ohm-cm slurry to the test arrester. Within 30 seconds, MCOV is applied for 15 minutes, during which time the surface leakage currents cause the surface to dry. Slurry applications are repeated for a total of 20 test cycles. After the 20th test cycle, MCOV is applied to the arrester for 30 to 60 minutes to demonstrate thermal stability. Surface leakage currents were measured at the end of the 5th, 10th, 15th and 20th test cycles.

explode causing porcelain fragmentation in the process; the one-piece fiberglass insulator will not experience puncture.

The polymer suitable for high voltage application consists of these materials:

1. Ethylene Propylene Monomer (EPM)
2. Ethylene Propylene Diene Monomer (EPDM)
3. Silicone Rubber (SR)

Both EPM and EPDM, jointly referred to as EP, are known for their inherent resistance to tracking and corrosion, and for their physical properties, SR offers good contamination performance and resistance to Ultra Violet (UV) sun rays. The result of combining these is a product that achieves the water repellent feature (hydrophobic) of silicone and the electromechanical advantages of EP rubber.

Different polymer materials may be combined to produce a polymer with special properties; for example, a silicone EPDM is highly resistant to industrial type pollution and ocean salt.

The advantageous strength to weight ratio of polymer as compared to porcelain makes possible lighter structures and overall costs as well as permitting more compact designs, resulting in narrower right-of-way requirements and smaller station layouts. The reduction in handling, shipping, packaging, storage, preparation and assembly, all with less breakage, are obvious—these, in addition to the superior electromechanical performance.

Fiberglass insulation with its polymer petticoats is supplanting porcelain in bushings associated with transformers, voltage regulators, capacitors, switchgear, circuit breakers, bus supports, instrument transformers, lightning or surge arresters, and other applications. The metallic rod or conductor inside the bushing body may be inserted in a fiberglass tube and sealed to prevent moisture or contamination entering between the conductor and the fiberglass tube around which the polymer petticoats are attached. More often, the fiberglass insulation is molded around the conductor, and the polymer petticoats attached in a similar fashion as in insulators. Figure A-5.

Lightning or surge arrester elements are enclosed in an insulated casing. Under severe operating conditions, or as a result of multiple operations, the pressure generated within the casing may rise to the point where pressure relief ratings are exceeded. The arrester then may fail, with or without external flashover, Figure A-6, exploding and violently expelling fragments of the casing as well as the internal components, causing possible injury to personnel and damage to surrounding structures. The action represents a race between pressures building up within the casing and an arcing or flashover outside the casing. The 'length' of the casing of the arrester limits its ability to vent safely. The use of polymer insulation for the casing permits puncturing to occur, without the fragmentation that may accompany breakdown and failure of porcelain.

Summarizing, the advantages of polymers over porcelain include:

Polymer insulation offers benefits in shedding rain water or sleet, particularly in contaminated environments.

Polymer products weigh significantly less than their porcelain counterparts, particularly line insulators, resulting in cost savings in structures, construction and installation costs. Table A-4.

Polymer insulators and surge arresters are resistant to damage resulting from installation and to damage from vandalism. The lack of flying fragments when a polymer insulator is shot deprives the vandal from his satisfaction with a spectacular event and should discourage insulators as convenient targets.

Polymer arresters allow for multiple operations (such as may result from station circuit reclosings), without violently failing. Figure A-6.

(Continued on page 379)

Figure A-5. Typical porcelain bushings that may be replaced with polymers.

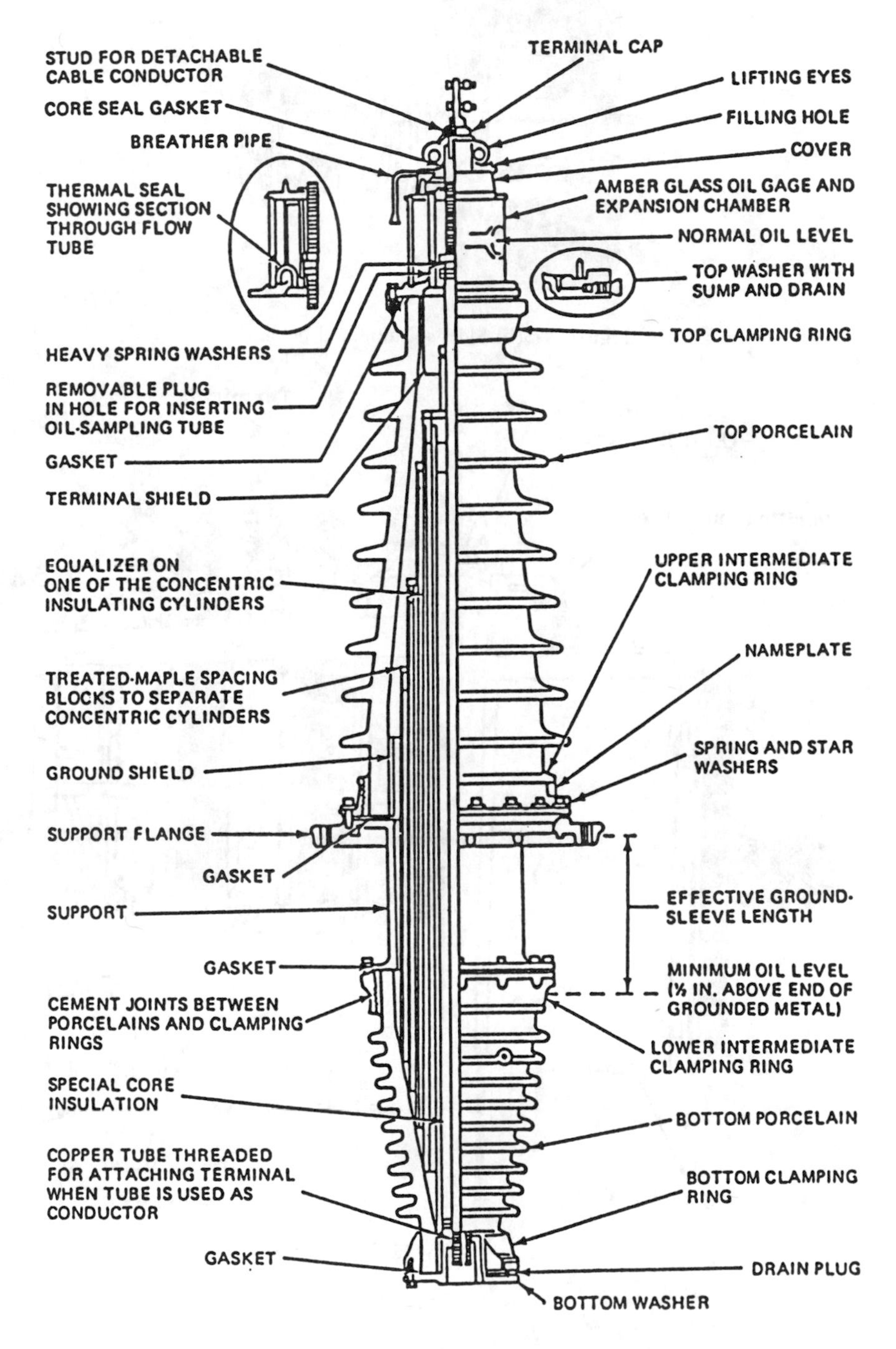

(a) Typical oil-filled bushing for 69 kV transformer.

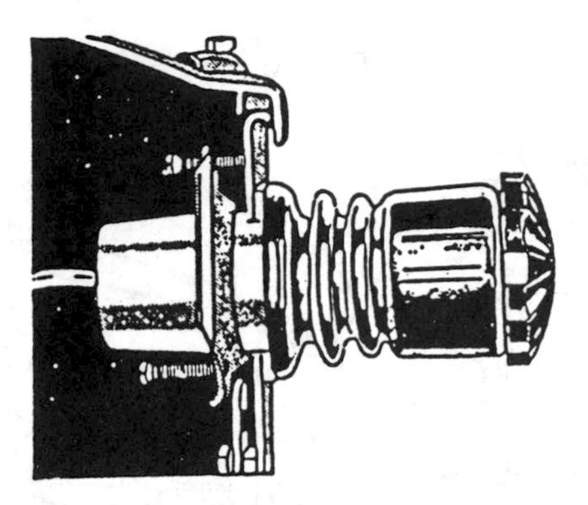

Figure A-5(b) Sidewall-mounted bushing

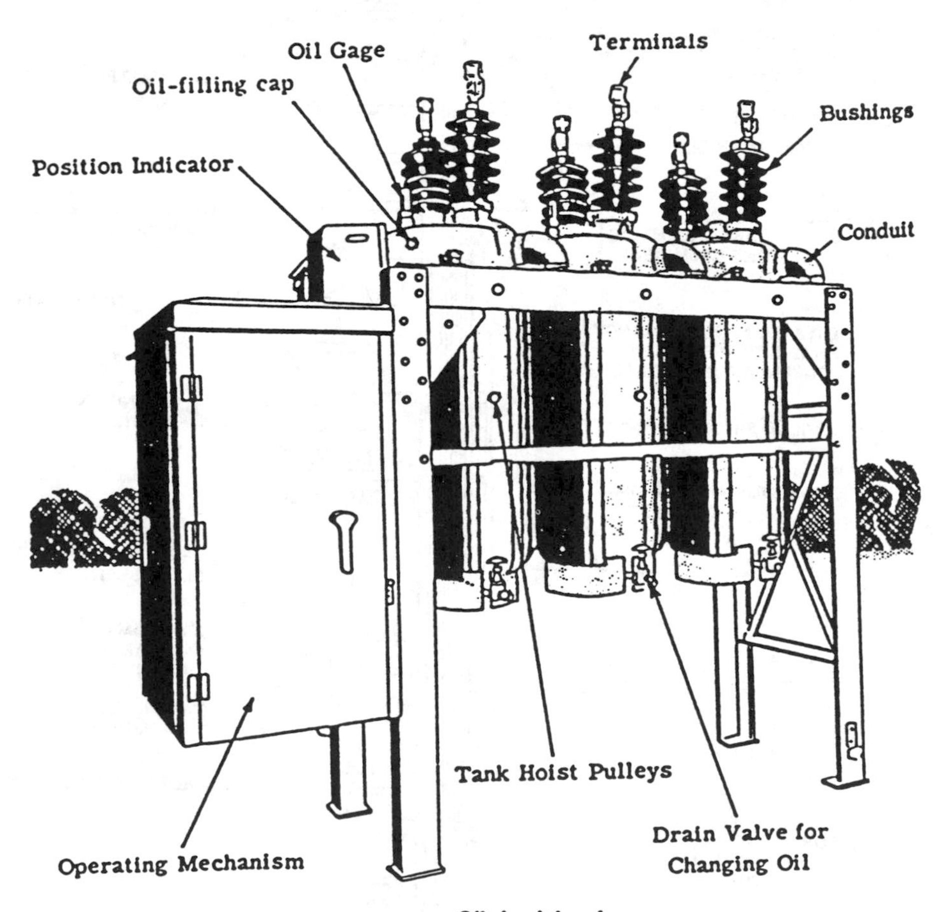

Oil circuit breaker.

Figure A-5(c)

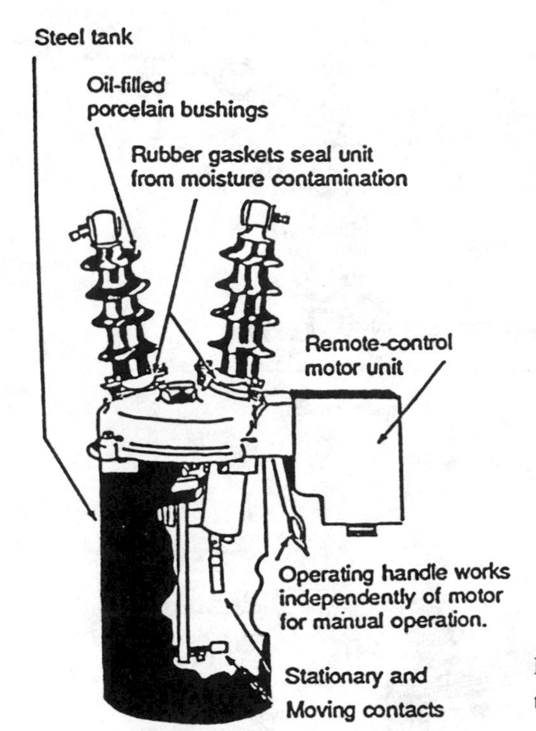

Figure A-5 (d) Bushing applications that may be replaced with polymers

Polymer insulators permit increased conductor (and static wire) line tensions, resulting in lower construction designs by permitting longer spans, fewer towers or lower tower heights.

Polymer one-piece insulators, lacking the flexibility of porcelain strings and the firm attachment of the conductor it support are said to produce a tendency to dampen galloping lines.

Although polymer insulation has become increasingly utilized over the past several decades, there are literally millions of porcelain insulated installations in this country alone; economics does not permit their wholesale replacement. Advantage is taken of maintenance and reconstruction of such facilities to make the change to polymers.

Much of the data and illustrations are courtesy of Hubbell Power systems, and is herewith duly acknowledged with thanks.

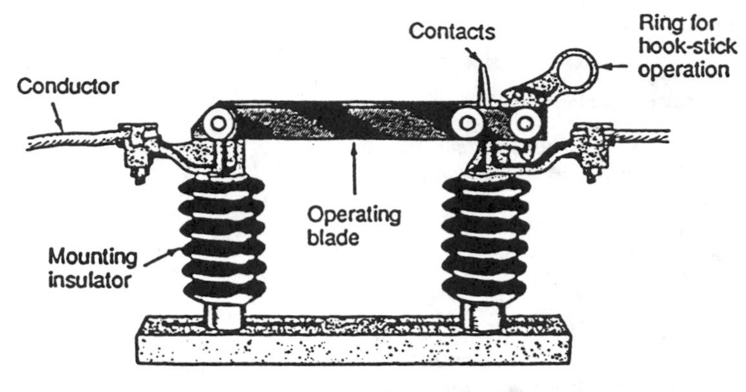

A disconnect switch.

Figure A-5(e)

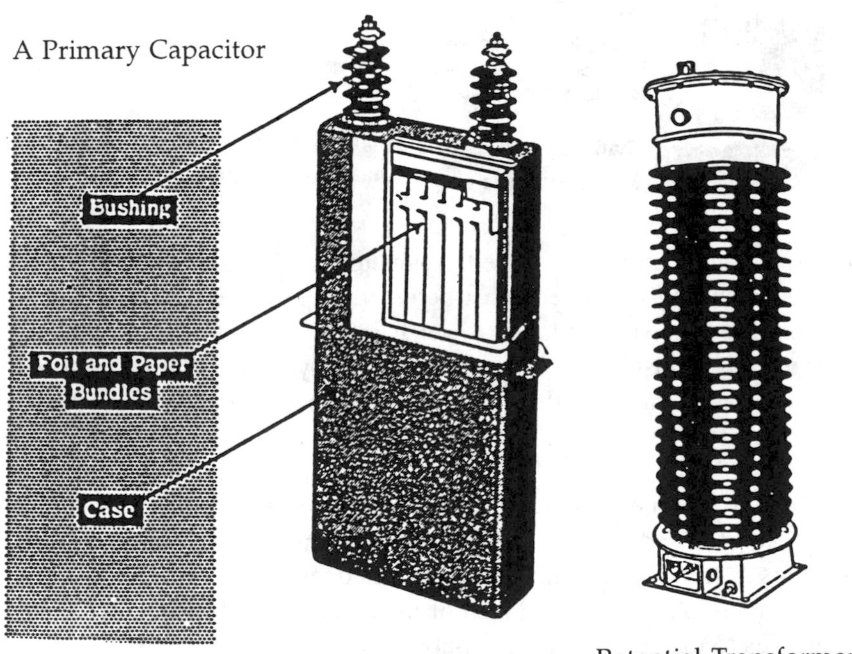

Figure A-5(f). More porcelain insulation that may be replaced with polymers.

Photographs show the pressure relief capability of surge arresters.

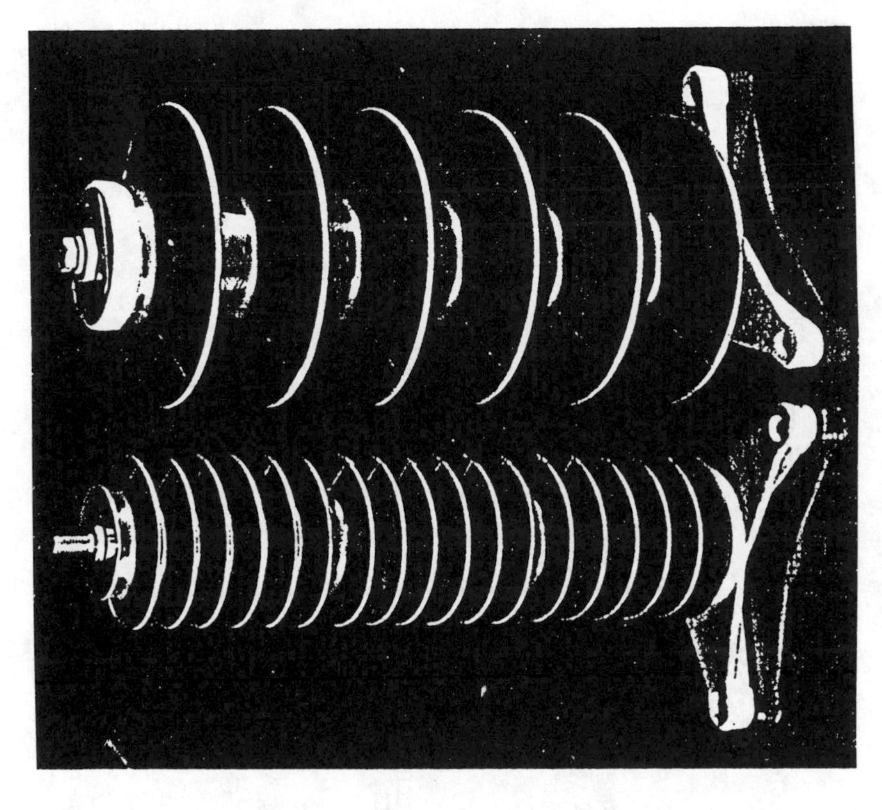

Polymer insulated arresters - station type

Figure A-6. Polymer and porcelain cased arresters

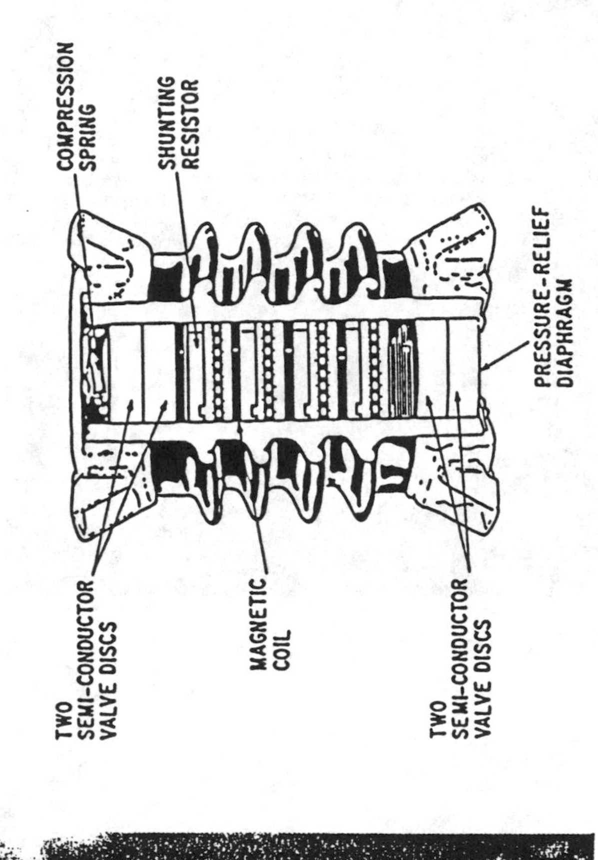

Cross-section of porcelain cased arrester

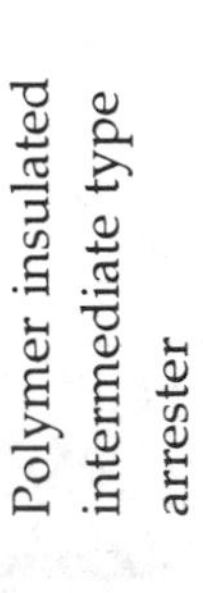

Polymer insulated intermediate type arrester

Figure A-6 (Cont'd). Polymer and porcelain cased arresters

Table A-4. Example - 10 Miles, 345 kV, 250 Strings of Insulators

Porcelain - 4500 bells, 52-3, 13.5 Ibs. ea., total 60,760 Ibs.
750 crates at 3.1 cu. ft. = 2,325 cu. ft.
Insulator cost = $51.760 ($11.60/bell)

Polymers - 250 units, 14.4 Ibs. ea., total 3 600 Ibs.
5 crates at 75 cu. ft. = 375 cu. ft.
Insulator cost = $51.750

	Savings	
1. Storage space at receiving point (3 mos.) porcelain - 580 sq. ft.; polymer - 100 sq. ft.	480 sq. ft.	$ 60.00
2. Off-load, re-load at receiving point porcelain - 10 man-hrs.; polymer - 2 man-hrs.	8 man-hrs.	$120.00
3. Breakage - off-loading, storage. re-load- porcelain - 1 percent; polymer - 0	1 percent	$517.50
4. Truck - receiving point to tower sites (5 miles) porcelain- 1.00/cwt.; polymer .50/cwt	$589.50	$589.50
5. Off load at tower site porcelain - 5 man-hrs.; polymer - 1 man-hr.	4 man-hrs.	$60.00
6. Unpack at tower site porcelain - 50 crates/hour, 25 man-hrs. polymer - 50 insulators/hour, 5 man-hrs.	20 man-hrs.	$300.00
7. Breakage - off-loading through string assembly & cleaning porcelain - 1 percent: polymer - 0	1 percent	$517.50
8. Assemble strings, attach blocks porcelain - 40 man-hrs.; polymer - 8 man-hrs.	32 man-hrs.	$480.00
9. Clean insulators porcelain-10 min./string; polymer - 3 min./string	29 man-hrs.	$435.00
10. Lift string into place (2 men) porcelain - 5 min./string; polymer - 2 min./string	25 man-hrs.	$375.00
11. Install & connect to tower (2 men) porcelain - 5 min./string polymer - 2 min./string	25 man-hrs.	$375.00
12. Breakage - lifting & installation porcelain - 0.5 percent; polymer - 0	0.5 percent	$258.75
13. Cleanup packaging materials at jobsite porcelain - 6 man-hrs.; polymer - 1.5 man-hrs.	4.5 man-hrs.	$67.50

(*Courtesy* Hubbell Power Systems)

INDEX

C

D

E

F

G

H

I

J

K

L

M

N

O

P

Q

R

S

T

V

W

X

Y

Z